Springer Proceedings in Mathematics & Statistics

Volume 159

Springer Proceedings in Mathematics & Statistics

This book series features volumes composed of selected contributions from workshops and conferences in all areas of current research in mathematics and statistics, including operation research and optimization. In addition to an overall evaluation of the interest, scientific quality, and timeliness of each proposal at the hands of the publisher, individual contributions are all refereed to the high quality standards of leading journals in the field. Thus, this series provides the research community with well-edited, authoritative reports on developments in the most exciting areas of mathematical and statistical research today.

More information about this series at http://www.springer.com/series/10533

Jozef Širáň · Robert Jajcay
Editors

Symmetries in Graphs, Maps, and Polytopes

5th SIGMAP Workshop, West Malvern, UK, July 2014

Springer

Editors
Jozef Širáň
The Open University
Milton Keynes
UK

Robert Jajcay
Comenius University
Bratislava
Slovakia

ISSN 2194-1009 ISSN 2194-1017 (electronic)
Springer Proceedings in Mathematics & Statistics
ISBN 978-3-319-30449-6 ISBN 978-3-319-30451-9 (eBook)
DOI 10.1007/978-3-319-30451-9

Library of Congress Control Number: 2016934436

Mathematics Subject Classification: 05E15, 05E18, 05C25, 05C10, 22E45

Printed on acid-free paper

This Springer imprint is published by Springer Nature
The registered company is Springer International Publishing AG Switzerland

Preface

The Symmetries In Graphs, Maps, And Polytopes Workshop 2014 was the fifth in a series of workshops, the first of which was organized by Steve Wilson in Flagstaff, Arizona, in 1998, under the acronym SIGMAC (with the original 'C' standing for Complexes). The initial workshop was followed in 2002 and 2006 by two meetings held in Aveiro, Portugal, organized by Antonio Breda d'Azevedo, and the fourth workshop, the first under the name SIGMAP, held in Oaxaca, Mexico, organized by Isabel Hubard in 2010. The aim of the workshops is to give the worldwide community of researchers in symmetries of discrete objects and structures the opportunity to gather together, exchange information and present their newest findings and advances.

The SIGMAP 2014 Workshop took place during 7–11 July 2014, in the idyllic environment of the ELIM Conference Centre in the beautiful area of Malvern, UK. It brought together a total of 62 researchers including a number of Ph.D. students. The list of invited plenary lecturers consisted of:

- Marston Conder, University of Auckland, New Zealand;
- Shaofei Du, Capital Normal University, Beijing, China;
- Gareth Jones, University of Southampton, UK;
- Roman Nedela, Matej Bel University, Slovakia;
- Primož Potočnik, University of Ljubljana, Slovenia;
- David Singerman, University of Southampton, UK;
- Asia Ivic Weiss, York University, Toronto, Canada;
- Jürgen Wolfart, J.W. Goethe University, Frankfurt, Germany.

The emphasis of the scientific program was on connections between maps, Riemann surfaces and dessins d'enfants. Gareth Jones, David Singerman and Jürgen Wolfart jointly delivered a mini-course on these connections. Beside the mini-course, the daily program consisted of plenary lectures, 34 contributed paper presentations and many informal discussions held in a collegial and encouraging atmosphere. Everybody merrily joined in congratulating Jozef Širáň on the occasion of his 60th birthday. The Wednesday conference trip took us to the Worcester Cathedral and Library and included a short organ recital.

All participants are to be thanked for their valuable contributions and for making SIGMAP 2014 a successful and memorable event. This volume contains 17 selected papers based on the talks delivered at the workshop, and it represents only a part of the workshop's rich scientific program. Despite that, it is representative of the variety of topics considered and the interactions between them.

Milton Keynes
Bratislava
December 2015

Jozef Širáň
Robert Jajcay

Acknowledgements

The organizer gratefully acknowledges support from a Conference Grant from the London Mathematical Society which enabled to cover accommodation and subsistence for the non-UK invited speakers, as well as support from the British Combinatorial Committee covering accommodation and subsistence for the UK-based invited speakers. The Open University is to be thanked for providing technical and organizational support.

We also wish to thank all our anonymous referees for their selfless and timely cooperation, without which the publication of these proceedings would be impossible.

Contents

Powers of Skew-Morphisms

Martin Bachratý and Robert Jajcay

Abstract Skew-morphisms have important applications in the classification of regular Cayley maps, and have also been shown to be fundamental in the study of complementary products of finite groups AB with B cyclic and $A \cap B = \{1\}$. As natural generalizations of group automorphisms, they share many of their properties but proved much harder to classify. Unlike automorphisms, not all powers of skew-morphisms are skew-morphisms again. We study and classify the powers of skew-morphisms that are either skew-morphisms or group automorphisms and consider reconstruction of skew-morphisms from such powers. We also introduce a new class of skew-morphisms that generalize the widely studied t-balanced skew-morphisms and which we call *coset-preserving skew-morphisms*. We show that, in certain cases, all skew-morphisms have powers that belong to this class and can therefore be reconstructed from these.

1 Introduction

The history of skew-morphisms started with their introduction in [8] a little more than a decade ago in the context of regular Cayley maps. From their very start, they have been seen as a very natural generalization of group automorphisms with which they share a number of important properties, and as such they constitute an essentially algebraic concept. After a number of applications in topological graph theory, the algebraic nature of skew-morphisms has been further underlined by discovering the fundamental role they play in the theory of *complementary products of finite groups*, specifically, with regard to products of the form AB, $A \cap B = \{1\}$, with B cyclic [2, 10].

M. Bachratý · R. Jajcay (✉)
Faculty of Mathematics, Physics and Computer Science, Comenius University, Bratislava, Slovakia
e-mail: robert.jajcay@fmph.uniba.sk

M. Bachratý
e-mail: bachraty@math.sk

J. Širáň and R. Jajcay (eds.), *Symmetries in Graphs, Maps, and Polytopes*, Springer Proceedings in Mathematics & Statistics 159,
DOI 10.1007/978-3-319-30451-9_1

A *skew-morphism* of a group G is a permutation $\varphi: G \to G$ of the elements of G that fixes the identity of G and satisfies the identity

$$\varphi(ab) = \varphi(a)\varphi^{\pi(a)}(b) \text{ for all } a, b \in G,$$

where the function $\pi : G \to \mathbb{Z}$ is called the *power function* of the skew-morphism φ and $\varphi^{\pi(a)}(a)$ is the image of the element $a \in G$ under $\pi(a)$ applications of φ. Clearly, every group automorphism of G is 'trivially' a skew-morphism of G with the constant power function $\pi(a) = 1$, for all $a \in G$. This justifies the claim that skew-morphisms constitute a generalization of group automorphisms. The similarities go even further. The value $\pi(1_G)$ is necessarily equal to 1 for the identity of G. The set of the elements of G for which $\pi(a) = 1$ forms a subgroup of G, called the *kernel* of φ and denoted by $\ker \varphi$ [8]. The kernel was shown to be non-trivial for all finite groups G and their skew-morphisms in [2]. This means that every skew-morphism of a finite group G has a restriction to a non-trivial subgroup of G (its kernel) that is a group isomorphism. It is an automorphism of the kernel if the kernel is preserved by φ, i.e., $\varphi(\ker \varphi) = \ker \varphi$; which is always true for abelian groups G. In addition, several results from [2] suggest that kernels of the majority of skew-morphisms are even considerably larger than the lower bound of 2.

Historically, first classes of skew-morphisms which were not automorphisms were skew-morphisms called anti-automorphisms [19] and their generalizations, the t-balanced skew-morphisms [1, 14]. The t-balanced skew-morphisms are skew-morphisms that possess a generating orbit $\mathcal{O}$ closed under inverses and that have the property that their power functions π are constant on $\mathcal{O}$: $\pi(a) = t$, for all $a \in \mathcal{O}$. One of the most important features of the t-balanced skew-morphisms is the large size of their kernels, which are always subgroups of index 2, and are preserved under the action of their skew-morphisms (unless $t = 1$, in which case the skew-morphism is a group automorphism and the kernel covers the entire group). Consequently, t-balanced skew-morphisms are equal to group automorphisms on all but a half of the underlying group. Thanks to their closeness to automorphisms, these are the only skew-morphisms characterized for finite cyclic groups $\mathbb{Z}_n$ [12].

One of the main sources of motivation for the present paper is based on the simple observations that all powers of t-balanced skew-morphisms are skew-morphisms again [1] and that all t-balanced skew-morphisms admit powers which are group automorphisms [4]. Translating these observations into the language of permutation actions on the elements of the underlying group yields that the orbits of t-balanced skew-morphisms are formed by merging orbits of group automorphisms (the intuitively clear concept of merging orbits is made precise in [7]). This suggested to us the possibility of constructing all t-balanced skew-morphisms from the orbits of automorphisms of their underlying groups. Another source of inspiration came from the paper [7] that contains a characterization of the skew-morphisms whose second powers are skew-morphisms again. This lead us toward considering the general question of which powers of skew-morphisms are necessarily skew-morphisms. The resulting classification (included in this paper and also derived in [2]) resulted finally in the discovery of a generalization of t-balanced skew-morphisms we call *coset-preserving skew-morphisms*.

Coset-preserving skew-morphisms constitute a very interesting class with perhaps the most important property being that all skew-morphisms of abelian groups that have a generating orbit possess non-trivial powers that are coset-preserving skew-morphisms. This, in analogy with the case of the t-balanced skew-morphisms, yields that the orbits of all skew-morphisms of abelian groups that possess a generating orbit are merges of orbits of the coset-preserving skew-morphisms of these groups. Thus, a complete list of coset-preserving skew-morphisms of a finite abelian group A provides a 'basis' for constructing all skew-morphisms of A. In view of this observation, and the obvious fact that every skew-morphism of a cyclic group has at least one generating orbit, the study of coset-preserving skew-morphisms of finite cyclic groups constitutes a very important step toward the classification of all skew-morphisms of cyclic groups. While classification of the skew-morphisms of cyclic groups giving rise to regular Cayley maps can be deduced from the recent paper [3], classification of *all* skew-morphisms of cyclic groups still awaits completion.

It should be noted that coset-preserving skew-morphisms are only useful for reconstructing skew-morphisms that preserve their kernels. Since finite simple groups do not admit skew-morphisms that preserve their kernels (except for group automorphisms) [21], these classes of groups, and non-abelian groups in general, will require different techniques.

Our paper is organized around the above outlined ideas. Section 2 consists of a summary of results concerning general properties of skew-morphisms and their connections to Cayley maps. After that comes Sect. 3 that contains our first generalization of the t-balanced skew-morphisms which is also our first example of the class of coset-preserving skew-morphisms—the main topic of our paper. Section 4 after that is devoted to the general problem of powers of skew-morphisms, and the paper is concluded with Sect. 5 about powers of skew-morphisms that are coset-preserving and Sect. 6 dealing with coset-preserving skew-morphisms of cyclic groups.

2 Basic Properties of Skew-Morphisms and Their Relation to Cayley Maps

Let us begin this section with a quick description of the concept of a Cayley map; further details can be found in [17]. Given a finite group G together with a generating set $X \subset G$ that is closed under taking inverses ($x^{-1} \in X$, for all $x \in X$) and does not contain the identity 1_G, the vertex set of the *Cayley graph* $C(G, X)$ consists of the elements of G and the edge set contains the pairs $\{\{g, gx\} \mid g \in G, \ x \in X\}$. Any cyclic permutation P of X determines a Cayley map $CM(G, X, P)$ which is a 2-cell embedding of the Cayley graph $C(G, X)$ in an orientable surface satisfying the property that the local ordering of the arcs emanating from any vertex $g \in G$ agrees with P: the counterclockwise neighbor of the arc (g, x) on the surface is the arc $(g, P(x))$, for all $g \in G$ and $x \in X$. Cayley maps proved repeatedly useful in many different contexts of topological graph theory—most importantly due to the fact that the permutations of G induced by left multiplications by the elements of G all give rise to distinct automorphisms of the map $CM(G, X, P)$. In terms of the action on the set of the arcs of the map, the permutation associated with an element $g \in G$ is the permutation $\sigma_g(h, x) = (gh, x)$, where $h \in G$ and $x \in X$. Since the left multiplication action of G is necessarily transitive on the elements of G, the group $G_L = \{ \sigma_g \mid g \in G \}$ is a vertex-transitive automorphism subgroup of $Aut(CM(G, X, P))$, for any Cayley map $CM(G, X, P)$. The action of the orientation preserving automorphism group of any orientable map on the set of the arcs of the map is well-known to be semi-regular (having trivial arc stabilizers). If the action of the full automorphism group of the map is also transitive (and therefore regular) on the arc set of the map, the map is called *regular*. Thus, regular orientable maps possess the highest possible level of symmetry, and as such constitute a central concept in the theory of highly symmetric orientable maps.

Since all Cayley maps $CM(G, X, P)$ admit a vertex-transitive automorphism group, Cayley maps are regular if and only if they also admit a map automorphism Φ mapping $(1_G, x)$ to its counterclockwise neighbor $(1_G, P(x))$. This follows easily from the fact that Φ preserves the surface around 1_G and thus maps all the arcs emanating from 1_G to their immediate neighbors, $\Phi(1_G, x) = (1_G, P(x))$, for all $x \in X$. Consequently, the vertex stabilizer of 1_G acts transitively on the set of arcs emanating from 1_G which together with the vertex-transitivity of G_L yields the arc-transitivity of the full automorphism group. It only takes a technical calculation to show that the permutation φ induced by Φ on the vertices of $CM(G, X, P)$ (i.e., on the group G itself) is a skew-morphism of G. Therefore, a Cayley map $CM(G, X, P)$ is regular if and only if G admits a skew-morphism φ preserving X and whose restriction to X is equal to P. Since X is assumed to generate G and be closed under inverses, the skew-morphism induced by Φ always has an orbit that generates G and is closed under inverses. However, not all skew-morphisms have such orbit, and only the skew-morphisms that have such orbit give rise to regular Cayley maps (with X consisting of the elements of the orbit and P induced by the cyclic action of the skew-morphism on X).

The *distribution of inverses* with respect to X and P is the function $\chi(x)$ defined to be equal to the smallest non-negative integer i with the property $P^i(x) = x^{-1}$ for all $x \in X$. This function determines many of the properties of the corresponding skew-morphism φ. In case when $\chi(x) = 0$ for all $x \in X$ (i.e., X consists of involutions) or when $|X|$ is even of the form $2k$ and $\chi(x) = k$ for all $x \in X$, the corresponding map is called *balanced* and the corresponding skew-morphism (in case of a regular map) is a group automorphism of G [18]. Naturally, this is the best understood case. Even though the question which group automorphisms possess an orbit that generates G and is closed under inverses turns out to be more complicated than one might expect, balanced regular Cayley maps have been already classified for a number of classes of finite groups [15, 18, 20].

The case $P(x^{-1}) = (P^{-1}(x))^{-1}$, for all $x \in X$, is another well studied case, and Cayley maps having this property are called *antibalanced* [19]. More generally, Cayley maps with the property $P(x^{-1}) = (P^{t}(x))^{-1}$, for all $x \in X$, are called *t-balanced*, and, except for the case $t = 1$ which gives rise to automorphisms, all have the property that the kernel of the corresponding skew-morphism is of index 2 in the underlying group G, the power function of the skew-morphism has two values 1 and t, and the parameter t must be a square root of 1 modulo the order of the skew-morphism (as a permutation) [1]. Regular t-balanced Cayley maps have also been classified for a number of classes of groups [1, 4, 11–14, 16].

As for classifying all regular Cayley maps $CM(G, X, P)$ for a fixed group G, classification results are very rare. Outside the case of cyclic groups of prime order which only admit automorphisms [8], only the recent paper [2] contains further classifications.

Skew-morphisms that give rise to regular Cayley maps have several important characteristics that are not necessarily shared by all skew-morphisms. This has sometimes caused confusion as many in the topological graph theory community only consider the skew-morphisms that give rise to regular maps. For example, it is easy to see that the kernels of the skew-morphisms that give rise to regular Cayley maps must be non-trivial [8]. This result was eventually proved for general skew-morphisms only using relatively strong results from permutation group theory [2]. Similarly, the order of a skew-morphism giving rise to a regular Cayley map must be equal to the size of one of its orbits (namely, the size of the set X) [8]. The order of these skew-morphisms is therefore always smaller than the order of the underlying group G. An orbit of a skew-morphism φ whose length is equal to the order of φ is called a *precise* orbit, and general skew-morphisms do not have to have a precise orbit. In fact, as proved by Horoševskiĭ [5], there are even group automorphisms that do not have a precise orbit. However, the second result mentioned above, i.e., the claim that the order of a skew-morphism cannot exceed the order of the underlying group was also eventually extended to general skew-morphisms in [2].

Even though skew-morphisms play a fundamental role in the theory of regular Cayley maps, general skew-morphisms are more closely related to complementary products of groups. If G is a group and K and H are subgroups of G with the property $K \cap H = \{1_G\}$, whose product is equal to G, $K \cdot H = G$, we say that G is a *complementary product* of K and H. In case when $H = \langle y \rangle$ is cyclic, the left multiplication of elements of K by y gives rise to a skew morphism φ of K such that $yk = \varphi(k)y^{\pi(k)}$ for all $k \in K$ [2]. Conversely, any skew-morphism φ of a group K gives rise to a complementary product $K \cdot \langle \varphi \rangle$ called *skew product* that is a generalization of the split (semidirect) product [2, 6, 10]. In general, if $G = K \cdot \langle y \rangle$ is a complementary product of finite groups, the order of the associated skew-morphism φ does not have to be equal to the order of y. The orders of φ and y are equal if and only if $\langle y \rangle$ does not contain a non-trivial normal subgroup of the product G, in which case G is isomorphic to the skew product $K \cdot \langle \varphi \rangle$ [2].

3 Generalization of t-Balanced Skew-Morphisms

One of the reasons the class of t-balanced skew-morphisms is the best understood class of skew-morphisms which are not group automorphisms lies in their 'closeness' to group automorphisms: every t-balanced skew-morphism is equal to a group automorphism on a subgroup of index 2. Also, t-balanced skew-morphisms share many properties of the automorphisms, and always have at least one power equal to a group automorphism (if t is not equal to -1, this power is non-trivial) [4]. In this section, we introduce two generalizations of t-balanced skew-morphisms, both of which prove extremely useful throughout the rest of our paper. First, we introduce a generalization of the t-balanced skew-morphisms to skew-morphisms that do not possess a generating orbit closed under taking inverses, and then we introduce a further generalization we will call coset-preserving skew-morphisms.

In accordance with [2], let Skew(G) denote the set of all skew-morphisms of G. Unlike the case of automorphisms, this set rarely forms a group under composition. This makes it necessary to consider the smallest subgroup of Sym(G) that contains (is generated by) Skew(G); we denote it by SkewGroup(G) and call it the *skew-morphism group* of G. Of the class of all finite abelian groups, only the cyclic group of order 4, the cyclic groups of order n where $\gcd(n, \phi(n)) = 1$, and the elementary abelian 2-groups have the property SkewGroup(G) $=$ Skew(G) $=$ Aut(G) [2]. The only known non-abelian class satisfying these equalities consists of the dihedral groups of prime degree $p > 3$ [2, 9], however, the theory of skew-morphisms of non-abelian groups is much less developed at this time. To give examples of the opposite kind, consider $C_3 \times C_3$, which has 48 automorphisms and 64 skew morphisms, which generate a group of order 40320, isomorphic to S_8 [2], and the group D_3 which has 12 skew-morphisms, only 6 of which are automorphisms, and the skew-morphisms generate a subgroup of order 120 [2].

As is well-known, every power of a group automorphism is a group automorphism. While this is not the case for skew-morphisms, it is interesting to note that there exist skew-morphisms with certain powers equal to group automorphisms. Recall that a t-balanced skew-morphism of G is a skew-morphism that has a generating orbit X closed under taking inverses and a power function equal to t on all of X. Then $t^2 \equiv 1 \pmod{|X|}$, and the power function has only two values 1 and t, and hence a kernel of index 2 [1]. Thus, t-balanced skew-morphisms are the skew-morphisms that are in a sense the closest to automorphisms: they have the largest possible kernel not equal to the whole group and only two power values. The kernel of a t-balanced skew-morphism φ is always preserved by φ, and therefore any t-balanced skew-morphisms is *equal* to a group automorphism on half of the underlying group. Moreover, the $(t+1)$-st power φ^{t+1} of a t-balanced skew-morphism φ is always a group automorphisms of the underlying group for all finite abelian groups [4], and every power of a t-balanced skew-morphism is a t-balanced skew-morphism (or a group automorphism) [1].

The class of t-balanced skew-morphisms is a good example of a class that until now has only been investigated within the context of regular Cayley maps. Specifically,

t-balanced skew-morphisms were introduced only in the context of the t-balanced Cayley maps: maps $CM(G, X, P)$ satisfying the property $P(x^{-1}) = (P^t(x))^{-1}$, for all $x \in X$. When trying to generalize t-balanced skew-morphisms to skew-morphisms that do not possess a generating orbit closed under taking inverses, one faces an important decision. While in the case of the t-balanced skew-morphisms *all* generating orbits closed under inverses are contained in the coset of the kernel assigned the power function value t, and each orbit is contained in the kernel or its coset, in case of a general skew-morphism φ whose power function only attains two values 1 and t, φ may possess orbits that intersect with both the kernel and its coset. Thus, skew-morphisms whose power functions only assume two values 1 and t naturally split into two subclasses: those that preserve their kernels and those that do not. As we will see, these two classes significantly differ. This is why we choose to restrict the name t-balanced only to those skew-morphisms whose power functions assume two values and which preserve their kernels. With a slight abuse of the previously used terminology, from now on, a skew-morphism φ (with or without a generating orbit closed under inverses) will be called t-balanced if *the power function of* φ *assumes only the values* 1 *and* t, *and* φ *preserves* $\ker \varphi$ *setwise* (if $t = 1$, the skew-morphism is a group automorphism). These more general t-balanced skew-morphisms share all the important properties of the t-balanced skew-morphisms that have a generating orbit closed under taking inverses.

Theorem 1 *Let G be a finite group and let φ be a t-balanced skew-morphism of G, $t \neq 1$. Then $t^2 \equiv 1 \pmod{|\varphi|}$, all powers of φ are skew-morphisms, and φ^{t+1} is a group automorphism of G.*

Proof Let φ be a t-balanced skew-morphism of G, $t \neq 1$. Then $\ker \varphi$ is a subgroup of G of index 2, preserved by φ, and the power function π of φ assumes the value 1 on all of $\ker \varphi$, and the value t on the rest of G. To prove the first claim of the theorem, we employ a formula from [8]:

$$\pi(ab) \equiv \sum_{0 \le i < \pi(a)} \pi(\varphi^i(b)) \pmod{|\varphi|}, \tag{1}$$

for all $a, b \in G$. Let $h \notin \ker \varphi$ and $\pi(h) = t$. Necessarily, $h^2 \in \ker \varphi$, as otherwise we would have $h^2 = bh$, for some $b \in \ker \varphi$, leading to $h = b$, which contradicts the choice of h. Thus,

$$1 = \pi(hh) \equiv \sum_{0 \le i < t} \pi(\varphi^i(h)) \equiv t \cdot t \pmod{|\varphi|},$$

since φ is assumed to preserve $\ker \varphi$, and therefore also its complement, and all the images of h under φ must belong to the complementary coset of $\ker \varphi$.

To prove the other two claims, we will use another formula from [8]—one that deals with powers of skew-morphisms. Like formula (1), it will be used repeatedly throughout this paper:

$$\varphi^j(gh) = \varphi^j(g)\varphi^{\pi(j,g)}(h), \tag{2}$$

where

$$\pi(j, g) = \sum_{0 \leq i < j} \pi(\varphi^i(g)) \pmod{|\varphi|}, \tag{3}$$

for all $g, h \in G$ and all $j \in \mathbb{N}$. Applying formula (2) yields the claims almost immediately:

$$\varphi^j(gh) = \varphi^j(g)\varphi^{\pi(j,g)}(h) = \varphi^j(g)\varphi^{j\pi(g)}(h) = \varphi^j(g)(\varphi^j)^{\pi(g)}(h)$$

and

$$\varphi^{t+1}(gh) = \varphi^{t+1}(g)\varphi^{(t+1)\pi(g)}(h) = \varphi^{t+1}(g)\varphi^{t+1}(h),$$

for all $g, h \in G$, with the first identity based on the fact that all elements within a single orbit have the same value of π, and the second identity obvious when $\pi(g) = 1$ and relying on the identity $(t+1)\pi(g) \equiv (t+1)t \equiv t^2 + t \equiv 1 + t \pmod{|\varphi|}$ when $\pi(g) = t$. □

Skew-morphisms of abelian groups always preserve their kernels [1]. The parameter t of a skew-morphism with two power function values that does not preserve its kernel must satisfy different arithmetic restrictions. Namely, suppose that G is finite, $[G : \ker \varphi] = 2$, and suppose that φ *does not preserve* $\ker \varphi$. Then there exists an orbit of φ that contains elements from both cosets of $\ker \varphi$, i.e., there exist elements $g, h \in G$, $\varphi(g) = h$, and $\pi(g) = 1$, $\pi(h) = t$; g belongs to $\ker \varphi$ while h does not. Being of index 2 in G, $\ker \varphi$ is normal in G, and thus $a(\ker \varphi) = (\ker \varphi)a$, for all $a \in G$. Hence, hg and gh belong to the same coset of $\ker \varphi$; the complement of $\ker \varphi$ in G. At the same time $h^2 \in \ker \varphi$ again. Applying formula (1) yields:

$$t = \pi(hg) \equiv \sum_{0 \leq i < t} \pi(\varphi^i(g)) \pmod{|\varphi|},$$

while

$$1 = \pi(hh) \equiv \sum_{0 \leq i < t} \pi(\varphi^i(h)) \pmod{|\varphi|}.$$

However,

$$\sum_{0 \leq i < t} \pi(\varphi^i(h)) = \sum_{0 \leq i < t} \pi(\varphi^i(g)) - \pi(g) + \pi(\varphi^{t-1}(h)),$$

and therefore $1 \equiv t - 1 + \pi(\varphi^{t-1}(h)) \pmod{|\varphi|}$. If $\pi(\varphi^{t-1}(h))$ were equal to 1, then we would get a contradiction $t \equiv 1 \pmod{|\varphi|}$. Thus, $\pi(\varphi^{t-1}(h)) \equiv t \pmod{|\varphi|}$, and therefore

$$2 \equiv 2t \pmod{|\varphi|}. \tag{4}$$

This also means that 2 must divide the order of φ. The following example of a skew-morphism with the above properties has been kindly provided to us by Marston Conder.

Example 1 Let G be the (non-abelian) direct product $\mathbb{D}_3 \times \mathbb{Z}_2$ of the dihedral group $\mathbb{D}_3$ of order 6 and the cyclic group $\mathbb{Z}_2$ of order 2. Denote the generators of $\mathbb{D}_3$ by a and b, of orders 3 and 2, respectively, and let c be the order 2 generator of $\mathbb{Z}_2$. Let

$$\varphi = (a, a^2)(b, bc, c)(ab, a^2bc, ac, a^2b, abc, a^2c).$$

Then φ is a skew-morphism of order 6, with kernel K of order 6 generated by a and b, and with power function value 4 on Kc (and 1 on K). While the order 3 subgroup of K generated by a is preserved by φ, K itself is not. The orbit $(ab, a^2bc, ac, a^2b, abc, a^2c)$ consists of elements with alternating power function values and is of length 6; which is an even number. Also, the parameter $t = 4$ satisfies the identity (4) which takes the form $2 \equiv 2t = 2 \cdot 4 \pmod 6$.

As mentioned repeatedly already, group automorphisms and t-balanced skew-morphisms constitute specific examples of a much wider class that is in many ways the main focus of our paper. We say that a skew-morphism φ of a finite group G is a *coset-preserving skew-morphism*, if every orbit of φ is contained within a single coset of $\ker\varphi$, or equivalently, $\pi(a) = \pi(b)$ for any elements $a, b \in G$ that belong to the same orbit of φ.

We close the present section with two results concerning conjugates of skew-morphisms by automorphisms. The following observation has also been made by others, but was first brought to the attention of the second author by Kan Hu.

Theorem 2 *Let φ be a skew-morphism of a finite group G, and let ψ be a group automorphism of G. Then the conjugate $\psi\varphi\psi^{-1}$ is a skew-morphism of G again.*

Proof The calculation is an easy exercise:

$$\begin{aligned}(\psi\varphi\psi^{-1})(gh) &= (\psi\varphi)\psi^{-1}(g)\psi^{-1}(h) = \psi(\varphi(\psi^{-1}(g))\varphi^{\pi(\psi^{-1}(g))}(\psi^{-1}(h)))\\ &= (\psi\varphi\psi^{-1})(g)(\psi\varphi^{\pi(\psi^{-1}(g))}\psi^{-1})(h) = (\psi\varphi\psi^{-1})(g)(\psi\varphi\psi^{-1})^{\pi(\psi^{-1}(g))}(h). \quad \square\end{aligned}$$

The above observation allows us to restate and generalize one of the implications of Lemma 2.4 from [13] concerning isomorphic regular Cayley maps with the same underlying group and the same distribution of inverses. Let $\mathcal{M}_1 = CM(G_1, X_1, P_1)$ and $\mathcal{M}_2 = CM(G_2, X_2, P_2)$ be Cayley maps, and let λ_i, ρ_i, $i \in \{1, 2\}$, be the corresponding arc-reversing involutions and rotations of the maps defined on their dart sets by the formulas: $\lambda_i(g, x) = (gx, x^{-1})$, $\rho_i(g, x) = (g, P_i(x))$. The *maps $\mathcal{M}_1$ and $\mathcal{M}_2$ are isomorphic* if there exists a bijection $\Phi : D(\mathcal{M}_1) \to D(\mathcal{M}_2)$ satisfying the property $\Phi\lambda_1 = \lambda_2\Phi$ and $\Phi\rho_1 = \rho_2\Phi$.

Theorem 3 *Let $\mathcal{M}_1 = CM(G_1, X_1, P_1)$ and $\mathcal{M}_2 = CM(G_2, X_2, P_2)$ be two Cayley maps of the same order and valency. If there exists a group isomorphism $\psi : G_1 \to G_2$ that maps X_1 to X_2, $\psi(X_1) = X_2$, and whose restriction to X_1 and X_2 commutes with P_1 and P_2, $P_2\psi = \psi P_1$, then the two maps are isomorphic.*

Moreover, if φ_1 is the skew-morphism corresponding to the map automorphism of $\mathcal{M}_1$ mapping $(1_{G_1}, x)$ to $(1_{G_1}, P_1(x))$, for all $x \in X_1$, and φ_2 is the skew-morphism corresponding to the map automorphism of $\mathcal{M}_2$ mapping $(1_{G_2}, x)$ to $(1_{G_2}, P_2(x))$, for all $x \in X_2$, then $\varphi_2 = \psi\varphi_1\psi^{-1}$.

Proof Let us assume the existence of a group isomorphism $\psi : G_1 \to G_2$ with the property $P_2\psi = \psi P_1$. The mapping $\Phi : D(\mathcal{M}_1) \to D(\mathcal{M}_2)$ defined via the formula $\Phi(g, x) = (\psi(g), \psi(x))$, for all $g \in G_1, x \in X_1$, is a well defined bijection from $D(\mathcal{M}_1)$ to $D(\mathcal{M}_2)$. The following calculations show that Φ is a map automorphism:

$$\Phi\lambda_1(g, x) = \Phi(gx, x^{-1}) = (\psi(gx), \psi(x^{-1}))$$
$$= (\psi(g)\psi(x), \psi(x^{-1})) = \lambda_2(\psi(g), \psi(x)) = \lambda_2\Phi(g, x),$$

and

$$\Phi\rho_1(g, x) = \Phi(g, P_1(x)) = (\psi(g), \psi(P_1(x)))$$
$$= (\psi(g), P_2(\psi(x))) = \rho_2(\psi(g), \psi(x)) = \rho_2\Phi(g, x).$$

If φ_1 and φ_2 are the two skew-morphisms described in the statement of the theorem, the restriction of φ_1 to X_1 is equal to P_1, and the restriction of φ_2 to X_2 is equal to P_2. Since $P_2\psi = \psi P_1$, the restriction of φ_2 is a conjugate of the restriction of φ_1 via the restriction of ψ to X_1. The rest of the proof follows from an induction on the length of the products of the generators in X_1 along the lines of the original proof in [13]. □

4 Powers of Skew-Morphisms

The set Skew(G) of the skew-morphisms of a finite group G rarely forms a group under the operation of composition. Moreover, the smallest subgroup of the full symmetric group acting on the elements of G that contains Skew(G), SkewGroup(G), is generally considerably bigger than Skew(G); often close or equal to the largest group stabilizing the identity of G—the stabilizer of 1_G in Sym(G). This does not necessarily imply that the powers of a single specific skew-morphism cannot be skew-morphisms again, however, computational evidence suggests that most powers of general skew-morphisms are not skew-morphisms again.

In the present section, we develop a general theory of powers of skew-morphisms with the aim of understanding the recursive possibilities of our approach in classifying skew-morphisms of finite groups.

In what follows, the ith *power of a skew-morphism* φ of a group G is the permutation $\varphi^i \in \mathrm{Sym}(G)$ obtained by composing φ with itself i times. Necessarily, all powers of a skew-morphism φ fix the identity 1_G, and thus φ^i belongs to the stabilizer of 1_G in $\mathrm{Sym}(G)$, for all i. The ith *root of a skew-morphism* φ is a permutation ψ with the property $\psi^i = \varphi$. It is important to note that a permutation can have many distinct ith roots, and the roots of a skew-morphism do not necessarily fix 1_G (to mention an extreme case, all permutations of order i constitute an ith root of the identity mapping).

In the first theorem of this section, we resolve the fundamental question which powers of skew-morphisms are skew-morphisms themselves. A slightly different version of this theorem that appears in [2] was discovered independently at about the same time we proved ours. The theorem is a generalization of a theorem from [7] which only addresses the second powers of skew-morphisms.

Theorem 4 *Let φ be a skew-morphism of order $|\varphi| = n$ of a finite group G, and let i be a positive integer. The power $\psi = \varphi^i$ is a skew-morphism if and only if, for every $g \in G$, the equation*

$$i \cdot x \equiv \pi(i, g) \pmod{n} \tag{5}$$

admits a solution.

If $\psi = \varphi^i$ is a skew-morphism and π_ψ is its power function, the value $\pi_\psi(g)$ is the smallest positive solution x of (5).

Proof Assume that $\psi = \varphi^i$ is a skew-morphism with the power function π_ψ. Then, using (2),

$$\psi(gh) = \varphi^i(gh) = \varphi^i(g)\varphi^{\pi(i,g)}(h),$$

while on the other hand,

$$\psi(gh) = \psi(g)\psi^{\pi_\psi(g)}(h) = \varphi^i(g)(\varphi^i)^{\pi_\psi(g)}(h) = \varphi^i(g)\varphi^{i\pi_\psi(g)}(h).$$

Therefore,

$$i\pi_\psi(g) \equiv \pi(i, g) \pmod{n},$$

for all $g \in G$, and $n = |\varphi|$; $\pi_{\psi(g)}$ is a solution of (5). Recall that $\pi_{\psi(g)}$ is the smallest positive integer with the property $\psi(gh) = \psi(g)\psi^{\pi_\psi(g)}(h)$ [8], and suppose that $0 < j < \pi_\psi(g)$ also satisfies $i \cdot j \equiv \pi(i, g) \pmod{n}$. Then

$$\psi(gh) = \varphi^i(g)\varphi^{\pi(i,g)}(h) = \varphi^i(g)\varphi^{ij}(h) = \psi(g)\psi^j(h),$$

for all $h \in G$. This would contradict $\pi_\psi(g)$ being the smallest with this property.

The opposite implication follows along similar lines. If each of the equations $i \cdot x \equiv \pi(i, g) \pmod{n}$ admits a solution, denote the smallest of these solutions by $\pi_\psi(g)$, and note that

$$\begin{aligned}\psi(gh) &= \varphi^i(gh) = \varphi^i(g)\varphi^{\pi(i,g)}(h) = \varphi^i(g)\varphi^{i\pi_\psi(g)}(h) = \varphi^i(g)(\varphi^i)^{\pi_\psi(g)}(h)\\ &= \psi(g)\psi^{\pi_\psi(g)}(h),\end{aligned}$$

hence $\psi = \varphi^i$ is a skew-morphism of G with power function π_ψ. □

If φ preserves its kernel set-wise and $g \in \ker\varphi$, then $\pi(i, g) = i$ for all i. Hence the smallest solution to $i \cdot x \equiv \pi(i, g) \pmod{n}$ is $x = 1$. If $\psi = \varphi^i$ is a skew-morphism, this yields that $\ker\varphi \leq \ker\psi$, an observation already made in [2].

Theorem 1 asserts that skew-morphisms that preserve their kernels of index 2 have the property that all of their powers are skew-morphisms again. This is also an easy consequence of the above Theorem 4 and, more importantly, this result can be extended to the much wider class of coset-preserving skew-morphisms we have defined in the previous section.

Theorem 5 *Let φ be a coset-preserving skew-morphism of a group G, and let i be a positive integer. Then φ^i is a coset-preserving skew-morphism for each i.*

Proof The property of being a coset-preserving skew-morphism yields that $\pi_\varphi(g) = \pi_\varphi(\varphi^i(g))$, for all $i > 0$ and $g \in G$. Thus, $\pi_\varphi(i, g) = i\pi_\varphi(g)$, for all $i > 0$ and $g \in G$, which means that i divides $\pi_\varphi(i, g)$ for all $i > 0$ and $g \in G$. Therefore, $\psi = \varphi^i$ is a skew-morphism of G for all $i > 0$ by Theorem 4.

If g and g' belong to the same orbit of ψ, they also belong to the same orbit of φ, and therefore $\pi_\varphi(g) = \pi_\varphi(g')$. Consequently, $\pi_\varphi(i, g) = \pi_\varphi(i, g')$, and therefore, $\pi_\psi(g) = \pi_\psi(g')$, since both values are the smallest solution to the same equation. Hence, ψ is coset-preserving. □

Another corollary of Theorem 4 is concerned with general skew-morphisms.

Corollary 1 *Let φ be a skew-morphism of order n of a group G, and let i be a positive integer. If n and i are relatively prime, φ^i is a skew-morphism of order n.*

Proof If n and i are relatively prime, each of the equations $i \cdot x \equiv \pi(i, g) \pmod{n}$ has solution regardless of the value of $\pi(i, g)$. Thus, due to Theorem 4, φ^i is a skew-morphism of G. A well-known formula yields that the order of φ^i is $\frac{n}{(n,i)}$, and $(n, i) = 1$ when n and i are relatively prime. □

Based on the above corollary, a skew-morphism φ of order n gives rise to at least $\phi(n)$ distinct skew-morphisms, all of them of order n, where $\phi(n)$ is the value of the Euler totient function at n. In the case when n and i are relatively prime, there exists a j such that $ji \equiv 1 \pmod{n}$. Hence, $\varphi = (\varphi^i)^j$, and therefore φ is a power of each of these $\phi(n)$ companion skew-morphisms. Since every power of a group automorphism is an automorphism, if φ is not an automorphism of the underlying group, none of these powers is a group automorphism either. Note that n and $n-1$ are relatively prime for all $n \geq 2$, and hence the inverse permutation to a skew-morphism is always a skew-morphism. We illustrate the above ideas in a series of examples. All our examples come from a list of skew-morphisms of cyclic groups maintained online by M. Conder.

We begin with a simple example of a skew-morphism whose power is not a skew-morphism.

Example 2 Let φ be the skew-morphism

$$\varphi = (1, 7, 9, 23, 16, 12, 19, 18, 6, 17, 4, 13, 21, 22, 14, 8, 11, 2, 24, 3)(5, 10, 20, 15)$$

of $\mathbb{Z}_{25}$ of order 20 with the corresponding power function values

$$[9, 17, 13, 5, 9, 17, 13, 5, 9, 17, 13, 5, 9, 17, 13, 5, 9, 17, 13, 5][1, 1, 1, 1].$$

Due to Theorem 4, the fifth power

$$\varphi^5 = (1, 12, 4, 8)(7, 19, 13, 11)(9, 18, 21, 2)(23, 6, 22, 24)(16, 17, 14, 3)$$

is not a skew-morphism since: $\pi(5, 1) = \pi(1) + \pi(7) + \pi(9) + \pi(23) + \pi(16) = 9+17+13+5+9 = 53$, and there is no solution to the equation: $5x \equiv 53 \pmod{20}$.

Example 3 The permutation $\varrho = (1, 7, 13, 3, 9, 15, 5, 11)$ is a 7-balanced skew-morphism of $\mathbb{Z}_{16}$ of order 8 with the corresponding power function values

$$[\pi(1), \pi(7), \pi(13), \pi(3), \pi(9), \pi(15), \pi(5), \pi(11)] = [7, 7, 7, 7, 7, 7, 7, 7].$$

All powers ϱ^i, $1 \leq i \leq 7$ are distinct skew-morphisms by Theorem 5.

The next example shows that the power function of a skew-morphism φ and the power function of its power φ^i, i relatively prime to the order n of φ, are not necessarily the same.

Example 4 The skew-morphism $\varsigma = (1, 11, 15, 13, 5, 9, 7, 17, 3)(2, 14, 8)(4, 10, 16)$ of $\mathbb{Z}_{18}$ of order 9 has the power function

$$[8, 2, 5, 8, 2, 5, 8, 2, 5][4, 4, 4][7, 7, 7]$$

while its second power $\varsigma^2 = (1, 15, 5, 7, 3, 11, 13, 9, 17)(2, 8, 14)(4, 16, 10)$ is a skew-morphism with the power function

$$[5, 2, 8, 5, 2, 8, 5, 2, 8][4, 4, 4][7, 7, 7].$$

Our last example answers in negative the question whether each skew-morphism must have a power that is a group automorphism.

Example 5 The skew-morphism

$$\zeta = (1, 10, 19, 28, 37, 7, 16, 25, 34, 4, 13, 22, 31)(2, 38, 35, 32, 29, 26, 23, 20, 17, 14, 11, 8, 5)$$

of $\mathbb{Z}_{39}$ has the power function

$$[3, 3, 3, 3, 3, 3, 3, 3, 3, 3, 3, 3, 3][9, 9, 9, 9, 9, 9, 9, 9, 9, 9, 9, 9, 9]$$

and order 13. Hence, every power ζ^i, $1 \leq i \leq 13$, is a skew-morphism of order 13. As pointed out in the discussion preceding the above examples, if any of these powers were a group automorphism, so would have to be ζ.

Clearly, the distribution of the values of the power function throughout the orbits of a skew-morphism φ is very important with regard to the question which powers of φ are skew-morphisms again. It has been shown in [1], that any skew-morphism φ of an abelian group A possessing a generating orbit X closed under taking inverses is either coset-preserving or X takes the form

$$\begin{array}{l} x, \varphi(x), \varphi^2(x), \dots, \varphi^{\ell-1}(x), \\ h_1x,\ \varphi(h_1)\varphi(x),\ \varphi^2(h_1)\varphi^2(x),\ \dots,\ \varphi^{\ell-1}(h_1)\varphi^{\ell-1}(x), \\ h_2x,\ \varphi(h_2)\varphi(x),\ \varphi^2(h_2)\varphi^2(x),\ \dots,\ \varphi^{\ell-1}(h_2)\varphi^{\ell-1}(x), \\ \dots \\ h_{k-1}x,\ \varphi(h_{k-1})\varphi(x),\ \varphi^2(h_{k-1})\varphi^2(x),\ \dots,\ \varphi^{\ell-1}(h_{k-1})\varphi^{\ell-1}(x), \end{array}$$

where $h_j = \varphi^\ell(h_{j-1})h_1$, $1 < j < k$, $k\ell = |X|$, and $\varphi^\ell(h_{k-1})h_1 = 1_A$. Moreover, at least the first two rows of this list are necessary, so that $k \geq 2$. Each h_j is a non-trivial element of H (for $1 \leq j < k$), and the ℓ elements $x, \varphi(x), \varphi^2(x), \dots, \varphi^{\ell-1}(x)$ belong to mutually distinct cosets of ker φ. This yields, in particular, that the values π takes on X for a skew-morphism φ of an abelian A that gives rise to a regular Cayley map are either all the same (if φ is coset-preserving) or 'periodic'—repeating in (at least two) sequences of equal length ℓ. For example, consider the skew-morphisms φ and ς from Examples 2 and 4. It is easy to see that the periodicity of π on X yields the equality $\pi(\ell, x) \equiv \pi(\ell, y) \pmod{|X|}$, for all $y \in X$. Moreover, $\sum_{i=0}^{|X|-1} \pi(\varphi^i(x)) \equiv 0 \pmod{|X|}$ [1], and thus $k\pi(\ell, x) \equiv 0 \pmod{|X|}$, which finally leads us to the conclusion that φ^ℓ is a coset-preserving skew-morphism of A.

The preceding arguments can be summarized in observing that the skew-morphisms of abelian groups that give rise to regular Cayley maps are either coset-preserving or have a non-trivial power that is coset-preserving. In this sense, coset-preserving skew-morphisms are the building stones of all skew-morphisms of abelian groups that give rise to regular Cayley maps. In the forthcoming paragraphs, we show that the same must also be true for skew-morphisms of abelian groups that have a generating orbit (which, however, is no longer required to be closed under taking inverses). We build our argument through a series of technical lemmas.

Lemma 1 *Let A be an abelian group, and let φ be a skew-morphism of A with the power function π. Then the following hold:*

(i) *if $a \in A$ and $\pi(a) = \pi(\varphi^k(a))$ for a positive integer k, then $\pi(\varphi^{ik}(a)) = \pi(a)$ for all $i \geq 0$;*
(ii) *if $a \in A$ and $\pi(a) = \pi(\varphi^k(a))$ for a positive integer k, then $\pi(b) = \pi(\varphi^k(b))$ for all b belonging to the orbit $\mathcal{O}_a$ of a under φ;*
(iii) *if $a, b \in A$ and $\pi(a) = \pi(b)$, then $\pi(\varphi^j(a)) = \pi(\varphi^j(b))$ for all $j \geq 0$;*
(iv) *if $a, b \in A$, $\pi(a) = \pi(\varphi^k(a))$, and $\pi(b) = \pi(\varphi^l(b))$, then $\pi(ab) = \pi(\varphi^{kl}(ab))$.*

Proof The best way to visualize the statements in this lemma is to compare its claims to the structure of the generating set X discussed in the paragraph preceding the lemma.

The identity $\pi(a) = \pi(\varphi^k(a))$ yields that a and $\varphi^k(a)$ belong to the same right coset of the kernel $\ker\varphi$ in A. Hence, $a = h_1 g$ and $\varphi^k(a) = h_2 g$ for some $g \in A$ and $h_1, h_2 \in \ker\varphi$.

(i) We proceed by induction on $i \geq 2$. First assume $i = 2$. The calculation $h_2 g = \varphi^k(a) = \varphi^k(h_1 g) = \varphi^k(h_1)\varphi^k(g)$ yields $\varphi^k(g) = (\varphi^k(h_1))^{-1}h_2 g$. Thus, $\varphi^{2k}(a) = \varphi^k(\varphi^k(a)) = \varphi^k(h_2 g) = \varphi^k(h_2)\varphi^k(g) = \varphi^k(h_2)(\varphi^k(h_1))^{-1}h_2 g$. As A is abelian, $\ker\varphi$ is preserved by φ (a well-known result contained for example in [1]), and we obtain $\varphi^k(h_2)(\varphi^k(h_1))^{-1}h_2 \in H$. It follows that $\varphi^{2k}(a)$ and a belong to the same right coset $(\ker\varphi)g$, and therefore $\pi(\varphi^{2k}(a)) = \pi(a)$. The general induction step proceeds along the same kind of calculations.
(ii) As b belongs to the orbit of a, $b = \varphi^j(a)$ for some positive integer j. Hence $b = \varphi^j(h_1 g) = \varphi^j(h_1)\varphi^j(g)$. Furthermore, $\varphi^k(b) = \varphi^{k+j}(a) = \varphi^j(\varphi^k(a)) = \varphi^j(h_2 g) = \varphi^j(h_2)\varphi^j(g)$. Since $\varphi^j(h_1)$ and $\varphi^j(h_2)$ are both in $\ker\varphi$, the two calculations show that b and $\varphi^k(b)$ belong to the same right coset of $\ker\varphi$, and thus, $\pi(b) = \pi(\varphi^k(b))$.
(iii) If $\pi(a) = \pi(b)$ we have $a = h_1'g'$ and $b = h_2'g'$ for some $h_1', h_2' \in \ker\varphi$ and $g' \in A$. Hence

$$\varphi^j(a) = \varphi^j(h_1)\varphi^j(g) \text{ and } \varphi^j(b) = \varphi^j(h_2)\varphi^j(g).$$

Since φ preserves $\ker\varphi$, the statement follows.
(iv) If $\pi(a) = \pi(\varphi^k(a))$, a and $\varphi^k(a)$ belong to the same right coset of $\ker\varphi$, and $\varphi^k(a) = ha$ for some $h \in \ker\varphi$. Then, $\varphi^k(ab) = \varphi^k(a)\varphi^{\pi(k,a)}(b) = ha\varphi^{\pi(k,a)}(b)$. Furthermore,

$$\begin{aligned}\varphi^{2k}(ab) &= \varphi^k(\varphi^k(ab)) = \varphi^k(\varphi^k(a)\varphi^{\pi(k,a)}(b)) = \varphi^k(ha\varphi^{\pi(k,a)}(b)) = \\ &= \varphi^k(h)\varphi^k(a\varphi^{\pi(k,a)}(b)) = \varphi^k(h)\varphi^k(a)\varphi^{\pi(k,a)}(\varphi^{\pi(k,a)}(b)) = \\ &= \varphi^k(h)\varphi^k(a)\varphi^{2\pi(k,a)}(b) = \varphi^k(h)ha\varphi^{2\pi(k,a)}(b).\end{aligned}$$

By induction, $\varphi^{lk}(ab) = h'a\varphi^{l\pi(k,a)}(b)$, for some $h' \in \ker\varphi$. Applying (i) to the equation $\pi(b) = \pi(\varphi^l(b))$ yields $\pi(b) = \pi(\varphi^{l\pi(k,a)}(b))$, and thus, $\varphi^{l\pi(k,a)}(b) = h''b$ for some $h'' \in \ker\varphi$. Hence, $\varphi^{lk}(ab) = h'ah''b = h'h''ab$. Therefore $\pi(ab) = \pi(\varphi^{lk}(ab))$.

□

The case (iii) of the above lemma has an interesting consequence. Namely, for any two elements $a, b \in A$, the set of the power function values assigned to the elements of $\mathcal{O}_a$ and the set of the power function values assigned to the elements of $\mathcal{O}_b$ must be either equal or disjoint. This is once again clearly exhibited in Examples 2 and 4, but is nevertheless a somewhat unexpected result.

Examples 2 and 4 also show that the length of the repeated sequence of the values of π may differ from an orbit to an orbit. To reflect this fact, we introduce the following definition.

Let φ be a skew-morphism of an abelian group A, π be its power function, and $a \in A$. *The periodicity of the power function π at a* is the smallest positive integer p_a with the property $\pi(a) = \pi(\varphi^{p_a}(a))$.

Properties of p_a are summarized in the following lemma. Note that (i) means that all elements belonging to the orbit $\mathcal{O}_a$ have the same periodicity. It is therefore immaterial which element of $\mathcal{O}_a$ we choose and thus we can talk about the periodicity of the orbit $\mathcal{O}_a$.

Lemma 2 *Let φ be a skew-morphism of an abelian group A with the power function π. Then the following hold:*

(i) *$p_a = p_b$ for each a and b that belong to the same orbit of φ;*
(ii) *for each $a \in A$ the number p_a divides both $|\mathcal{O}_a|$ and $|\varphi|$;*
(iii) *if $\pi(a) = \pi(b)$, then $p_a = p_b$,*
(iv) *$\pi(a) = \pi(\varphi^i(a))$ if and only if $p_a \mid i$.*

Proof This lemma is a consequence of Lemma 1.

(i) This result follows from Lemma 1 (ii) applied to both a and b, and the minimality of p_a and p_b.
(ii) Using Lemma 1 (i), $\pi(a) = \pi(\varphi^{ip_a}(a))$, for each positive integer i, while $\pi(a) = \pi(\varphi^{|\mathcal{O}_a|}(a))$. The rest follows from the usual argument using the minimality of p_a.
(iii) Since $\pi(a) = \pi(b)$, Lemma 1 (iii) yields $\pi(\varphi^k(a)) = \pi(\varphi^k(b))$, for each positive integer k. Hence p_a and p_b are necessarily the same.
(iv) This final claim can be proved using the arguments from (ii).

□

5 Coset-Preserving Powers

In the previous section, we have focused on the distribution of the power function values throughout the orbits of a skew-morphism. In this section, we use the information gained to show that each skew-morphism of an abelian group that gives rise to a regular Cayley map or has a generating orbit, and thus specifically each skew-morphism of a cyclic group, admits a non-trivial power that is a coset-preserving skew-morphism.

In view of the power values distribution, coset-preserving skew-morphisms are the second simplest skew-morphisms after the automorphisms—each orbit of a coset-preserving skew-morphism consists entirely of elements of the same power value.

We begin the section with a 'globalization' of the concept of the periodicity of an element to the entire skew-morphism. Let p_φ denote the least common multiple of all $p_a, a \in A$. Then, p_φ is the smallest positive integer, such that $\pi(a) = \pi(\varphi^{p_\varphi}(a))$ for each $a \in A$. In view of Lemma 2 (ii), p_φ necessarily divides the order of φ. In the next few paragraphs, we prove some preliminary results concerning the value p_φ followed by one of the main results of our paper.

First, a simple result from number theory.

Lemma 3 *Let $c_1, c_2, \ldots, c_k$ be positive integers, such that $c_j - c_i \not\equiv j - i \pmod{k}$, for each $1 \le i < j \le k$. Then $k \mid c_1 + c_2 + \cdots + c_k$.*

Proof Suppose that $c_1, c_2, \ldots, c_k$ satisfy the conditions. An easy argument by contradiction shows that the numbers $c_1 - 1, , c_2 - 2, , \ldots, c_k - k$ are pairwise different modulo k. Hence, $\{c_1 - 1, c_2 - 2, \ldots, c_k - k\} = \{0, 1, 2, \ldots, k - 1\} \pmod{k}$ and therefore $c_1 - 1 + c_2 - 2 + \cdots + c_k - k \equiv 1 + 2 + 3 + \cdots + k \pmod{k}$, which yields $c_1 + c_2 + \cdots + c_k \equiv 2 \cdot \sum_{i=0}^{k-1} i \equiv 0 \pmod{k}$. □

The next lemma is a generalization of a result from [1], where it has been shown for any skew-morphism φ having a generating orbit X closed under inverses that

$$\sum_{i=0}^{|\varphi|-1} \pi(\varphi^i(g)) \equiv 0 \pmod{|\mathcal{O}_h|},$$

for all $g, h \in G$.

Lemma 4 *Let φ be a skew-morphism of order n of an abelian group A, and let a be an arbitrary element of A. Then $p_a \mid \pi(p_a, a')$ for all $a' \in \mathcal{O}_a$.*

Proof We have already argued that $\pi(p_a, a') = \pi(p_a, a)$, for all $a' \in \mathcal{O}_a$, and by the minimality of p_a, the values $\pi(a), \pi(\varphi(a)), \ldots, \pi(\varphi^{p_a-1}(a))$ are all different. Assume $0 \le i < j < p_a$, and consider the following two-way calculation. First, $\varphi(\varphi^i(a)\varphi^j(a)) = \varphi^{i+1}(a)\varphi^{\pi(\varphi^i(a))+j}(a)$. Similarly, $\varphi(\varphi^j(a)\varphi^i(a)) = \varphi^{j+1}(a)\varphi^{\pi(\varphi^j(a))+i}(a)$. Since $\varphi(\varphi^i(a)\varphi^j(a)) = \varphi(\varphi^j(a)\varphi^i(a))$ by commutativity, we conclude that

$$\varphi^{i+1}(a)\varphi^{\pi(\varphi^i(a))+j}(a) = \varphi^{j+1}(a)\varphi^{\pi(\varphi^j(a))+i}(a). \tag{6}$$

We claim that the above equality yields that $\pi(\varphi^j(a)) - \pi(\varphi^i(a)) \not\equiv j - i \pmod{p_a}$. To prove our claim via contradiction, assume on the contrary that $\pi(\varphi^j(a)) - \pi(\varphi^i(a)) \equiv j - i \pmod{p_a}$, or that $\pi(\varphi^j(a)) \equiv \pi(\varphi^i(a)) + j - i \pmod{p_a}$. Plugging in into (6) forces

$$\varphi^{i+1}(a)\varphi^{\pi(\varphi^i(a))+j}(a) = \varphi^{j+1}(a)\varphi^{\pi(\varphi^i(a))+j-i+i}(a),$$

consequently, $\varphi^{i+1}(a) = \varphi^{j+1}(a)$, contrary to $0 \le i < j < p_a$. Thus, we can apply Lemma 3, and conclude that $p_a \mid \pi(p_a, a')$, for all $a' \in \mathcal{O}_a$. □

We now have at hand all the necessary ingredients for proving the main theorem of this section.

Theorem 6 *Let φ be a skew-morphism of an abelian group A. Then $\psi = \varphi^{p_\varphi}$ is a coset-preserving skew-morphism of A.*

Furthermore, if a and a' belong to the same orbit of φ, and π_ψ is the power function of ψ, then $\pi_\psi(a) = \pi_\psi(a')$.

Proof Let a be an arbitrary element of A. Due to our definition of p_φ, $p_a \mid p_\varphi$, and thus, $p_\varphi = rp_a$ for some positive integer r. By Lemma 4, we have $p_a \mid \pi(p_\varphi, a)$. Furthermore, $\pi(p_\varphi, a) = r \cdot (\pi(a) + \cdots + \pi(\varphi^{p_a-1}(a))) = r\pi(p_a, a)$. Hence $p_\varphi = rp_a$ divides $r\pi(p_a, a) = \pi(p_\varphi, a)$, for each $a \in A$. Theorem 4 asserts that $\psi = \varphi^{p_\varphi}$ is a skew-morphism of A. Clearly, ψ preserves the cosets of $\ker\varphi$, and since the kernel of ψ contains the kernel of φ, ψ preserves the cosets of $\ker\psi$ as well. If a' belongs to the orbit of a, $\pi(p_\varphi, a') = \pi(p_\varphi, a)$, and hence, $\pi_\psi(a') = \pi_\psi(a)$, as both are equal to the smallest positive solution of the equation $p_\varphi x \equiv \pi(p_\varphi, a) \pmod{|\varphi|}$ (Theorem 4). □

This is a good place to summarize and reiterate our results. First, a skew-morphism φ is coset-preserving if and only if $p_\varphi = 1$. As for the skew-morphisms that are not coset-preserving, we have shown that every skew-morphism φ of an abelian group A that is not coset-preserving must have a very specific structure: Any orbit of φ that is not entirely contained in a single coset of $\ker\varphi$ consists of elements of several different cosets which are visited one by one in a uniquely defined order, and any other orbit of φ that contains at least one element from these cosets has this very same properties. The skew-morphism φ^{p_φ} discussed in the above theorem not only never leaves the cosets of φ, but the second part of Theorem 6 asserts that *all* cosets of $\ker\varphi$ touched by some orbit of φ will become a part of a single coset of $\ker\varphi^{p_\varphi}$; a much stronger result than the one in which we have argued that the kernel of φ^{p_φ} must be a supergroup of the kernel of φ. This observation correlates with our comment following Lemma 1 where we pointed out that the power function values of distinct orbits of a skew-morphism φ are either identical or disjoint (as sets).

Unlike the case of the skew-morphisms that give rise to regular Cayley maps (i.e., have a generating orbit closed under inverses), general skew-morphisms do not have to have the property that φ^{p_φ} is necessarily non-trivial: If φ does not have a precise orbit, it may contain orbits of relatively prime periodicities, which in turn may cause p_φ become too big (equal to $|\varphi|$), and therefore φ^{p_φ} may turn out to be trivial.

Our next goal is to find additional assumptions under which we can prove that the coset-preserving skew-morphism ψ constructed in the previous theorem is a non-trivial skew-morphism. We will rely on the following lemma.

Lemma 5 *Let φ be a skew-morphism of an abelian group A. Then for any $a, b \in A$, $p_{ab} \mid lcm(p_a, p_b)$, and, in particular, $p_{a_1 a_2 \ldots a_l} \mid lcm(p_{a_1}, p_{a_2}, \ldots, p_{a_l})$ for any $a_1, \ldots, a_l \in A$.*

Proof Denote $k = lcm(p_a, p_b)$ and note that $k = ip_a$ and $k = jp_b$ for some positive integers i and j. For any positive integer l there exist $h, h' \in \ker\varphi$ such that $\varphi^{lp_a}(a) = ha$ and $\varphi^{lp_b}(b) = h'b$. Furthermore, $\pi(k, a) = \pi(ip_a, a) = i\pi(p_a, a)$, and $p_a \mid \pi(p_a, a)$ (Theorem 4), therefore $k = ip_a \mid \pi(k, a)$, $jp_b \mid \pi(k, a)$, and $\pi(k, a) = j' p_b$ for some positive j'. It follows that

$$\varphi^k(ab) = \varphi^k(a)\varphi^{\pi(k,a)}(b) = \varphi^{ip_a}(a)\varphi^{j'p_b}(b) = h_1 a h_2 b = h_1 h_2 ab,$$

for some $h_1, h_2 \in \ker\varphi$. Thus, $\pi(\varphi^k(ab)) = \pi(ab)$, and by Lemma 2, $p_{ab} \mid k = lcm(p_a, p_b)$. The general claim of the lemma follows by induction. □

Consider now the case when a skew-morphism φ of an abelian group A has a generating orbit $\mathcal{O}$ (not necessarily closed under inverses). Since $p_a = p_b$, for all $a, b \in \mathcal{O}$, and every element of A is a product of elements from $\mathcal{O}$, Lemma 5 yields that $p_c \mid p_a$ for all $c \in A$, and therefore $p_\varphi = p_a$, where a is any element of $\mathcal{O}$. It has been shown in [7], that the order of a skew-morphism φ with a generating orbit $\mathcal{O}$ is necessarily equal to the size of $\mathcal{O}$. Thus, in this case, $p_\varphi \mid |\mathcal{O}| = |\varphi|$. As discussed above, Theorem 6 is only meaningful if $p_\varphi < |\varphi|$, in which case $\psi = \varphi^{p_\varphi}$ is a non-trivial coset-preserving skew-morphism of A. In the next theorem, we show that this is always the case when φ possesses a generating orbit.

Theorem 7 *Let A be an abelian group, let φ be a non-identity skew-morphism of A with a generating orbit $\mathcal{O}$, and let $a \in \mathcal{O}$. Then φ^{p_a} is a non-identity coset-preserving skew-morphism of A.*

Proof As argued prior to the statement of the theorem, the assumptions yield $p_a = p_\varphi$, and thus φ^{p_a} is a coset-preserving skew-morphism of A by Theorem 6.

To prove that φ^{p_a} is not the identity mapping, we need to show that $p_a < |\varphi|$. Once again, by the results of [7], we observe that $|\mathcal{O}| = |\varphi|$. The rest of the proof is a simple consequence of the Pigeonhole Principle. Since φ preserves $\ker\varphi$, which is a proper subgroup of A, $\mathcal{O} \cap \ker\varphi = \emptyset$, and therefore the power function values of the elements in $\mathcal{O}$ must all belong to the set $\{2, 3, \ldots, |\varphi| - 1 = |\mathcal{O}| - 1\}$. It follows that at least two elements of $\mathcal{O}$ are assigned the same power function value, and hence $\mathcal{O}$ visits each coset of $\ker\varphi$ at least twice, each time in the same order, and hence $p_a \le \frac{|\mathcal{O}|}{2} < |\mathcal{O}| = |\varphi|$. □

While not every skew-morphism of an abelian group has necessarily a generating orbit, in the case of a single element generated group (i.e., the cyclic group $\mathbb{Z}_n$), every skew-morphism must have at least one generating orbit; for example, the orbit containing the generator 1. Thus, based on Theorem 7, every non-trivial skew-morphism of a cyclic group must have a non-trivial coset-preserving power. This easy consequence is nevertheless fundamental for our further considerations. For that reason, we state it in the form of a corollary.

Corollary 2 *Let φ be a non-identity skew-morphism of a cyclic group $\mathbb{Z}_n$. Then for every generator $a \in \mathbb{Z}_n$, φ^{p_a} is a non-trivial coset-preserving skew-morphism of $\mathbb{Z}_n$.*

Thus, *every* skew-morphism of a cyclic group is either coset-preserving or is a root of a non-trivial coset-preserving skew-morphism. In the forthcoming paragraphs, we show that the same must be true for all skew-morphisms of abelian groups that give rise to regular Cayley maps (i.e., skew-morphisms with a generating orbit closed under inverses that preserve their kernels).

First note that the coset-preserving skew-morphisms of abelian groups giving rise to regular Cayley maps have an even more restricted structure.

Lemma 6 *If φ is a coset-preserving skew-morphism of an abelian group A that gives rise to a regular Cayley map $CM(A, X, P)$, then φ is either t-balanced or a group automorphism of A.*

Proof If a coset-preserving skew-morphism φ possesses a generating orbit X that is closed under inverses, all elements of X belong to the same coset of $\ker \varphi$, and have therefore the same power function value t. On the other hand, for any $x \in X$, x^{-1} is also in X, and therefore $1 = \pi(xx^{-1}) = t^2$ by formula (1), and consequently $\pi(xy) = 1$, for all $x, y \in X$. Thus, $xy \in \ker \varphi$, for all $x, y \in X$. Finally, since X generates all of A, every element of A is a product of elements in X. Since all even length products of elements in X belong to $\ker \varphi$, and all odd length products belong to the coset containing X, the index of $\ker \varphi$ in A is at most 2. The lemma follows. □

Very similar arguments lead to the following strong result concerning skew-morphisms of abelian groups giving rise to regular Cayley maps; the original topic of interest within the context of Cayley maps.

Theorem 8 *Let φ be a skew-morphism of an abelian group A that gives rise to a regular Cayley map $CM(A, X, P)$. Then, φ^{p_x} is either a non-trivial t-balanced skew-morphism or a non-trivial group automorphism of A, for all $x \in X$.*

In the case when A is of odd order, φ^{p_x} is necessarily a group automorphism of A.

Proof If φ gives rise to a regular Cayley map, it possesses a generating orbit X that is closed under inverses. Then, $p_\varphi = p_x$, for all $x \in X$, and Theorem 6 yields that $\psi = \varphi^{p_\varphi}$ is a non-trivial coset-preserving skew-morphism of A. The second part of this very same theorem yields further that all elements of X, which may not constitute a single orbit of ψ anymore, belong nevertheless to the same coset of $\ker \psi$. Thus, a coset of ψ contains a subset that generates A and is closed under inverses, and, following the same line of argument as in the proof of the previous lemma, $\ker \psi$ has at most two distinct cosets in A. The non-trivial skew-morphism ψ is therefore either a t-balanced skew-morphism or a group automorphism of A. If A is of odd order, the index of $\ker \psi$ in A cannot be 2, and ψ has to be an automorphism. □

Both of the previous corollaries underscore the significance of the coset-preserving skew-morphisms of abelian groups. For example, all t-balanced skew-morphisms of cyclic groups giving rise to regular Cayley maps have already been classified in [12]. Once we complete the classification of *all* t-balanced skew-morphisms of cyclic groups, the complete list of skew-morphisms of cyclic groups giving rise to regular Cayley maps will be a subset of the set of the roots of the t-balanced skew-morphisms. In the last section of our paper, we investigate the class of coset-preserving skew-morphisms of cyclic groups in detail. As t-balanced skew-morphisms form a subset of the set of coset-preserving skew-morphisms, our results appear to suggest that classification of skew-morphisms of cyclic groups giving rise to regular Cayley maps might be within our reach. Classification of the coset-preserving skew-morphisms of general abelian groups appears however much less likely.

We close this section with a further discussion of properties of coset-preserving skew-morphisms.

Lemma 7 *Let φ be a coset-preserving skew-morphism of an abelian group A, and let π be its power function. Then π is a homomorphism from A into the multiplicative group $\mathbb{Z}^{\#}_{|\varphi|}$.*

Proof This is an easy-to-prove result that further underlines that coset-preserving skew-morphisms have a very restrictive structure. Applying formula (1) yields

$$\pi(ab) \equiv \sum_{0 \le i < \pi(a)} \pi(\varphi^i(b)) \equiv \pi(a)\pi(b) \pmod{|\varphi|},$$

since $\pi(\varphi^i(b)) = \pi(b)$, for all i. □

Lemma 8 *Let φ be a coset-preserving skew-morphism of an abelian group A, let π be its power function, and let $a \in A$. If $\pi(a) = \pi(a^{-1})$, then $\pi(a)$ is a square root of 1 in $\mathbb{Z}^{\#}_{|\varphi|}$.*

Furthermore, if A is the cyclic group $\mathbb{Z}_n$ and a is a generator of A, then $\pi(a) = \pi(a^{-1})$ implies that φ is a t-balanced skew-morphism with $t = \pi(a)$.

Proof Employing Lemma 7 gives

$$1 = \pi(aa^{-1}) \equiv \pi(a)\pi(a^{-1}) \equiv \pi(a)^2 \pmod{|\varphi|}.$$

The second claim follows from the structure of $\mathbb{Z}_n$: If $\pi(a) = \pi(-a) = 1$, φ is necessarily an automorphism (and therefore t-balanced with $t = 1$), and if $\pi(a) = \pi(-a) \neq 1$, a and $-a$ belong to the same right coset of $\ker \varphi$ in $\mathbb{Z}_n$, hence $2a \in \ker \varphi$, and therefore the index of $\ker \varphi$ in $\mathbb{Z}_n$ is 2. Since skew-morphisms of abelian groups always preserve their kernels, the orbit of a has no points in common with $\ker \varphi$, and consists of elements whose power is equal to $\pi(a)$. □

6 Coset-Preserving Skew-Morphisms of Cyclic Groups

As stated in the introduction, the two main articles concerning classification of skew-morphisms of cyclic groups are the fundamental paper [3] that contains a very indirect classification of the skew-morphisms of the cyclic groups that give rise to regular Cayley maps and the paper [12] that classifies regular t-balanced Cayley maps of cyclic groups. Due to their interest in regular maps, neither paper deals with general skew-morphisms, i.e., skew-morphisms that do not possess generating orbits closed under inverses. In the present section, we focus on the class of coset-preserving skew-morphisms of cyclic groups regardless of whether they do or do not possess a generating orbit closed under inverses. As we have shown in the previous section, every skew-morphism of a cyclic group is a root of some non-trivial coset-preserving skew-morphism, and thus, we see our efforts as a significant step toward the classification of all skew-morphisms of cyclic groups.

The key to the classification of coset-preserving skew-morphisms of cyclic groups lies in finding necessary and sufficient sets of parameters that uniquely determine such skew-morphisms. Intuitively, every coset-preserving skew-morphism φ of $\mathbb{Z}_n$ is uniquely determined by its kernel $\ker\varphi$, the action of φ on $\ker\varphi$, the image $\varphi(1)$, and the value of the power function $\pi(1)$.

Theorem 9 *Let φ be a non-identity coset-preserving skew-morphism of $\mathbb{Z}_n$. Let b be the smallest non-zero element of $\ker\varphi$, let t be the smallest solution of the equation $\varphi(b) = t \cdot b$, let $h = \varphi(1) - 1 \pmod{n}$, and let $p = \pi(1)$. The five-tuple of parameters $(n; b, t, h, p)$ satisfies the following conditions:*

(i) *the parameters $(n; b, t, h, p)$ uniquely determine φ; if φ and φ' are coset-preserving skew-morphisms of $\mathbb{Z}_n$, then $\varphi = \varphi'$ if and only if their corresponding five-tuples are equal;*
(ii) *$b \mid n$ and $\ker\varphi = \langle b\rangle$;*
(iii) *$1 \le t \le \frac{n}{b}$, $(t, \frac{n}{b}) = 1$, and the restriction of φ to $\ker\varphi$ is the right multiplication by t, i.e., $\varphi(g) = tg$, for each $g \in \ker\varphi$;*
(iv) *$h \in \ker\varphi$ and $h \neq 0$;*
(v) *the order of p in $\mathbb{Z}^{\#}_{|\varphi|}$ is b;*
(vi) *$tb = \varphi(1) + \varphi^{p}(1) + \varphi^{p^2}(1) + \cdots + \varphi^{p^{b-1}}(1)$;*
(vii) *φ^{p-1} restricted to $\ker\varphi$ is an identity; $\varphi^{p-1}(g) = g$, for each $g \in \ker\varphi$.*

Proof By choosing the smallest parameter in cases where there might exist more than one solution, we made sure that the parameters $(n; b, t, h, p)$ are uniquely determined by the choice of the skew-morphism φ. We now prove the seven properties listed in the theorem, but leave the proof of the most important property (i) for the last.

(ii) The kernel $\ker\varphi$ is a subgroup of $\mathbb{Z}_n$ and every subgroup of $\mathbb{Z}_n$ is generated by its smallest non-zero element which in turn must divide n.
(iii) It is well-known (and easy to see) that, if φ preserves $\ker\varphi$, the restriction of φ to $\ker\varphi$ is a group automorphism of $\ker\varphi$. Thus $\varphi(b) = tb$, for some $1 \le t \le \frac{n}{b}$ relatively prime to the order $\frac{n}{b}$ of $\ker\varphi$.

(iv) As 1 and $\varphi(1)$ belong to the same right coset of $\ker\varphi$, we have $\varphi(1)-1 \in \ker\varphi$. If $\varphi(1)$ were equal to 1, φ would be an identity automorphism.

(v) Lemma 7 asserts that π is a homomorphism from $\mathbb{Z}_n$ into $\mathbb{Z}^{\#}_{|\varphi|}$, and therefore $\pi(\mathbb{Z}_n)$ is a cyclic multiplicative subgroup of $\mathbb{Z}^{\#}_{|\varphi|}$. The order of $\pi(\mathbb{Z}_n)$ is the number of different values of π on $\mathbb{Z}_n$, i.e., the index b of $\ker\varphi$ in $\mathbb{Z}_n$. Since $\pi(1)$ is a generator of $\pi(\mathbb{Z}_n)$, the order of $\pi(1)$ is also b.

(vi) Formulas (1) and (2) imply, for any non-zero element $a \in \mathbb{Z}_n$,

$$\varphi(a) = \varphi(1) + \varphi^{p}(1) + \varphi^{p^2}(1) + \cdots + \varphi^{p^{a-1}}(1). \tag{7}$$

On the other hand, part (iii) of our theorem asserts that $\varphi(b) = tb$.

(vii) Let $g \in \ker\varphi$, and consider the following calculation:

$$\varphi(g) + \varphi(1) = \varphi(g+1) = \varphi(1+g) = \varphi(1) + \varphi^{p}(g).$$

It follows that $\varphi^{p}(g) = \varphi(g)$, or equivalently, $\varphi^{p-1}(g) = g$.

(i) The orbit of the element 1 under φ is uniquely determined by the parameters h and t by means of the following equation:

$$\varphi^{m}(1) = 1 + h + th + t^2h + \cdots + t^{m-1}h, \tag{8}$$

that holds for all $m \geq 1$. We prove the equation by induction on m. Since $\varphi(1) = 1+h$, the basis of our induction holds true. Now suppose that $\varphi^{m}(1) = 1 + h + th + t^2h + \cdots + t^{m-1}h$, for some $m \geq 1$. Then

$$\begin{aligned} &\varphi^{m+1}(1) = \\ &= \varphi(1 + (h + th + t^2h + \cdots + t^{m-1}h)) = \varphi((h + th + t^2h + \cdots + t^{m-1}h) + 1) = \\ &= \varphi(h + th + t^2h + \cdots + t^{m-1}h) + \varphi(1) = th + t^2h + \cdots + t^{m}h + 1 + h = \\ &= 1 + h + th + t^2h + \cdots + t^{m}h, \end{aligned}$$

which proves the induction claim. Plugging the parameters $(n; b, t, h, p)$ into the equations (8) and (7) allows for a deterministic calculation of $\varphi(a)$, for all $a \in \mathbb{Z}_n$. Thus, φ is uniquely determined by the parameters $(n; b, t, h, p)$. Finally, if the five-tuples associated with two skew-morphisms φ, φ' differ in at least one number, it is easy to see that the two skew-morphisms differ as permutations.

□

We refer to the five-tuple $(n; b, t, h, p)$ as the *parameters of the coset-preserving skew-morphism* φ. While the conditions from Theorem 9 are necessarily satisfied by every five-tuple of parameters of a coset-preserving skew-morphism, unless we prove that the above conditions are also sufficient, the set of all five-tuples satisfying the conditions (ii) through (vii) may contain five-tuples that do not belong to any skew-morphisms. It is not hard to see, however, that a five-tuple $(n; b, t, h, p)$ corresponds to a skew-morphism if and only if the mapping $\varphi : \mathbb{Z}_n \to \mathbb{Z}_n$ built in accordance with formulas (8) and (7):

$$\begin{aligned} &\varphi(0) = 0; \\ &\varphi^m(1) = 1 + h + th + t^2h + \cdots + t^{m-1}h, \qquad \text{for all } 1 \leq m < r; \\ &\varphi(a) = \varphi(1) + \varphi^p(1) + \varphi^{p^2}(1) + \cdots + \varphi^{p^{a-1}}(1), \text{ for all } a \in \mathbb{Z}_n, a \neq 0, 1; \end{aligned}$$

is a well-defined coset-preserving skew-morphism of $\mathbb{Z}_n$. This means that one way to construct all coset-preserving skew-morphisms of $\mathbb{Z}_n$ is to create the (finite) list of all five-tuples $(n; b, t, h, p)$ satisfying conditions (ii) through (vii) together with their corresponding mappings defined as above, and then choosing those mappings from the list that are skew-morphisms of $\mathbb{Z}_n$.

Acknowledgments The second author acknowledges support from the projects VEGA 1/0577/14, VEGA 1/0474/15, NSFC 11371307, and Project: Mobility—ITMS code: 26110230082. We also wish to express our thanks to M. Conder who generously allowed us to use his lists of skew-morphisms of cyclic groups.

References

1. M. Conder, R. Jajcay, and T. Tucker, Regular t-balanced Cayley maps, J. Combin. Theory Ser. B **97**, 453–473 (2007).
2. M. Conder, R. Jajcay, and T. Tucker, *Cyclic complements and skew morphisms of groups*, J. Algebra **453**, 68–100 (2016).
3. M. Conder and T. Tucker, Regular Cayley maps for cyclic groups, Trans. Amer. Math. Soc. **366** no. 7, 3585–3609 (2014).
4. R. Feng, R. Jajcay and Y. Wang, Regular t-balanced Cayley maps for abelian groups, Discrete Math. **311**, 2309–2316 (2011).
5. M.V. Horoševskiĭ, Automorphisms of finite groups, Math. USSR Sbornik **22**, 584–594 (1974).
6. R. Jajcay, On a new product of groups, European J. Combin. **15**, 251–252 (1994).
7. R. Jajcay and R. Nedela, *Half-regular Cayley maps*, Graphs and Combinatorics **31**, 1003–1018 (2015).
8. R. Jajcay and J. Širáň, Skew morphisms of regular Cayley maps, Discrete Math. **244**, 167–179 (2002).
9. I. Kovács, D. Marušič and M. Muzychuk On G-arc-regular dihedrants and regular dihedral maps, J Algebr. Comb. **38**, 437–455 (2013).
10. I. Kovács and R. Nedela, Decomposition of skew-morphisms of cyclic groups, Ars Math. Contemp. **4**, 329–349 (2011).
11. J.H. Kwak and J.-M. Oh, A classification of regular t-balanced Cayley maps on dicyclic groups, European J. Combin. **29**, no. 5, 1151–1159 (2008).
12. Y.-S. Kwon, A classification of regular t-balanced Cayley maps for cyclic groups, Discrete Math. **313**, no. 5, 656–664 (2013).
13. J.H. Kwak, Y.-S. Kwon and R. Feng, A classification of regular t-balanced Cayley maps on dihedral groups, European J. Combin. **27** (3), 382–392 (2006).
14. Ľ. Lišková, M. Mačaj and M. Škoviera, Regular maps from Cayley graphs, Discrete Math. **307**, no. 3-5, 517–533 (2007).
15. M. Muzychuk, On balanced Cayley maps over abelian groups, The International Conference on Topological and Geometric Graph Theory, 115–118, Electron. Notes Discrete Math. **31**, Elsevier Sci. B. V., Amsterdam, 2008.

16. J.-M. Oh, Regular t-balanced Cayley maps on semi-dihedral groups, J. Combin. Theory Ser. B **99**, no. 2, 480–493 (2009).
17. R.B. Richter, R. Jajcay, J. Širáň, T.W. Tucker, and M.E. Watkins, Cayley maps, J. Combin. Theory Ser. B **95**, 189–245 (2005).
18. M. Škoviera and J. Širáň, Regular Maps from Cayley Graphs, Part I. Balanced Cayley Maps, Discrete Math. **109**, 265–276 (1992).
19. J. Širáň and M. Škoviera, Regular Maps from Cayley Graphs II. Antibalanced Cayley Maps, Discrete Math. **124**, 179–191 (1994).
20. Y. Wang, R.-Q. Feng, Regular balanced Cayley maps for cyclic, dihedral and generalized quaternion groups, Acta Math. Sin. (Engl. Ser.) **21**, no. 4, 773–778 (2005).
21. J.-Y. Zhang, A classification of regular Cayley maps with trivial Cayley-core for dihedral groups, to appear in *Discrete Math.*

Census of Quadrangle Groups Inclusions

António Breda d'Azevedo, Domenico A. Catalano, Ján Karabáš and Roman Nedela

Abstract In a classical result of 1972 Singerman classifies the inclusions between triangle groups. We extend the classification to a broader family of triangle and quadrangle groups forming a particular subfamily of Fuchsian groups. With two exceptions, each inclusion determines a finite bipartite map (hypermap) on a 2-dimensional spherical orbifold that encodes the complete information and gives a graphical visualisation of the inclusion. A complete description of all the inclusions is contained in the attached tables.

1 Introduction

The search for inclusions between triangle groups, and more generally between Fuchsian groups, was motivated by the theory of Riemann surfaces and algebraic geometry. Triangle and quadrangle groups are particular instances of Fuchsian groups, which are finitely generated discrete subgroups of $PSL(2, \mathbb{R})$, the group of conformal automorphisms of the upper-half plane. Inclusions of Fuchsian groups played an important rôle in the investigation of Teichmüller spaces, see for instance Greenberg [10, Theorem 1]. Later Singerman extended some of Greenberg's results and obtained a complete list of normal inclusions between Fuchsian groups having the

A.B. d'Azevedo · D.A. Catalano
Departamento de Matemática, Universidade de Aveiro, 3810-193 Aveiro, Portugal
e-mail: breda@ua.pt

D. A. Catalano
e-mail: domenico@ua.pt

J. Karabáš (✉)
Department of Computer Science, Faculty of Natural Sciences, Matej Bel University, 97401 Banská Bystrica, Slovakia
e-mail: karabas@savbb.sk

R. Nedela
University of West Bohemia, Pilsen, Czech Republic
e-mail: nedela@ntis.zcu.cz

J. Širáň and R. Jajcay (eds.), *Symmetries in Graphs, Maps, and Polytopes*, Springer Proceedings in Mathematics & Statistics 159,
DOI 10.1007/978-3-319-30451-9_2

same Teichmüller space dimension. In addition, he gives all non-normal inclusions between triangle groups [14, 15]. Another motivation for looking at inclusions of Fuchsian groups in triangle groups comes from the connection between algebraic curves over complex numbers, Riemann surfaces and dessins d'enfant, established explicitly by a result of Belyĭ [1], see also [17]. It follows that every hypermap endows its underlying closed orientable surface with a complex structure by lifting the complex structure of the Riemann sphere via a Belyĭ function, a meromorphic function ramified above at most three points (located at 0, 1 and ∞). A natural question arises: *Under which conditions do two hypermaps determine the same Riemann surface?* In certain circumstances, inclusions of Fuchsian groups in triangle groups with spherical quotients correspond to Riemann-surface-preserving transformations of hypermaps, see [5, 16].

The main aim of this paper is to present a complete list of finite index inclusions $P < Q$, with both P and Q being either a triangle or a quadrangle group (with finite periods). In what follows, we give an outline of the proof followed by instructions how to read the attached census.

2 Generalised Quadrangle Groups and Constellations

Quadrangle groups. By a *generalised quadrangle group* we mean a Fuchsian group Q with presentation

$$Q(k,l,m,n) = \langle x, y, z, w \mid x^k = y^l = z^m = w^n = xyzw = 1 \rangle,$$

where k, l, m, n are positive integers satisfying $\frac{1}{k} + \frac{1}{l} + \frac{1}{m} + \frac{1}{n} < 2$. Clearly, at most one of k, l, m, n can be equal to one. Therefore a generalised quadrangle group is either a triangle or a quadrangle group. In what follows, we assume that the parameters k, l, m, and n are ordered in a non-decreasing order. This is motivated by the fact that a permutation of the parameters (or of the generators) in the above presentation gives an isomorphic copy of $Q(k, l, m, n)$. In particular, the group $Q(1, l, m, n)$ is just the triangle group $\Delta(l, m, n)$. Inclusions between triangle groups were classified by Singerman in [15] and they are listed in Appendix (see Tables 1 and 2).

Constellations. Let $P = Q(p, q, r, s)$ and $Q = Q(k, l, m, n)$ be two generalised quadrangle groups and let P be a subgroup of index N in Q. We write $P <_N Q$. The meaning of the parameters N, p, q, r, s, k, l, m, and n will be fixed throughout the whole paper. There is an induced action of Q on the (right) cosets of P represented by four permutations $\pi_x, \pi_y, \pi_z, \pi_w$ corresponding to the images of the four generators of Q in the natural homomorphism into the symmetric group $\mathrm{Sym}(N)$. In accordance with Lando and Zvonkin [13, Chap. 1], we call the four-tuple $\mathscr{C} = [\pi_x, \pi_y, \pi_z, \pi_w]$ a *constellation* (or a *4-constellation*) of degree N and the sequence $[\lambda_x, \lambda_y, \lambda_z, \lambda_w]$ of partitions of N, where each λ_a is the cycle structure of the permutation π_a, the *passport* of the constellation $\mathscr{C}$. The *monodromy group* $\mathrm{Mon}(\mathscr{C})$ of the constellation $\mathscr{C}$ is the group $\langle \pi_x, \pi_y, \pi_z, \pi_w \rangle \leq \mathrm{Sym}(N)$. By definition, the action of $\mathrm{Mon}(\mathscr{C})$ is transitive on the set $\{1, 2, \ldots, N\}$ and $\pi_x \pi_y \pi_z \pi_w = 1$. We write the cycle structure of a permutation in the exponential notation: for instance, the permutation $\varrho = (1,2,3)(4,5)(6,7,8)(9)$ has the cycle structure $[1.2.3^2]$. For convenience, each cycle structure in a passport in the census is ordered in a non-decreasing order.

Two constellations $\mathscr{C} = [\pi_x, \pi_y, \pi_z, \pi_w]$ and $\mathscr{C}' = [\pi'_x, \pi'_y, \pi'_z, \pi'_w]$ of degree N are *equivalent* if there exists $\alpha \in \mathrm{Sym}(N)$ such that the corresponding permutations are simultaneously conjugated by α. In particular, if $P < Q$ and $P' < Q$ are two inclusions of generalised quadrangle groups, then the associated constellations are equivalent if and only if the subgroups P and P' are conjugate in Q. In fact, an inclusion $P < Q$ determines a constellation $\mathscr{C} = [\pi_x, \pi_y, \pi_z, \pi_w]$ which corresponds to a Q-marked hypermap $\mathscr{P} = (Q/P;\ xP^*, yP^*, zP^*)$, where P^* is the core of P in Q and π_x, π_y, π_z are the actions of xP^*, yP^*, zP^* on the N cosets of P in Q respectively (see [3] for definitions). Replacing Θ by Q in [3, Theorem 19], we have that two Q-marked hypermaps $\mathscr{P}$ and $\mathscr{K} = (Q/K;\ xK^*, yK^*, zK^*)$, corresponding to the inclusions $P < Q$ and $K < Q$ respectively, are isomorphic if and only if P and K are conjugate in Q, say $K = P^g$, for some $g \in Q$; the isomorphism is the conjugation morphism $\iota^g : Q/P \to Q/K$, $Pq \mapsto Q^g q^g$. If $\psi_1 : Q/P \to \{1, 2, \ldots, N\}$ and $\psi_2 : Q/K \to \{1, 2, \ldots, N\}$ are the bijections to their transversals, then $\alpha = \psi_1^{-1}\iota^g\psi_2$ is the permutation that makes the constellations corresponding to $\mathscr{P}$ and $\mathscr{K}$ equivalent.

Our approach will follow the one outlined in Singerman's classification of inclusions between triangle groups [15]. The census is obtained in two steps. First, we find all admissible sets of parameters $N; p, q, r, s; k, l, m, n$ satisfying the Riemann-Hurwitz equation with additional numerical constrains. Each such numerical solution gives rise to a passport. Secondly, for each passport we either find all equivalency classes of constellations with that passport, or we show that such a constellation does not exist. Both steps are computer-assisted. As a byproduct, we confirm Singerman's classification of triangle group inclusions. More details follow.

Numerical solution. If we have an inclusion between generalised quadrangle groups $P <_N Q$ with parameters $N; p, q, r, s; k, l, m, n$, then the Riemann-Hurwitz formula holds true

$$N = \frac{2 - \left(\frac{1}{p} + \frac{1}{q} + \frac{1}{r} + \frac{1}{s}\right)}{2 - \left(\frac{1}{k} + \frac{1}{l} + \frac{1}{m} + \frac{1}{n}\right)}. \tag{1}$$

Our aim is to determine all possible solutions $N;\ p, q, r, s;\ k, l, m, n$ with their associated passports. To do this, the following two well known facts are useful:

- elements of finite order in $PSL(2, \mathbb{R})$ are elliptic,
- any element g of finite order in a Fuchsian group is conjugate to a power of an elliptic generator h; in symbols $g \vdash h$.

Let a, b, c, d be the generators of P of orders p, q, r, s, respectively. If $P < Q$, then one of the following four cases (up to a group isomorphism) happens:

Case 1. $a, b, c, d \vdash w$, which implies $p|n$, $q|n$, $r|n$ and $s|n$,
Case 2. $a \vdash z$ and $b, c, d \vdash w$, which implies $p|m$ and $q|n$, $r|n$ and $s|n$,

Case 3. $a, b \vdash z$, and $c, d \vdash w$, which implies $p|m$, $q|m$ and $r|n$, $s|n$
Case 4. $a \vdash y$, $b \vdash z$ and $c, d \vdash w$, which implies $p|l$, $q|m$ and $r|n$, $s|n$.

Remark It might appear that the case $a \vdash x$, $b \vdash y$, $c \vdash z$ and $d \vdash w$ is missing, however (1) implies $N = 1$, that is, $P = Q$ in this case.

Using an argument by Singerman [14, Theorem 1], if there is a constellation associated to one of the aforementioned cases, then it has one of the following passports, according to the case it belongs to:

$$\textbf{Case1}.\ \left[k^{\frac{N}{k}}, l^{\frac{N}{l}}, m^{\frac{N}{m}}, \frac{n}{p} \cdot \frac{n}{q} \cdot \frac{n}{r} \cdot \frac{n}{s} \cdot n^{\frac{N-\left(\frac{n}{p}+\frac{n}{q}+\frac{n}{r}+\frac{n}{s}\right)}{n}}\right],$$

$$\textbf{Case2}.\ \left[k^{\frac{N}{k}}, l^{\frac{N}{l}}, \frac{m}{p} \cdot m^{\frac{N-\frac{m}{p}}{m}}, \frac{n}{q} \cdot \frac{n}{r} \cdot \frac{n}{s} \cdot n^{\frac{N-\left(\frac{n}{q}+\frac{n}{r}+\frac{n}{s}\right)}{n}}\right],$$

$$\textbf{Case3}.\ \left[k^{\frac{N}{k}}, l^{\frac{N}{l}}, \frac{m}{p} \cdot \frac{m}{q} \cdot m^{\frac{N-\left(\frac{m}{p}+\frac{m}{q}\right)}{m}}, \frac{n}{r} \cdot \frac{n}{s} \cdot n^{\frac{N-\left(\frac{n}{r}+\frac{n}{s}\right)}{n}}\right],$$

$$\textbf{Case4}.\ \left[k^{\frac{N}{k}}, \frac{l}{p} \cdot l^{\frac{N-\frac{l}{p}}{l}}, \frac{m}{q} \cdot m^{\frac{N-\frac{m}{q}}{m}}, \frac{n}{r} \cdot \frac{n}{s} \cdot n^{\frac{N-\left(\frac{n}{r}+\frac{n}{s}\right)}{n}}\right],$$

where all of the fractions appearing in the above passports are integers. We adopt the convention that factors with zero exponent in passports are vacuous (and are not to be taken as equal to 1). For instance, $\frac{n}{r} \cdot \frac{n}{s} \cdot n^0$ should be interpreted as $\frac{n}{r} \cdot \frac{n}{s}$ and not as $\frac{n}{r} \cdot \frac{n}{s} \cdot 1$, which has a different meaning in a passport.

Each passport which belongs to one of the above four cases with parameters satisfying (1) will be called *admissible*. Admissible passports may or may not be passports of constellations.

By definition, the length of each cycle of π_x (resp. π_y, π_z and π_w) in a constellation is a divisor of k (l, m, and n, respectively). A cycle of a permutation π_x, π_y, π_z or π_w will be called *singular* if its length is strictly less then the order k, l , m or n of the corresponding generator x, y, z or w. If $p \neq 1$, that is, if P is not a triangle group, then each admissible passport has exactly four cycle lengths that are proper divisors of k, l, m or n. If P is a triangle group ($p = 1$), then by [15, Proposition 5 and Theorem 1], Q is a triangle group as well, or equivalently $k = 1$. In this case there are exactly three singular cycles in the constellation associated with the inclusion $P < Q$. The four types of passports are distinguished by their respective distributions of the singular cycles between the four permutations.

As we already mentioned, not every admissible passport can be realised by an inclusion $P <_N Q$. To determine the inclusions that realise admissible passports, we used computer algebra systems MAGMA [2] and GAP [9]. With the exception of two families described in Table 3, each inclusion $P <_N Q$ has the parameter k equal to 1, which means that Q is a triangle group. In this case Q can be viewed as a group of orientation preserving automorphisms of an infinite regular *hypermap* (or *bipartite map*) **U** on the hyperbolic plane with hypervertices (black vertices) of valency l, hyperedges (white vertices) of valency m and hyperfaces of valency n (faces of valency $2n$). Then P is a group of automorphisms of **U** and the quotient

$\mathbf{U}/P$ is a hypermap on the Riemann sphere (P is a Fuchsian group of signature $(0; \{p, q, r, s\})$), more precisely, the quotient hypermap lies on a spherical orbifold with exactly 4 or 3 singular points corresponding to the singular cycles of the constellation associated with the inclusion $P < Q$.

Dessins d'enfant. When Q is a triangle group, $k = 1$ and therefore $\pi_x = 1$. Then the inclusion $P <_N Q$ gives rise to a 4-constellation $\mathscr{C}$ that can be reduced to the 3-constellation $[\pi_y, \pi_z, \pi_w]$. This can be regarded as a spherical hypermap $\mathscr{H}$ on the set $\{1, 2, \ldots, N\}$, whose hypervertices and hyperedges are the orbits of π_y and π_z, respectively. The hyperfaces of $\mathscr{H}$ are the orbits of π_w.

There exists a regular branched covering from the universal infinite hypermap $\mathbf{U} = \mathbf{U}(l, m, n)$ on the hyperbolic plane onto the spherical hypermap $\mathbf{U}/P$ either with exactly four branched points with indexes p, q, r, s, or with exactly three branched points with indexes q, r, s, located at some hypervertices, hyperedges or hyperfaces. This hypermap (on the spherical orbifold) with the additional information about the branched points and their respective indices will be called a *dessin d'enfant* or simply a *dessin*. It is more precise to talk about a hypermap on an orbifold with signature $(0; \{p, q, r, s\})$, or with signature $(0; \{q, r, s\})$ when $p = 1$, rather than simply talking of a "hypermap". There is a one-to-one correspondence between the singular cycles of π_y, π_z, and π_w and the branched points of the associated dessin.

Recall that the monodromy group Mon($\mathscr{C}$) of the constellation associated to an inclusion $P <_N Q$ acts transitively on $\{1, 2, \ldots, N\}$. Moreover, Mon($\mathscr{C}$) acts regularly on $\{1, 2, \ldots, N\}$ if and only if $P \lhd_N Q$ is a normal inclusion. In this case the associated dessin is also called *regular*.

Families of inclusions. The inclusions may form *infinite families* parametrised by one, two or three integer parameters in the signatures (corresponding to the number of zero exponents in the factors of the respective passport). All the inclusions of an infinite family share the same (non-parametrised) passport. Note that each member of an infinite family is represented by the same hypermap in the census. An inclusion not belonging to an infinite family is called *sporadic*.

Although there are infinitely many inclusions, the number of admissible passports is finite. Each admissible passport gives rise to a finite number of constellations (or dessins), since the index N of any inclusion is finite and bounded by 84 (the Riemann-Hurwitz bound). Thus, the identification of all dessins associated to quadrangle group inclusions is a finite problem.

More details (and proofs) on the classification of inclusions of generalised quadrangle groups will be discussed in the forthcoming paper [6].

3 How to Read the Census

The attached tables contain the complete list of inclusions $P <_N Q$ between generalised quadrangle groups. Two inclusions $P <_N Q$ and $P' <_N Q$ are distinguished up to conjugation in Q; if $P' = P^g$ for some $g \in Q$, then $P <_N Q$ and $P' <_N Q$ give equivalent constellations and so the two inclusions are essentially the same.

Each conjugacy class of inclusions $P <_N Q$ of generalised quadrangle groups forms one entry in the census. Excluding Table 3 (explained below), the corresponding row displays the following data:

- the associated passport $\lambda = [\lambda_y, \lambda_z, \lambda_w]$ (up to a permutation of its entries),
- the number of realisations, which is equal to the number of non-conjugate subgroups of Q isomorphic to P,
- the non-isomorphic dessins with passport λ up to mirror images,
- the monodromy group, or a structure description of the monodromy group, or the prime factor decomposition of its size [8, 9].

If two dessins form a chiral pair, only one member of the pair is depicted. Thus the number of the depicted dessins may not match the displayed number of realisations in the row.

The three cycle structures in a passport describe the following:

1. first item in a passport gives the sequence of hypervertex valencies (degrees of black vertices),
2. the second item gives the sequence of hyperedge valencies (degrees of white vertices), and
3. the third item gives the sequence of hyperface valencies.

However, there are 6 possible passports formed by permuting the entries of any given passport and therefore, each dessin $\mathscr{D}$ may be in principle associated with 5 (or 11, if $\mathscr{D}$ is chiral) additional non-isomorphic dessins. There is no essential reason to prefer any particular choice of one of these dessins for the census. The criteria we took into account were "aesthetic"—to indicate some symmetry of a dessin—or "space constraint"—a dual image of a dessin sometimes fits better into the reserved space—or we chose a dessin that was "triangulation resembling".

In Table 3 (when Q is not a triangle group) the corresponding row displays:

- the associated passport $\lambda = [\lambda_x, \lambda_y, \lambda_z, \lambda_w]$ (up to a permutation of its entries),
- the associated constellation (since there is only one),
- an alternative picture based on Δ_2-marked hypermaps [4]; these have blue, green and white vertices whose valencies (number of incidences of pairs of blue and green coloured edges (b, g), in counter-clockwise order) give the first, the second and the third entry of the passport (the last entry corresponds to face-valencies),
- the monodromy group.

The entries of the census are organised into six tables. In each table, the inclusions (entries) are ordered by their indices in a non-decreasing order. The first two tables (Tables 1 and 2) include the case of normal and non-normal inclusions of triangle groups classified by Singerman [15]. Table 3 contains the two families of inclusions between (pure) quadrangle groups, while Table 4 gives a classification of normal inclusions of quadrangle groups in triangle groups. Table 5 lists the infinite families of non-normal inclusions of quadrangle groups in triangle groups. The longest table is Table 6, which contains the classification of the sporadic inclusions of quadrangle groups in triangle groups.

A lot of information on inclusions can be dug from the tables. For instance:

- the indices of the inclusions cover all the integers from 2 to 22, additional integers covered are 24, 26, 27, 28, 29, 30, 36, 37, 44, 45, 52, 60;
- the largest possible index is 60, realised by six inclusions of $Q(7,7,7,7)$ in $\Delta(2,3,7)$;
- the largest number of non-conjugate realisations (16) is achieved by the inclusion $Q(2,3,7,7) < \Delta(2,3,7)$ of index 44;
- the number of realisations of an inclusion varies from 1 to 16. There are inclusions such that each of their realisations is chiral, such that each of their realisation is reflexible, and those that have both chiral and reflexible realisations.

For the sake of completeness, in Table 7 we collect all the solutions of (1) with the respective admissible passports for which there is no inclusion. Additional information on the inclusions of the generalised quadrangle groups can be found at the web page [7].

Acknowledgments This work was supported by the Research and Development Cooperation project FCT (Fundação para a Ciência e a Tecnologia) Portugal—Slovakia.

The work of the first two authors was partially supported by Portuguese funds through the CIDMA—Center for Research and Development in Mathematics and Applications (University of Aveiro), and the Portuguese Foundation for Science and Technology FCT, within project PEst-OE/MAT/UI4106/2014.

The work of the third and the fourth author was supported by the Ministry of Education of the Slovak Republic, the grant VEGA 1/0150/14, by the project APVV SK-PT-0004-12, and by the project "Mobility-Enhancing Research, Science and Education," Matej Bel University (ITMS code 26110230082) under the Operational Programme of Education co-financed by the European Social Foundation.

The fourth author was supported by the project LO1506 of the Czech Ministry of Education, Youth and Sports.

The authors are very grateful to the anonymous referee for his/her careful and meticulous reading of the paper. The review was detailed and helpful to finalise the manuscript.

Appendix

Triangle Groups Inclusions

See Tables 1 and 2.

Table 1 Normal inclusions of triangle groups

Label	Inclusion	Passport	Realisation(s)
(a)	$\Delta(m,n,n) \lhd_2 \Delta(2,2m,n)$	$[2,2,1^2]$	1 realisation
	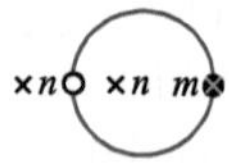	C_2	
(b)	$\Delta(n,n,n) \lhd_3 \Delta(3,3,n)$	$[3,3,1^3]$	1 realisation
	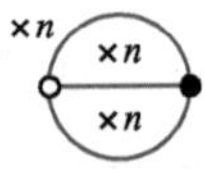	C_3	
(c)	$\Delta(n,n,n) \lhd_6 \Delta(2,3,2n)$	$[2^3,2^3,3^2]$	1 realisation
	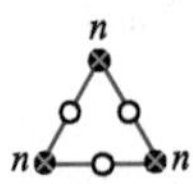	$S_3 \cong PSL(2,2)$	

Table 2 Non-normal inclusions of triangle groups

Label	Inclusion	Passport	Realisation(s)
(A)	$\Delta(7,7,7) <_{24} \Delta(2,3,7)$	$[1^3.7^3, 2^{12}, 3^8]$	1 realisation
		$PSL(3,2)$	
(B)	$\Delta(2,7,7) <_9 \Delta(2,3,7)$	$[1^2.7, 1.2^4, 3^3]$	1 realisation
	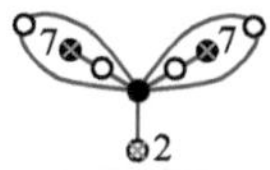	PSL(2,8)	
(C)	$\Delta(3,3,7) <_8 \Delta(2,3,7)$	$[1.7, 2^4, 1^2.3^2]$	1 realisation
	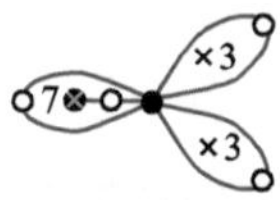	$PSL(3,2)$	
(D)	$\Delta(4,8,8) <_{12} \Delta(2,3,8)$	$[1^2.2.8, 2^6, 3^4]$	1 realisation
	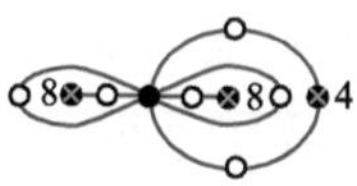	$(C_4{}^2 : C_3) : C_2$	

(continued)

Table 2 (continued)

Label	Inclusion	Passport	Realisation(s)
(E)	$\Delta(3,8,8) <_{10} \Delta(2,3,8)$	$[1^2.8, 2^5, 1.3^3]$	1 realisation
	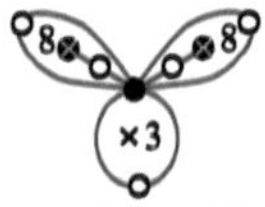	$A_6 : C_2 \cong PGL(2,9)$	
(F)	$\Delta(9,9,9) <_{12} \Delta(2,3,9)$	$[1^3.9, 2^6, 3^4]$	1 realisation
		$(((C_3{}^2 : C_2) : C_3) : C_2) : C_3$	
(G)	$\Delta(4,4,5) <_6 \Delta(2,4,5)$	$[1.5, 2^3, 1^2.4]$	1 realisation
	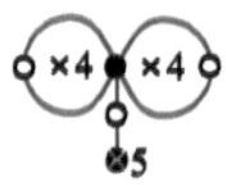	$S_5 \cong PGL(2,5)$	
(H)	$\Delta(n,4n,4n) <_6 \Delta(2,3,4n)$	$[1^2.4, 2^3, 3^2]$	1 realisation
	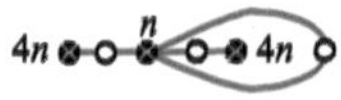	$S_4 \cong PGL(2,3)$	
(I)	$\Delta(n,2n,2n) <_4 \Delta(2,4,2n)$	$[1^2.2, 2^2, 4]$	1 realisation
	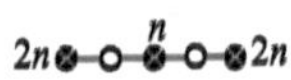	D_4	
(J)	$\Delta(3,n,3n) <_4 \Delta(2,3,3n)$	$[1.3, 2^2, 1.3]$	1 realisation
	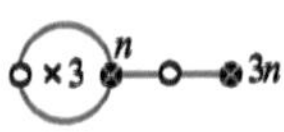	$A_4 \cong PSL(2,3)$	
(K)	$\Delta(2,n,2n) <_3 \Delta(2,3,2n)$	$[1.2, 1.2, 3]$	1 realisation
	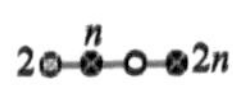	$S_3 \cong PSL(2,2)$	

Normal Inclusions of Quadrangle Groups

See Tables 3 and 4.

Table 3 Inclusions of quadrangle groups in quadrangle groups

Label	Inclusion	Passport	Constellation
(Q1)	$Q(m,m,n,n) \lhd_2 Q(2,2,m,n)$	$[1^2,1^2,2,2]$	$[(),(),(1\,2),(1\,2)]$
		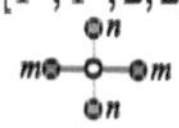 C_2	
(Q2)	$Q(n,n,n,n) \lhd_4 Q(2,2,2,n)$	$[1^4,2^2,2^2,2^2]$	$[(),(1\,2)(3\,4),(1\,3)(2\,4),(1\,4)(2\,3)]$
		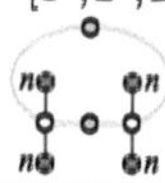 $C_2 \times C_2$	

Table 4 Infinite families of normal inclusions of quadrangle groups in triangle groups

Label	Inclusion	Passport	Realisation(s)
(f1)	$Q(l,m,n,n) \lhd_2 \Delta(2l,2m,n)$	$[2,2,1^2]$	1 realisation
	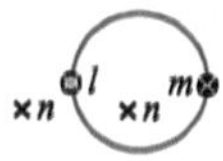	C_2 (a) of Table 1A for $l = 1$.	
(f2)	$Q(m,n,n,n) \lhd_3 \Delta(3,3m,n)$	$[3,3,1^3]$	1 realisation
	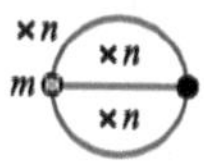	C_3 (b) of Table 1A for $m = 1$.	
(f3)	$Q(n,n,n,n) \lhd_4 \Delta(4,4,n)$	$[4,4,1^4]$	1 realisation
	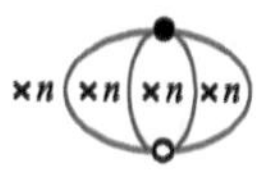	C_4	
(f4)	$Q(m,m,n,n) \lhd_4 \Delta(2,2m,2n)$	$[2^2,2^2,2^2]$	1 realisation
	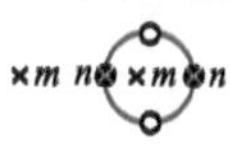	$C_2 \times C_2$	
(f5)	$Q(n,n,n,n) \lhd_8 \Delta(2,4,2n)$	$[2^4,2^4,4^2]$	1 realisation
		D_4	
(f6)	$Q(n,n,n,n) \lhd_{12} \Delta(2,3,3n)$	$[3^4,2^6,3^4]$	1 realisation
	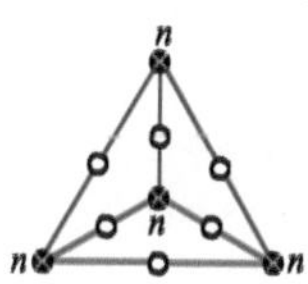	A_4	

Non-normal Inclusions of Quadrangle Groups

See Tables 5 and 6.

Table 5 Infinite families of non-normal inclusions of quadrangle groups in triangle groups

Label	Inclusion	Passport	Realisation(s)
(F1)	$Q(2,m,n,2n) <_3 \Delta(2,3m,2n)$	$[1.2,1.2,3]$	1 realisation
	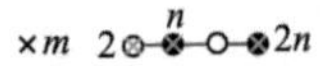	S_3 (K) of Table 1B for $m=1$.	
(F2)	$Q(m,2m,n,2n) <_3 \Delta(3,2m,2n)$	$[1.2,1.2,3]$	1 realisation
	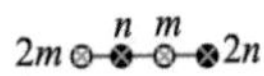	S_3 (K) of Table 1B for $m=1$.	
(F3)	$Q(3,n,2n,2n) <_4 \Delta(3,4,2n)$	$[4,1^2.2,1.3]$	1 realisation
	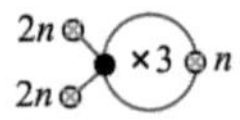	S_4	
(F4)	$Q(2,2,n,3n) <_4 \Delta(2,4,3n)$	$[4,1^2.2,1.3]$	1 realisation
	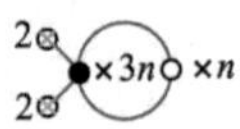	S_4	
(F5)	$Q(2,2,3,n) <_4 \Delta(2,3,4n)$	$[4,1^2.2,1.3]$	1 realisation
	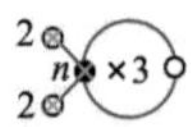	S_4	
(F6)	$Q(m,n,2n,2n) <_4 \Delta(2,4m,2n)$	$[1^2.2,2^2,4]$	1 realisation
	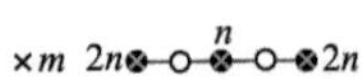	D_4 (I) of Table 1B for $m=1$.	
(F7)	$Q(2,2,n,n) <_4 \Delta(2,4,2n)$	$[1^2.2,2^2,4]$	1 realisation
	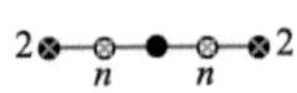	D_4	
(F8)	$Q(3,3,n,3n) <_4 \Delta(3,3,3n)$	$[1.3,1.3,1.3]$	1 realisation
	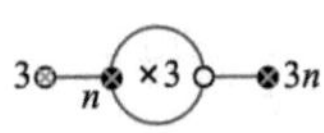	A_4	
(F9)	$Q(3,3,n,n) <_4 \Delta(3,3,2n)$	$[1.3,2^2,1.3]$	1 realisation
	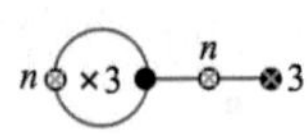	A_4	
(F10)	$Q(m,3m,n,3n) <_4 \Delta(2,3m,3n)$	$[1.3,2^2,1.3]$	1 realisation
	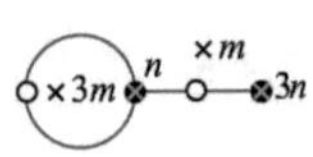	A_4 (J) of Table 1B for $m=1$.	
(F11)	$Q(2,n,3n,3n) <_5 \Delta(2,5,3n)$	$[5,1.2^2,1^2.3]$	1 realisation
	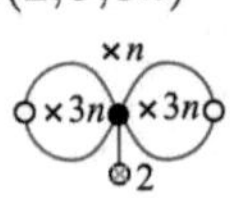	A_5	

(continued)

Table 5 (continued)

Label	Inclusion	Passport	Realisation(s)
(F12)	$Q(2,3,3,n) <_5 \Delta(2,3,5n)$	$[5, 1.2^2, 1^2.3]$	1 realisation
	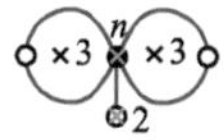	A_5	
(F13)	$Q(2,n,n,2n) <_5 \Delta(2,5,2n)$	$[5, 1.2^2, 1.2^2]$	1 realisation
	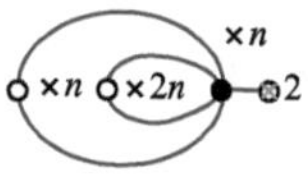	D_5	
(F14)	$Q(2,4,n,4n) <_5 \Delta(2,4,4n)$	$[1.4, 1.2^2, 1.4]$	2 realisations
	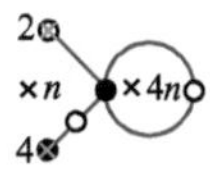	$C_5 : C_4$ chiral	
(F15)	$Q(2,4,2n,3n) <_5 \Delta(2,4,6n)$	$[1.4, 1.2^2, 2.3]$	1 realisation
	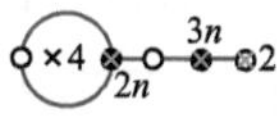	S_5	
(F16)	$Q(n,3n,3n,3n) <_6 \Delta(2,6,3n)$	$[6, 2^3, 1^3.3]$	1 realisation
	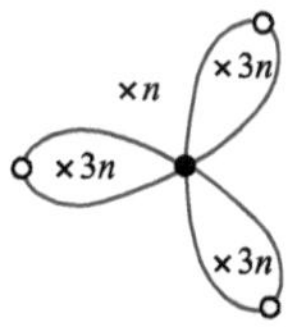	$S_3 \times C_3$	
(F17)	$Q(3,3,3,n) <_6 \Delta(2,3,6n)$	$[6, 2^3, 1^3.3]$	1 realisation
	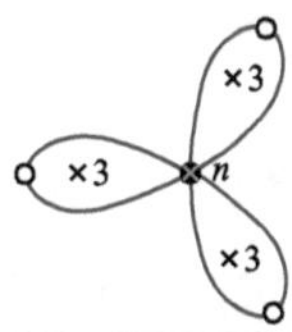	$S_3 \times C_3$	
(F18)	$Q(2,2,n,5n) <_6 \Delta(2,3,5n)$	$[1.5, 1^2.2^2, 3^2]$	1 realisation
	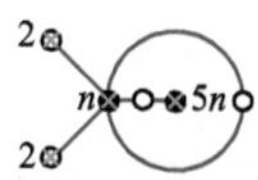	A_5	
(F19)	$Q(n,n,2n,2n) <_6 \Delta(2,6,2n)$	$[6, 2^3, 1^2.2^2]$	1 realisation
	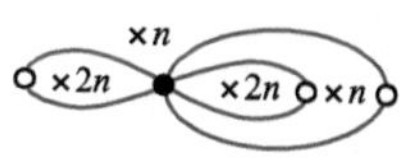	D_6	
(F20)	$Q(2,2,n,2n) <_6 \Delta(2,3,4n)$	$[2.4, 1^2.2^2, 3^2]$	1 realisation
	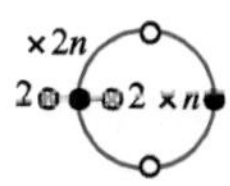	S_4	

(continued)

Table 5 (continued)

Label	Inclusion	Passport	Realisation(s)
(F21)	$Q(n,n,2n,2n) <_6 \Delta(3,3,2n)$	$[3^2, 1^2.2^2, 3^2]$	1 realisation
	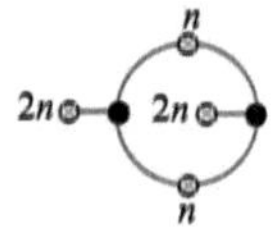	A_4	
(F22)	$Q(2,2,n,n) <_6 \Delta(2,3,3n)$	$[3^2, 1^2.2^2, 3^2]$	1 realisation
	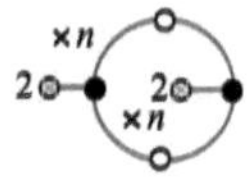	A_4	
(F23)	$Q(5,n,4n,4n) <_6 \Delta(2,5,4n)$	$[1.5, 2^3, 1^2.4]$	1 realisation
	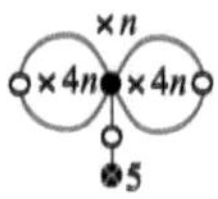	S_5 (G) of Table 1B for $n = 1$.	
(F24)	$Q(4,4,n,5n) <_6 \Delta(2,4,5n)$	$[1.5, 2^3, 1^2.4]$	1 realisation
	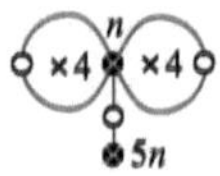	S_5 (G) of Table 1B for $n = 1$.	
(F25)	$Q(4,4,n,n) <_6 \Delta(2,4,3n)$	$[1^2.4, 2^3, 3^2]$	1 realisation
	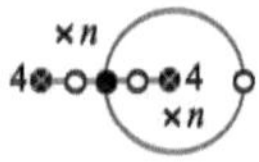	S_4	
(F26)	$Q(5,2n,3n,6n) <_6 \Delta(2,5,6n)$	$[1.5, 2^3, 1.2.3]$	1 realisation
	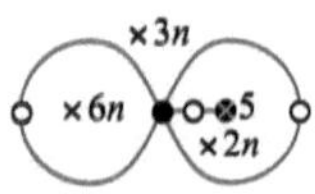	S_6	
(F27)	$Q(2,2n,3n,6n) <_6 \Delta(2,4,6n)$	$[2.4, 2^3, 1.2.3]$	1 realisation
	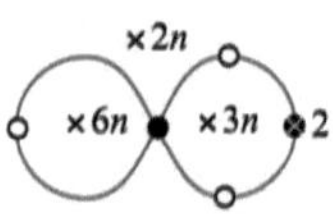	$S_3{}^2 : C_2$	
(F28)	$Q(2,3,n,6n) <_7 \Delta(2,3,6n)$	$[1.6, 1.2^3, 1.3^2]$	2 realisations
	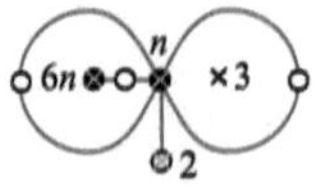	$(C_7 : C_3) : C_2$ chiral	
(F29)	$Q(2,3,2n,5n) <_7 \Delta(2,3,10n)$	$[2.5, 1.2^3, 1.3^2]$	1 realisation
	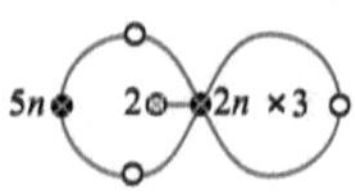	S_7	

(continued)

Table 5 (continued)

Label	Inclusion	Passport	Realisation(s)
(F30)	$Q(2,3,3n,4n) <_7 \Delta(2,3,12n)$	$[3.4, 1.2^3, 1.3^2]$	1 realisation
	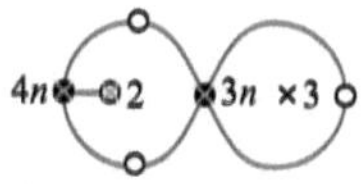	S_7	
(F31)	$Q(n,2n,4n,4n) <_8 \Delta(2,4,4n)$	$[4^2, 2^4, 1^2.2.4]$	1 realisation
	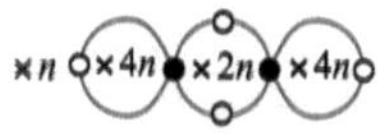	$((C_4 \times C_2) : C_2) : C_2$	
(F32)	$Q(3,3,n,7n) <_8 \Delta(2,3,7n)$	$[1.7, 2^4, 1^2.3^2]$	1 realisation
	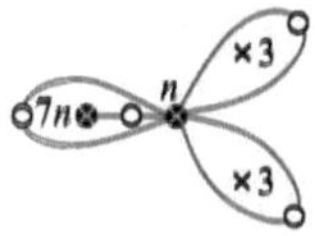	$PSL(3,2)$ (C) of Table 1B for $n = 1$.	
(F33)	$Q(3,3,n,3n) <_8 \Delta(2,3,6n)$	$[2.6, 2^4, 1^2.3^2]$	1 realisation
	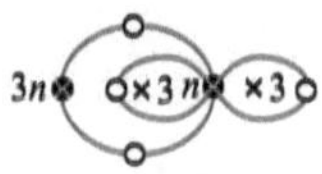	$A_4 \times C_2$	
(F34)	$Q(3,3,n,n) <_8 \Delta(2,3,4n)$	$[4^2, 2^4, 1^2.3^2]$	1 realisation
	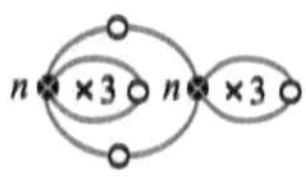	S_4	
(F35)	$Q(n,n,3n,3n) <_8 \Delta(2,4,3n)$	$[4^2, 2^4, 1^2.3^2]$	1 realisation
	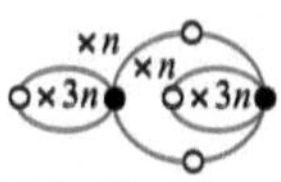	S_4	
(F36)	$Q(2,n,7n,7n) <_9 \Delta(2,3,7n)$	$[1^2.7, 1.2^4, 3^3]$	1 realisation
	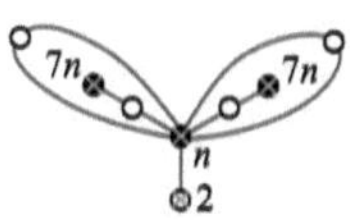	$PSL(2,8)$ (B) of Table 1B for $n = 1$.	
(F37)	$Q(2,n,3n,6n) <_9 \Delta(2,3,6n)$	$[1.2.6, 1.2^4, 3^3]$	1 realisation
	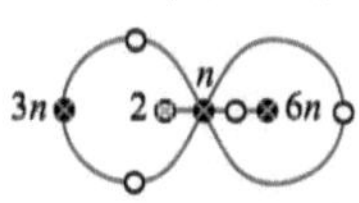	$(C_3{}^2 : C_3) : C_2$	
(F38)	$Q(2,3n,5n,15n) <_9 \Delta(2,3,15n)$	$[1.3.5, 1.2^4, 3^3]$	1 realisation
	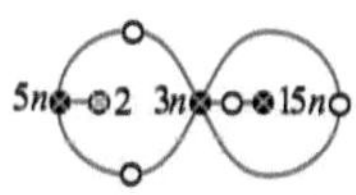	A_9	

(continued)

Table 5 (continued)

Label	Inclusion	Passport	Realisation(s)
(F39)	$Q(2,3n,4n,6n) <_9 \Delta(2,3,12n)$	$[2.3.4, 1.2^4, 3^3]$	1 realisation
	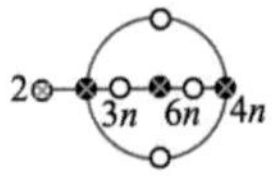	$(((C_3 \times (C_3{}^2 : C_2)) : C_2) : C_3) : C_2$	
(F40)	$Q(3,n,8n,8n) <_{10} \Delta(2,3,8n)$	$[1^2.8, 2^5, 1.3^3]$	1 realisation
	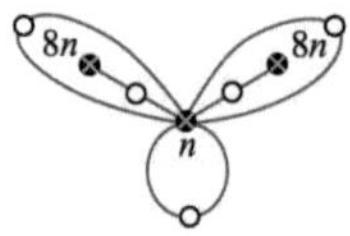	$A_6 : C_2 \cong PGL(2,9)$ (E) of Table 1B for $n = 1$.	
(F41)	$Q(3,2n,7n,14n) <_{10} \Delta(2,3,14n)$	$[1.2.7, 2^5, 1.3^3]$	1 realisation
	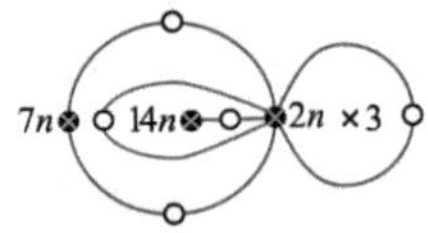	S_{10}	
(F42)	$Q(3,4n,5n,20n) <_{10} \Delta(2,3,20n)$	$[1.3^3, 2^5, 1.4.5]$	1 realisation
	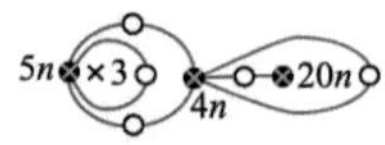	S_{10}	
(F43)	$Q(3,6n,10n,15n) <_{10} \Delta(2,3,30n)$	$[2.3.5, 2^5, 1.3^3]$	1 realisation
	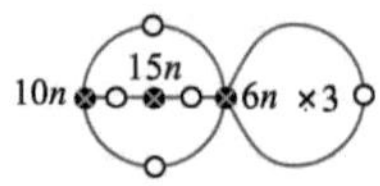	S_{10}	
(F44)	$Q(n,9n,9n,9n) <_{12} \Delta(2,3,9n)$	$[1^3.9, 2^6, 3^4)$	1 realisation
	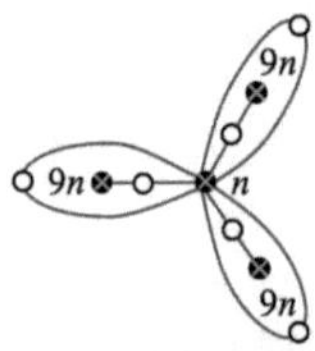	$(((C_3{}^2 : C_2) \times C_3) : C_2) : C_3$ (F) of Table 1B for $n = 1$.	
(F45)	$Q(n,4n,8n,8n) <_{12} \Delta(2,3,8n)$	$[1^2.2.8, 2^6, 3^4]$	1 realisation
	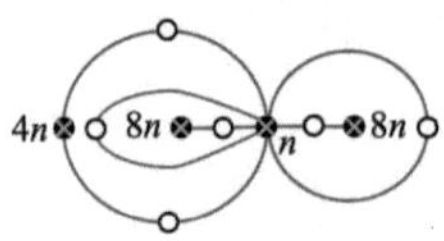	$(C_4{}^2 : C_3) : C_2$ (D) of Table 1B for $n = 1$.	
(F46)	$Q(n,n,5n,5n) <_{12} \Delta(2,3,5n)$	$[1^2.5^2, 2^6, 3^4]$	1 realisation
	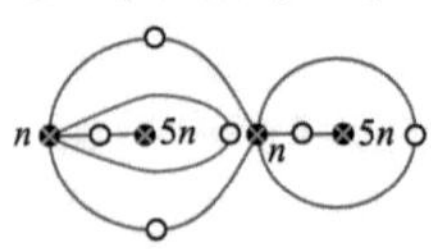	A_5	

(continued)

Table 5 (continued)

Label	Inclusion	Passport	Realisation(s)
(F47)	$Q(n,2n,3n,6n) <_{12} \Delta(2,3,6n)$	$[1.2.3.6, 2^6, 3^4]$	1 realisation
	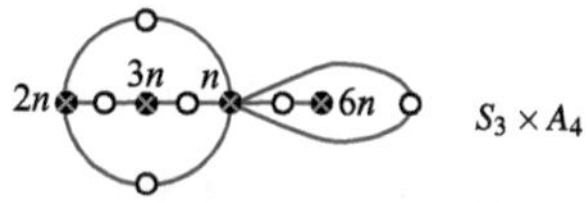		
(F48)	$Q(n,n,2n,2n) <_{12} \Delta(2,3,4n)$	$[2^2.4^2, 2^6, 3^4]$	1 realisation
	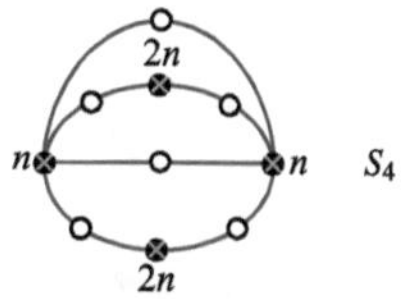		

Table 6 Sporadic non-normal inclusions of quadrangle groups in triangle groups. (There are no sporadic normal inclusions)

Label	Inclusion	Passport	Realisation(s)
(S1)	$Q(2,2,2,4) <_5 \Delta(2,4,5)$	$[5, 1^3.2, 1.4]$	1 realisation
	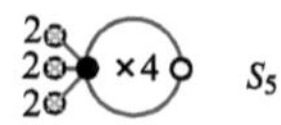		
(S2)	$Q(3,3,3,3) <_5 \Delta(3,3,5)$	$[5, 1^2.3, 1^2.3]$	1 realisation
	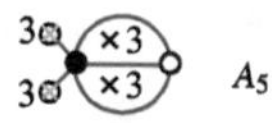		
(S3)	$Q(3,3,4,4) <_5 \Delta(3,4,4)$	$[1.4, 1.4, 1^2.3]$	1 realisation
	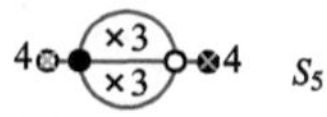		
(S4)	$Q(3,3,3,5) <_6 \Delta(3,3,5)$	$[1^3.3, 1.5, 3^2]$	1 realisation
	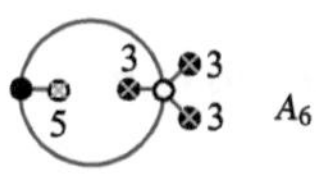		
(S5)	$Q(2,3,3,3) <_6 \Delta(3,3,4)$	$[1^3.3, 2.4, 3^2]$	1 realisation
	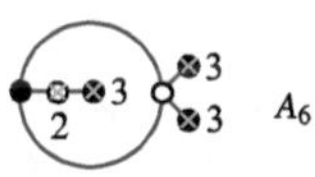		
(S6)	$Q(2,2,4,4) <_6 \Delta(2,4,6)$	$[1^2.4, 1^2.2^2, 6]$	2 realisations
	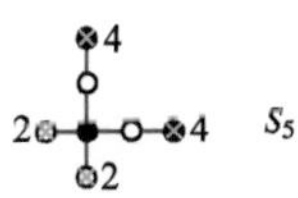	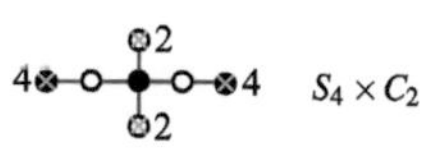	

(continued)

Table 6 (continued)

Label	Inclusion	Passport	Realisation(s)
(S7)	$Q(2,2,5,5) <_6 \Delta(2,5,5)$	$[1.5, 1^2.2^2, 1.5]$	3 realisations
	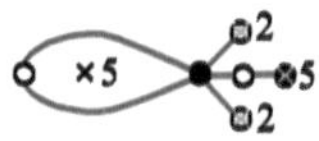 A_5 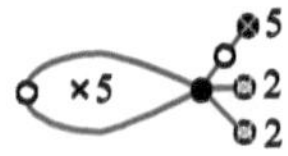 A_6 chiral		
(S8)	$Q(2,2,2,5) <_6 \Delta(2,4,5)$	$[1.5, 1^2.2^2, 2.4]$	2 realisations
	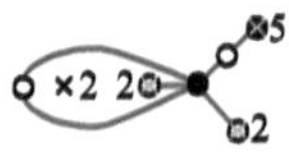 A_6 chiral		
(S9)	$Q(4,4,4,4) <_6 \Delta(3,4,4)$	$[1^2.4, 1^2.4, 3^2]$	2 realisations
	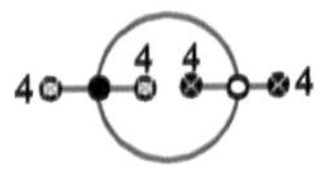 S_4 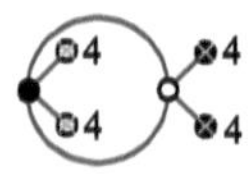 S_5		
(S10)	$Q(2,2,2,3) <_7 \Delta(2,3,7)$	$[7, 1^3.2^2, 1.3^2]$	2 realisations
	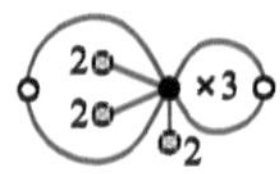 $PSL(3,2)$ chiral		
(S11)	$Q(2,4,4,4) <_7 \Delta(2,4,7)$	$[7, 1.2^3, 1^3.4]$	1 realisation
	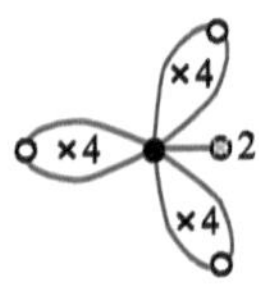 S_7		
(S12)	$Q(2,5,5,6) <_7 \Delta(2,5,6)$	$[1.6, 1.2^3, 1^2.5]$	3 realisations
	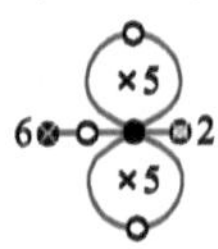 S_7 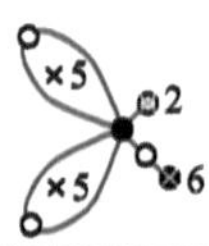 S_7 chiral		
(S13)	$Q(3,3,5,5) <_7 \Delta(3,3,5)$	$[1^2.5, 1.3^2, 1.3^2]$	1 realisation
	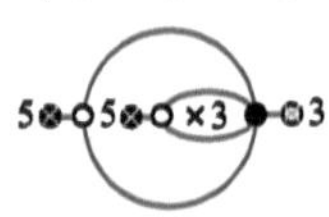 A_7		
(S14)	$Q(2,2,4,6) <_7 \Delta(2,4,6)$	$[1.6, 1.2^3, 1.2.4]$	4 realisations
	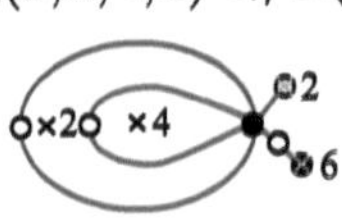 S_7 chiral 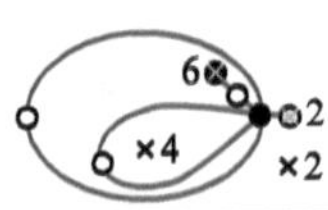 S_7 chiral		

(continued)

Table 6 (continued)

Label	Inclusion	Passport	Realisation(s)
(S15)	$Q(2,3,3,4) <_7 \Delta(3,3,4)$	$[1.2.4, 1.3^2, 1.3^2]$	4 realisations
	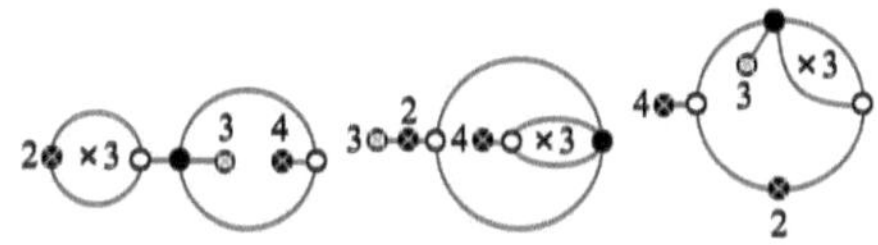 $PSL(3,2)$ $PSL(3,2)$ $PSL(3,2)$ chiral		
(S16)	$Q(4,4,4,4) <_8 \Delta(2,4,8)$	$[8, 2^4, 1^4.4]$	1 realisation
	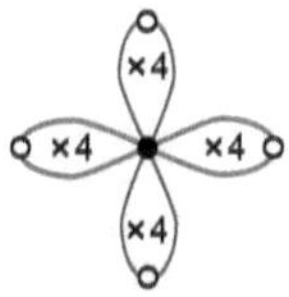 $C_4{}^2 : C_2$		
(S17)	$Q(5,5,5,7) <_8 \Delta(2,5,7)$	$[1.7, 2^4, 1^3.5]$	1 realisation
	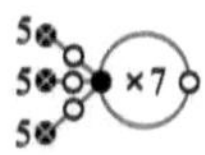 A_8		
(S18)	$Q(2,2,3,3) <_8 \Delta(2,3,8)$	$[8, 1^2.2^3, 1^2.3^2]$	4 realisations
	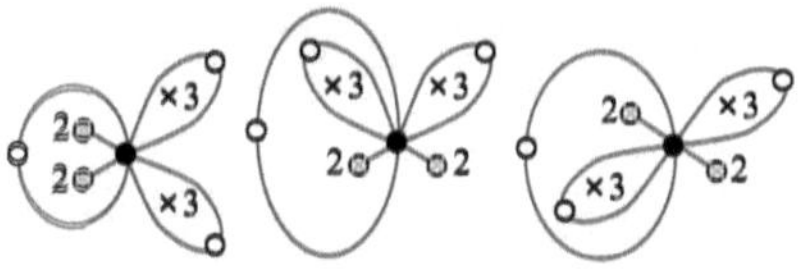 $PSL(3,2):C_2$ $PSL(3,2):C_2$ $GL(2,3)$ chiral		
(S19)	$Q(2,2,6,6) <_8 \Delta(2,4,6)$	$[1^2.6, 1^2.2^3, 4^2]$	3 realisations
	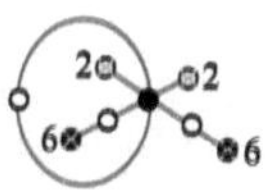 $PSL(3,2):C_2$ 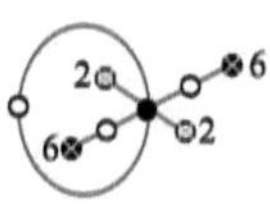 $((((D_4 \times C_2):C_2):C_3):C_2):C_3$ chiral		
(S20)	$Q(2,4,4,7) <_8 \Delta(2,4,7)$	$[1^2.2.4, 2^4, 1.7]$	2 realisations
	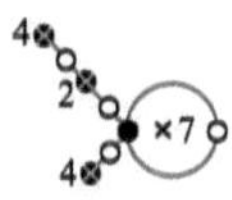 $C_2{}^3 : PSL(3,2)$ chiral		
(S21)	$Q(2,3,4,4) <_8 \Delta(2,4,6)$	$[1^2.2.4, 2^4, 2.6]$	1 realisation
	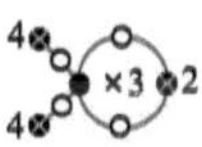 $((C_2{}^4 : C_3):C_2):C_2$		
(S22)	$Q(3,3,3,3) <_8 \Delta(3,3,4)$	$[4^2, 1^2.3^2, 1^2.3^2]$	3 realisations
	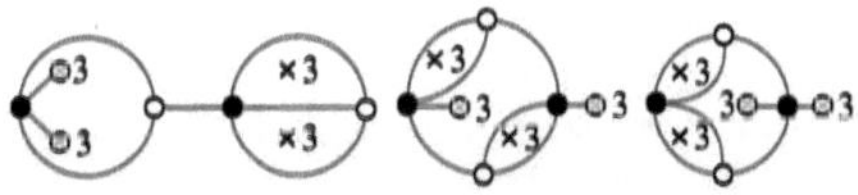 $PSL(3,2)$ $PSL(3,2)$ $SL(2,3)$		

(continued)

Table 6 (continued)

Label	Inclusion	Passport	Realisation(s)
(S23)	$Q(6,6,6,6) <_8 \Delta(2,6,6)$	$[1^2.6, 2^4, 1^2.6]$	2 realisations
	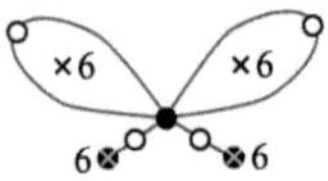 $PSL(3,2):C_2$	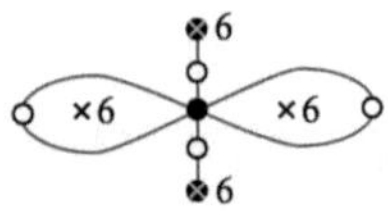	$(((D_4 \times C_2):C_2):C_3):C_2$
(S24)	$Q(2,2,2,8) <_9 \Delta(2,3,8)$	$[1.8, 1^3.2^3, 3^3]$	2 realisations
	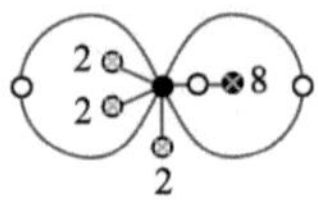	$((C_3{}^2:Q_4):C_3):C_2$ chiral	
(S25)	$Q(2,3,3,3) <_9 \Delta(2,3,9)$	$[9, 1.2^4, 1^3.3^2]$	2 realisations
	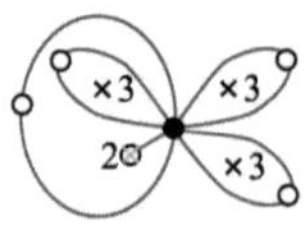	$PSL(2,8):C_3$ chiral	
(S26)	$Q(3,3,3,4) <_9 \Delta(3,3,4)$	$[1^3.3^2, 1.4^2, 3^3]$	1 realisation
	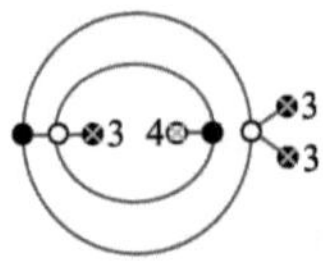	$(C_3{}^2:Q_4):C_3$	
(S27)	$Q(2,4,7,7) <_9 \Delta(2,4,7)$	$[1.4^2, 1.2^4, 1^2.7]$	4 realisations
	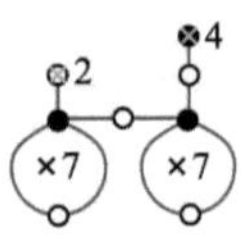 A_9 chiral	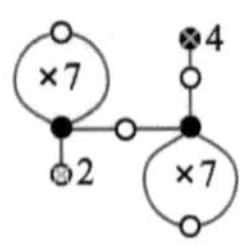	A_9 chiral
(S28)	$Q(2,3,4,6) <_9 \Delta(2,4,6)$	$[1.4^2, 1.2^4, 1.2.6]$	4 realisations
	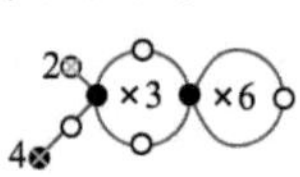 A_9 chiral	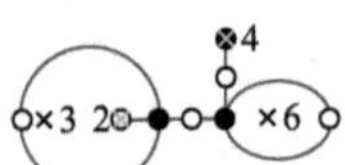	A_9 chiral
(S29)	$Q(3,3,3,3) <_{10} \Delta(2,3,10)$	$[10, 2^5, 1^4.3^2]$	1 realisation
	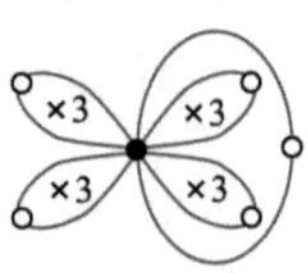	$A_5 \times C_2$	
(S30)	$Q(6,6,6,6) <_{10} \Delta(2,5,6)$	$[5^2, 2^5, 1^4.6]$	1 realisation
	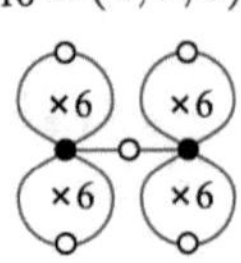	$(C_2{}^4:A_5) \times C_2$	

(continued)

Table 6 (continued)

Label	Inclusion	Passport	Realisation(s)
(S31i)	$Q(2,2,4,4) <_{10} \Delta(2,4,5)$	$[5^2, 2^5, 1^2.2^2.4]$	2 realisations
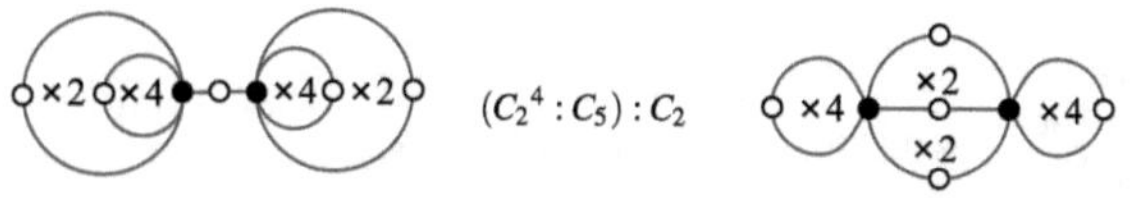 $(C_2{}^4 : C_5) : C_2$; $(C_2{}^4 : C_5) : C_2$			
(S31ii)	$Q(2,2,4,4) <_{10} \Delta(2,4,5)$	$[5^2, 1^2.2^4, 1^2.4^2]$	5 realisations
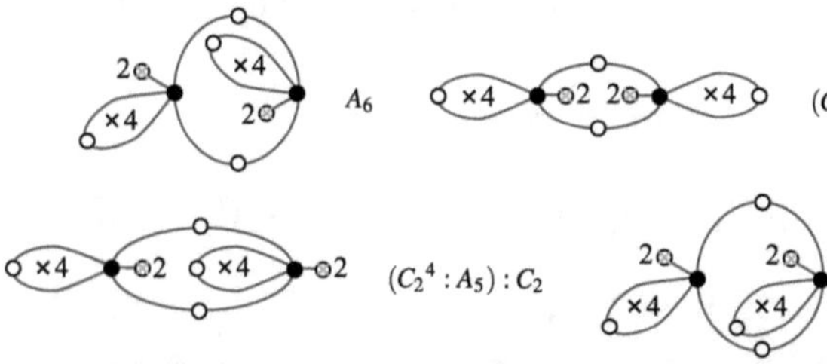 A_6; $(C_2{}^4 : C_5) : C_2$; $(C_2{}^4 : A_5) : C_2$; $(C_2{}^4 : A_5) : C_2$ chiral			
(S32)	$Q(2,2,3,9) <_{10} \Delta(2,3,9)$	$[1.9, 1^2.2^4, 1.3^3]$	6 realisations
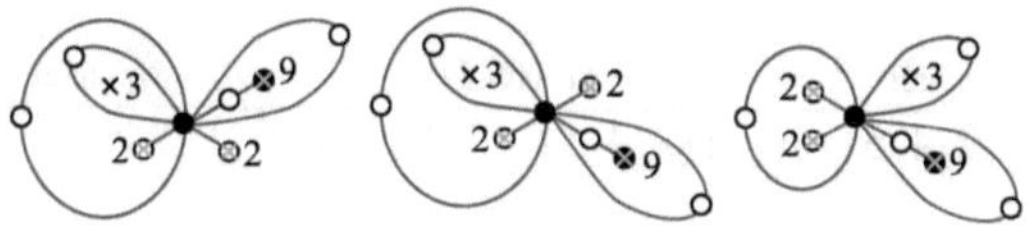 A_{10} chiral; A_{10} chiral; A_{10} chiral			
(S33)	$Q(2,2,3,4) <_{10} \Delta(2,3,8)$	$[2.8, 1^2.2^4, 1.3^3]$	3 realisations
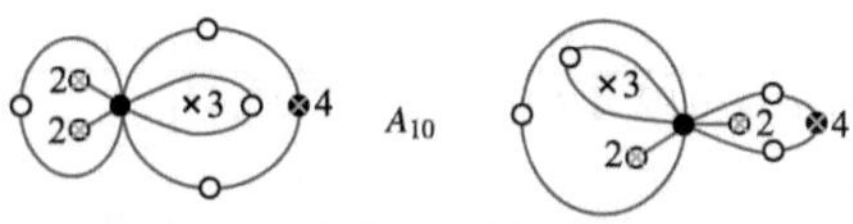 A_{10}; A_{10} chiral			
(S34)	$Q(4,4,8,8) <_{10} \Delta(2,4,8)$	$[1^2.4^2, 2^5, 1^2.8]$	3 realisations
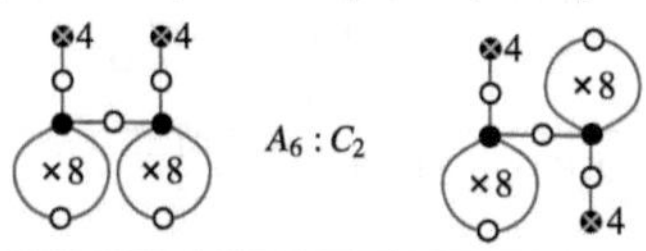 $A_6 : C_2$; $((C_2{}^4 : C_5) : C_4) : C_2$ chiral			
(S35)	$Q(3,3,4,4) <_{10} \Delta(3,3,4)$	$[1^2.4^2, 1.3^3, 1.3^3]$	4 realisations
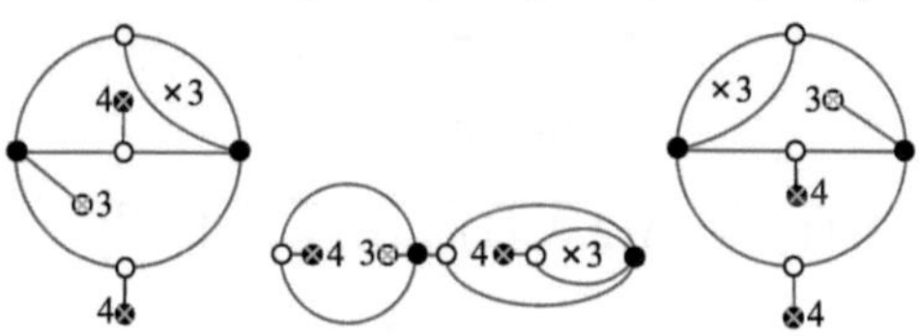 A_6; A_{10}; A_{10} chiral			

(continued)

Table 6 (continued)

Label	Inclusion	Passport	Realisation(s)
(S36)	$Q(2,4,4,6) <_{10} \Delta(2,4,6)$	$[1^2.4^2, 2^5, 1.3.6]$	3 realisations
	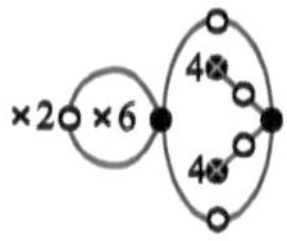 $(A_6 \cdot C_2) : C_2$ 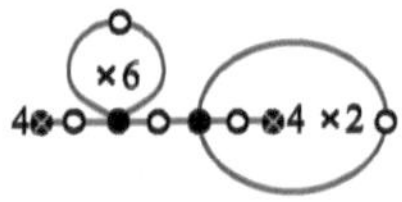 $(A_6 \cdot C_2) : C_2$ chiral		
(S37)	$Q(3,3,4,4) <_{10} \Delta(2,4,6)$	$[1^2.4^2, 2^5, 2^2.6]$	1 realisation
	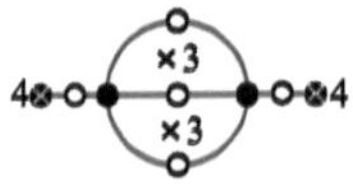 $S_5 \times C_2$		
(S38)	$Q(2,3,3,10) <_{11} \Delta(2,3,10)$	$[1.10, 1.2^5, 1^2.3^3]$	6 realisations
	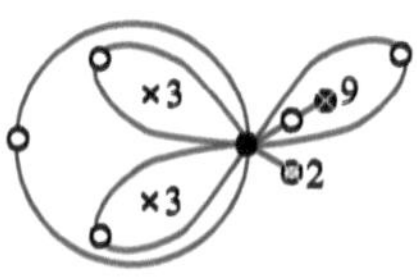 S_{11} chiral 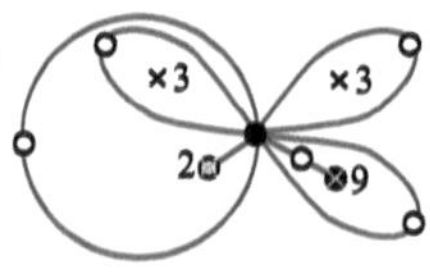 S_{11} chiral 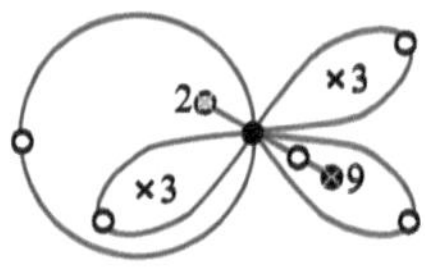 S_{11} chiral		
(S39)	$Q(2,2,4,5) <_{11} \Delta(2,4,5)$	$[1.5^2, 1.2^5, 1.2.4^2)$	5 realisations
	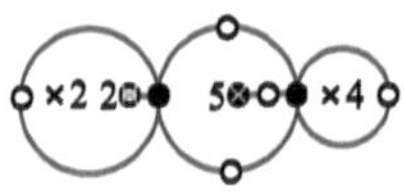 S_{11} 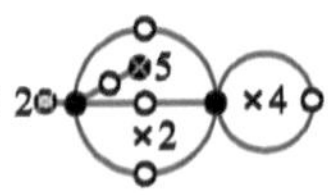 S_{11} chiral 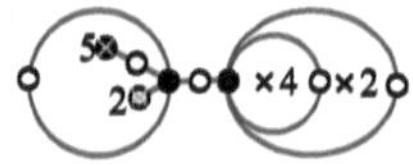 S_{11} chiral		
(S40)	$Q(4,4,4,4) <_{12} \Delta(2,4,6)$	$[6^2, 2^6, 1^4.4^2]$	2 realisations
	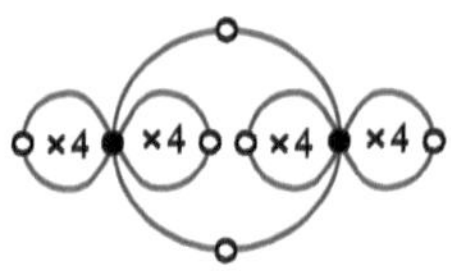 $S_4 \times C_2$ 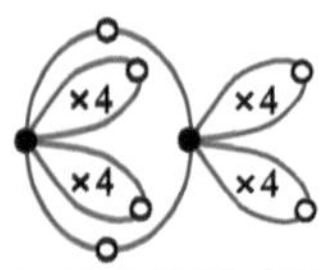 $S_5 \times C_2$		
(S41)	$Q(4,4,4,4) <_{12} \Delta(3,3,4)$	$[1^4.4^2, 3^4, 3^4]$	1 realisation
	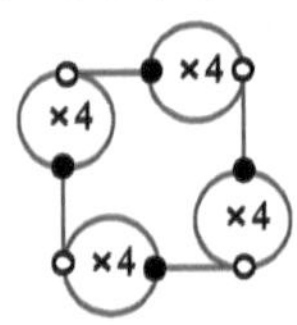 $C_4{}^2 : C_3$		
(S42)	$Q(3,3,3,11) <_{12} \Delta(2,3,11)$	$[1.11, 2^6, 1^3.3^3]$	2 realisations
	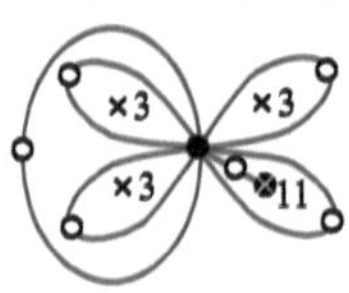 M_{12} chiral		

(continued)

Table 6 (continued)

Label	Inclusion	Passport	Realisation(s)
(S43)	$Q(3,3,3,5) <_{12} \Delta(2,3,10)$	$[2.10, 2^6, 1^3.3^3]$	1 realisation
	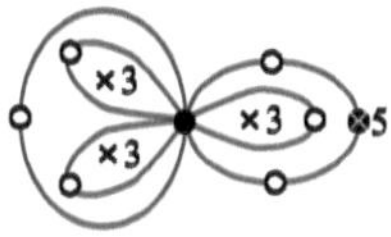	$PGL(2,9)$	
(S44)	$Q(3,3,3,3) <_{12} \Delta(2,3,9)$	$[3.9, 2^6, 1^3.3^3]$	1 realisation
	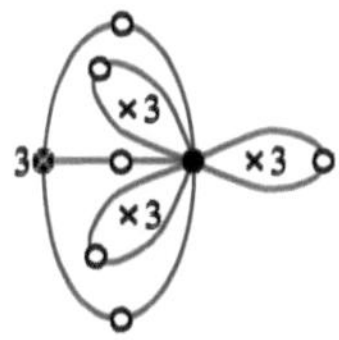	$(((C_3{}^2 : C_2) \times C_3) : C_2) : C_3$	
(S45)	$Q(2,3,3,3) <_{12} \Delta(2,3,8)$	$[4.8, 2^6, 1^3.3^3]$	1 realisation
	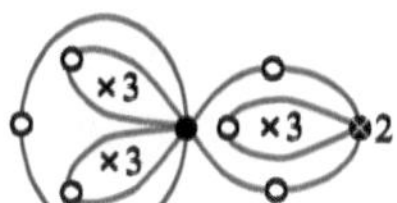	$PGL(2,9)$	
(S46)	$Q(2,6,6,6) <_{12} \Delta(2,4,6)$	$[1^3.3.6, 2^6, 4^3]$	1 realisation
	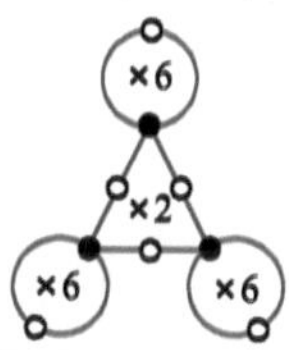	$(S_3{}^2 \times C_3) : C_2$	
(S47i)	$Q(2,2,5,5) <_{12} \Delta(2,4,5)$	$[1^2.5^2, 2^6, 2^2.4^2]$	3 realisations
	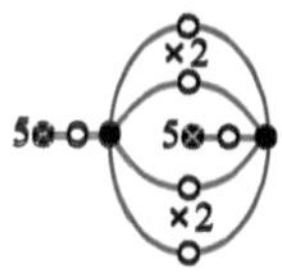 S_5	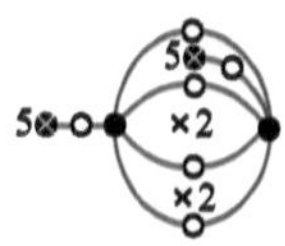 $A_6 \times C_2$ chiral	
(S47ii)	$Q(2,2,5,5) <_{12} \Delta(2,4,5)$	$[1^2.5^2, 1^2.2^5, 4^3]$	6 realisations
	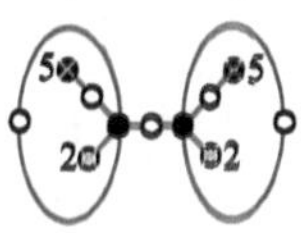 $PSL(2,11) : C_2$	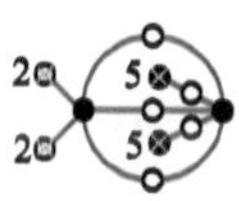 $PSL(2,11) : C_2$	
	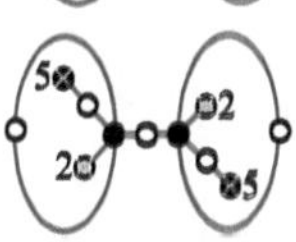 $(C_2{}^5 : A_6) : C_2$ chiral	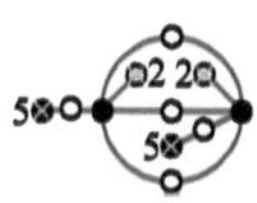 $(C_2{}^5 : A_6) : C_2$ chiral	

(continued)

Table 6 (continued)

Label	Inclusion	Passport	Realisation(s)
(S48)	$Q(2,2,10,10) <_{12} \Delta(2,3,10)$	$[1^2.10, 1^2.2^5, 3^4]$	4 realisations
	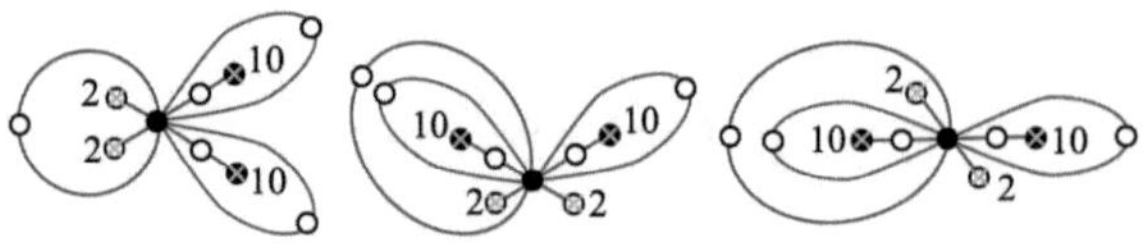 $PSL(2,11):C_2$; $PSL(2,11):C_2$; $(C_2{}^5:A_5):C_2$ chiral		
(S49)	$Q(2,2,4,4) <_{12} \Delta(2,3,8)$	$[2^2.8, 1^2.2^5, 3^4]$	1 realisation
	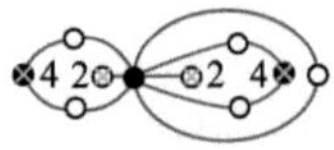 $(C_4{}^2:C_3):C_2$		
(S50)	$Q(3,3,6,6) <_{12} \Delta(2,4,6)$	$[4^3, 2^6, 1^2.2^2.6]$	1 realisation
	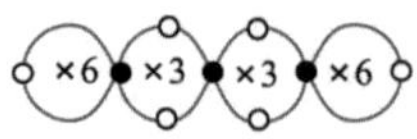 $S_3{}^2:C_2$		
(S51)	$Q(5,5,5,5) <_{12} \Delta(2,5,5)$	$[1^2.5^2, 2^6, 1^2.5^2]$	5 realisations
	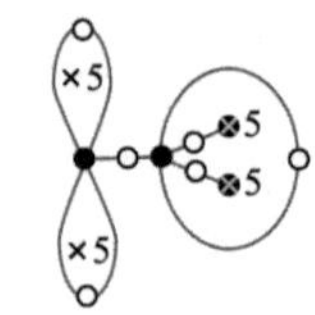 $PSL(2,11)$; 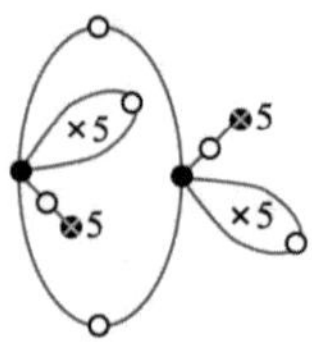 $PSL(2,11)$; 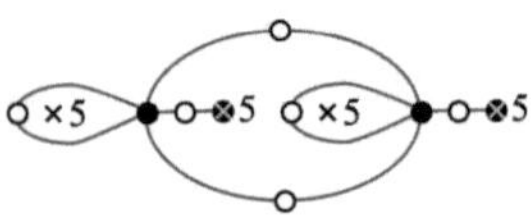 A_5; 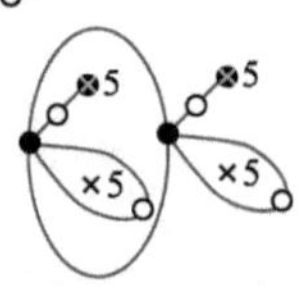 $C_2{}^5:A_6$ chiral		
(S52)	$Q(2,3,11,11) <_{13} \Delta(2,3,11)$	$[1^2.11, 1.2^6, 1.3^4]$	6 realisations
	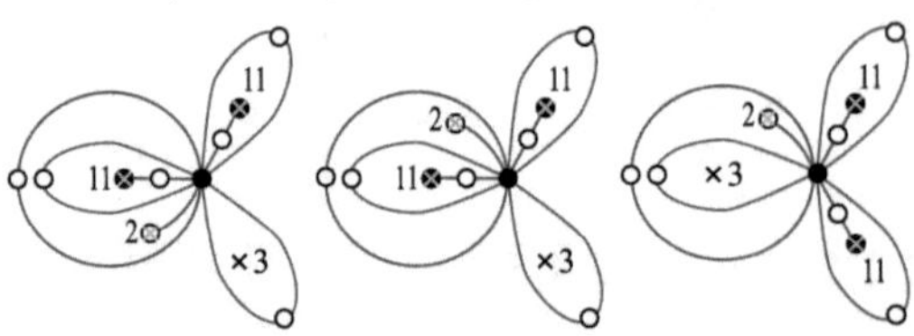 A_{13} chiral; A_{13} chiral; A_{13} chiral		
(S53)	$Q(2,3,5,10) <_{13} \Delta(2,3,10)$	$[1.2.10, 1.2^6, 1.3^4]$	6 realisations
	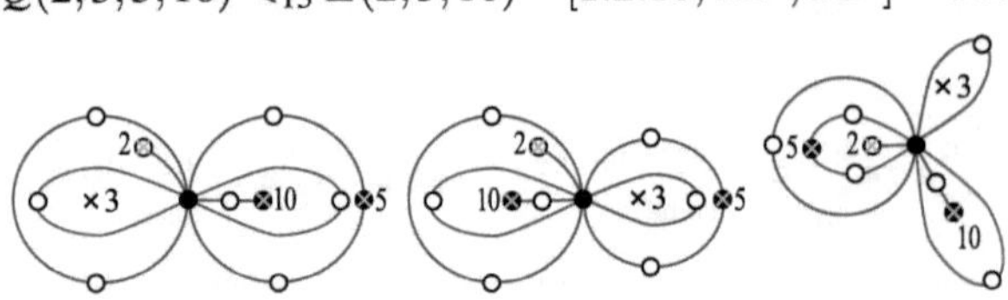 A_{13} chiral; A_{13} chiral; A_{13} chiral		

(continued)

Table 6 (continued)

Label	Inclusion	Passport	Realisation(s)
(S54)	$Q(2,3,3,9) <_{13} \Delta(2,3,9)$	$[1.3.9, 1.2^6, 1.3^4]$	4 realisations
	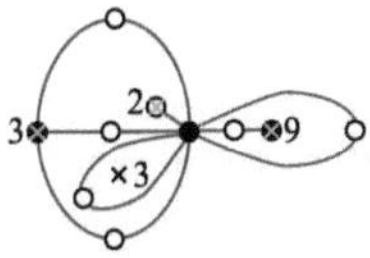 A_{13} chiral	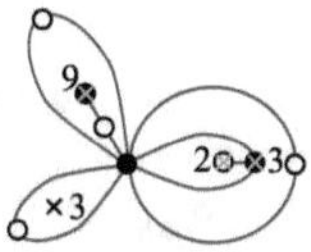 A_{13} chiral	
(S55)	$Q(2,2,3,8) <_{13} \Delta(2,3,8)$	$[1.4.8, 1.2^6, 1.3^4]$	4 realisations
	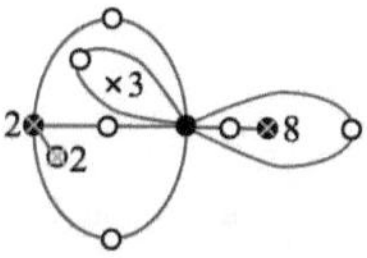 A_{13} chiral	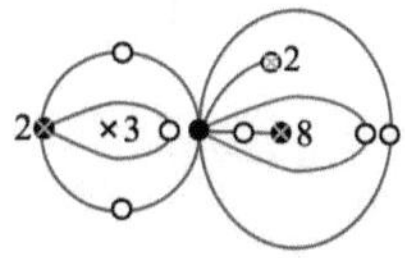 A_{13} chiral	
(S56)	$Q(2,2,3,3) <_{14} \Delta(2,3,7)$	$[7^2, 1^2.2^6, 1^2.3^4]$	9 realisations
	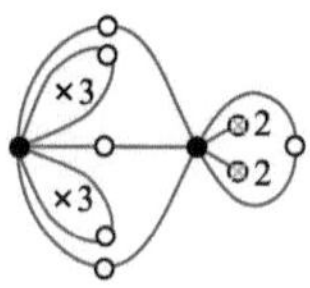 $PSL(2,13)$ 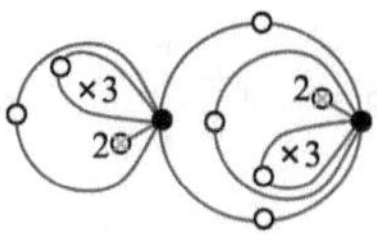 $PSL(2,13)$	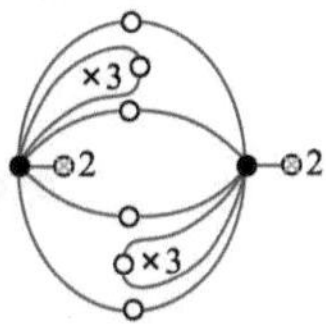 $PSL(2,13)$	
	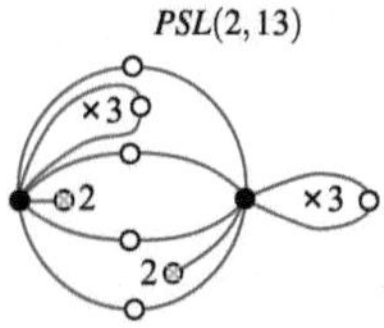 $PSL(3,2)$ chiral 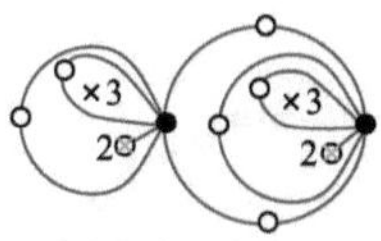 $C_2{}^3 \cdot PSL(3,2)$ chiral	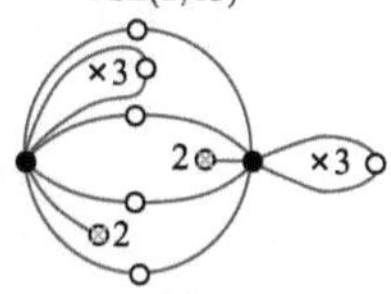 $C_2{}^3 \cdot PSL(3,2)$ chiral	
(S57)	$Q(3,3,12,12) <_{14} \Delta(2,3,12)$	$[1^2.12, 2^7, 1^2.3^4]$	4 realisations
	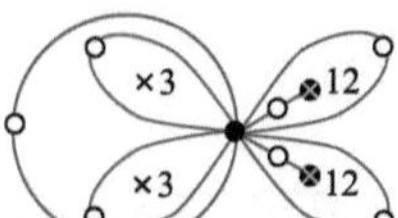 $PSL(2,13):C_2$ 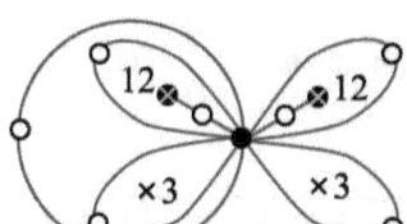 $PSL(2,13):C_2$	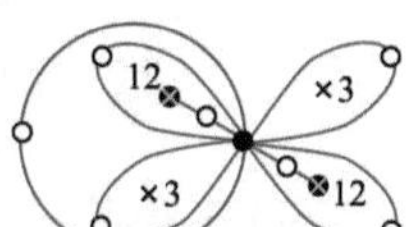 $((C_2{}^6:C_7):C_3):C_2$ chiral	
(S58)	$Q(3,3,5,5) <_{14} \Delta(2,3,10)$	$[2^2.10, 2^7, 1^2.3^4]$	1 realisation
	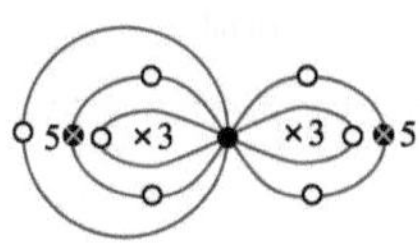 S_7		
(S59)	$Q(2,3,3,4) <_{14} \Delta(2,3,8)$	$[2.4.8, 2^7, 1^2.3^4]$	2 realisations
	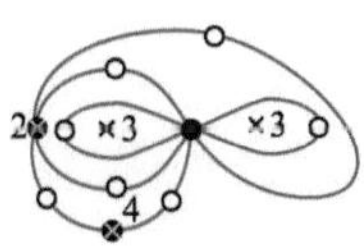 $PSL(3,2):C_2$	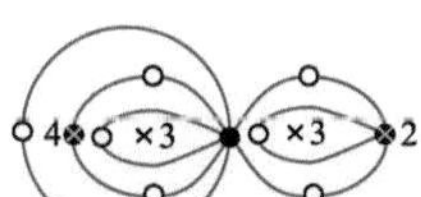 $PSL(3,2):C_2$	

(continued)

Table 6 (continued)

Label	Inclusion	Passport	Realisation(s)
(S60)	$Q(4,4,6,6) <_{14} \Delta(2,4,6)$	$[1^2.6^2, 2^7, 1^2.4^3]$	5 realisations
	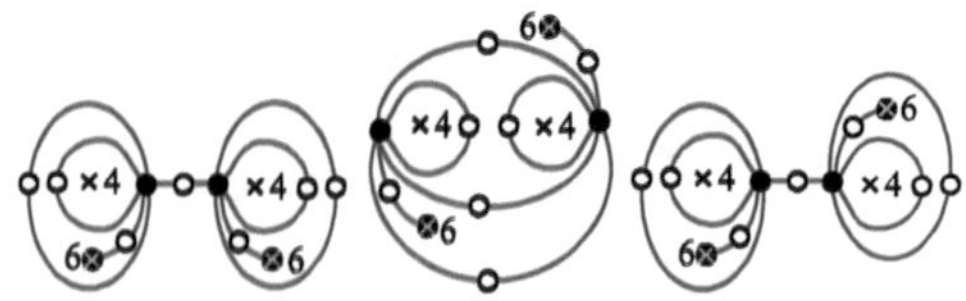		
(S61)	$Q(2,2,2,7) <_{15} \Delta(2,3,7)$	$[1.7^2, 1^3.2^6, 3^5]$	3 realisations
	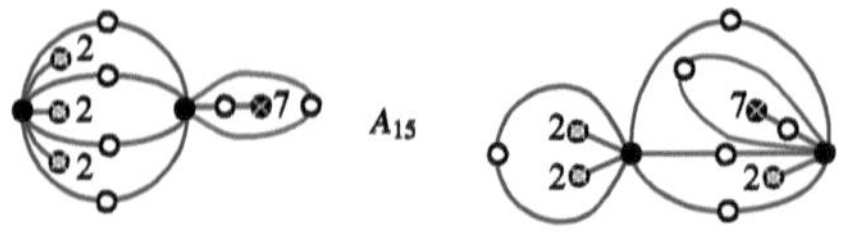 A_{15}; A_{15} chiral		
(S62)	$Q(2,4,4,4) <_{15} \Delta(2,4,5)$	$[5^3, 1.2^7, 1^3.4^3]$	1 realisation
	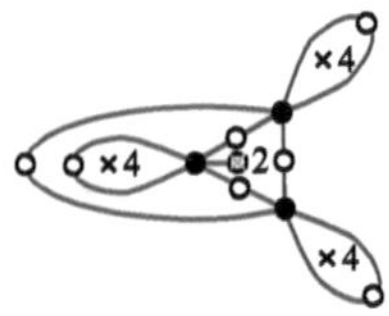 S_{15}		
(S63)	$Q(2,12,12,12) <_{15} \Delta(2,3,12)$	$[1^3.12, 1.2^7, 3^5]$	2 realisations
	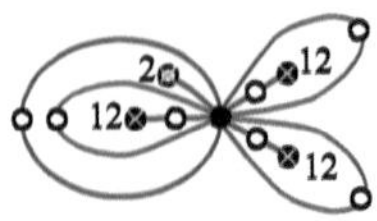 S_{15} chiral		
(S64)	$Q(2,5,5,10) <_{15} \Delta(2,3,10)$	$[1.2^2.10, 1.2^7, 3^5]$	1 realisation
	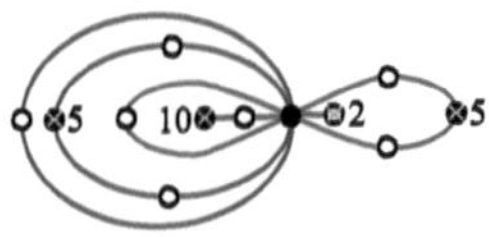 $(C_5{}^2 : C_3) : C_2$		
(S65)	$Q(2,2,4,8) <_{15} \Delta(2,3,8)$	$[1.2.4.8, 1.2^7, 3^5]$	2 realisations
	 $((((C_5{}^2 : C_4) \times C_5) : C_4) : C_3) : C_2$ chiral		

(continued)

Table 6 (continued)

Label	Inclusion	Passport	Realisation(s)
(S66)	$Q(3,3,3,3) <_{16} \Delta(2,3,8)$	$[8^2, 2^8, 1^4.3^4]$	3 realisations
	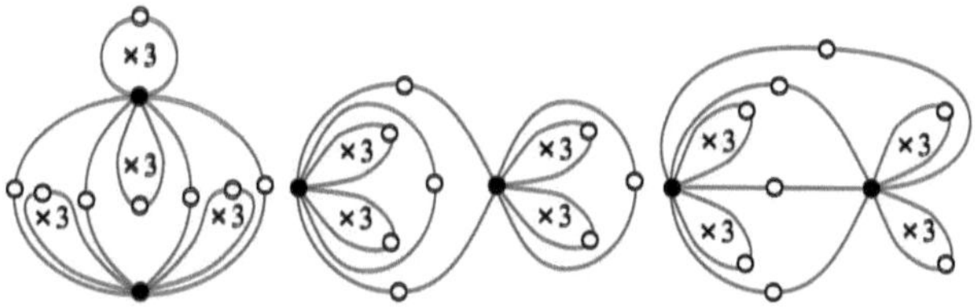 $GL(2,3)$; $PSL(3,2):C_2$; $PSL(3,2):C_2$		
(S67)	$Q(6,6,6,6) <_{16} \Delta(2,4,6)$	$[4^4, 2^8, 1^4.6^2]$	2 realisations
	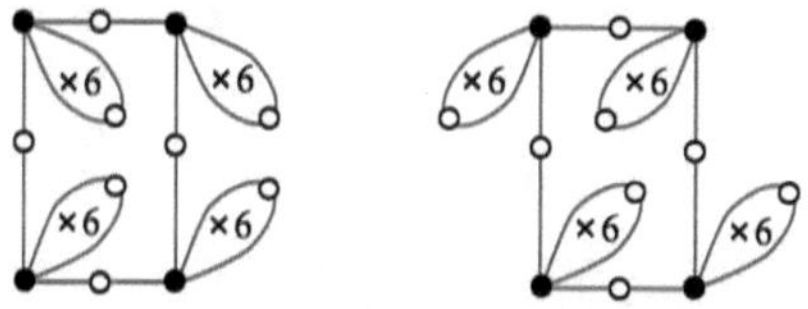 $(PSL(3,2):C_2)\times C_2$; $((((D_4\times C_2):C_2):C_3):C_2):C_2$		
(S68)	$Q(3,13,13,13) <_{16} \Delta(2,3,13)$	$[1^3.13, 2^8, 1.3^5]$	2 realisations
	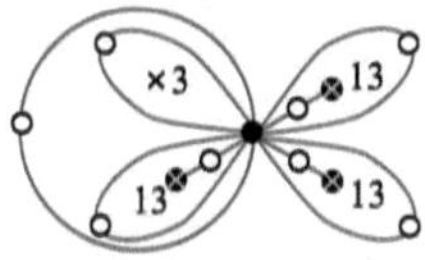 A_{16} chiral		
(S69)	$Q(2,4,4,5) <_{16} \Delta(2,4,5)$	$[1.5^3, 2^8, 1^2.2.4^3]$	3 realisations
	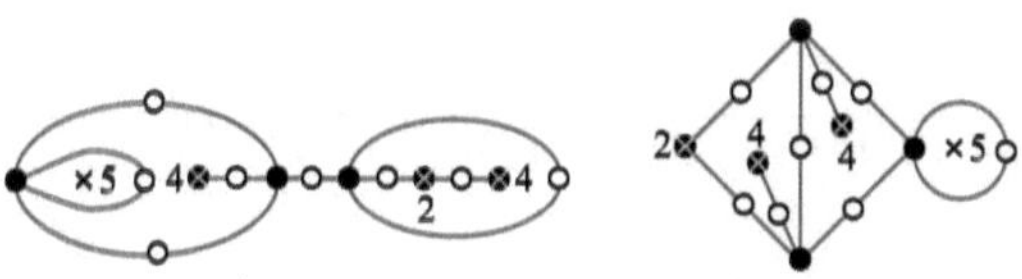 $(C_2{}^4:A_5):C_2$; $(C_2{}^4:A_5):C_2$ chiral		
(S70)	$Q(3,6,12,12) <_{16} \Delta(2,3,12)$	$[1^2.2.12, 2^8, 1.3^5]$	3 realisations
	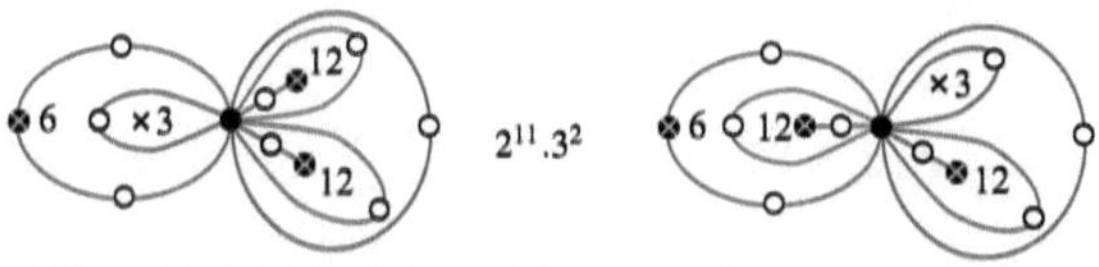 $2^{11}.3^2$; $2^{14}.3^6.5^3.7^2.11.13$ chiral		
(S71)	$Q(3,3,3,9) <_{16} \Delta(2,3,9)$	$[1.3^2.9, 2^8, 1.3^5]$	1 realisation
	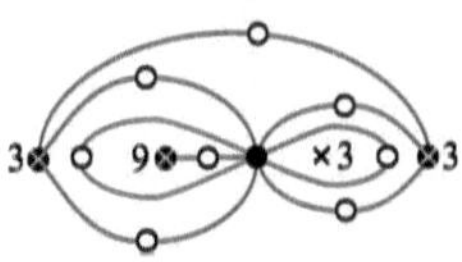 $2^{10}.3^4$		

(continued)

Table 6 (continued)

Label	Inclusion	Passport	Realisation(s)
(S72)	$Q(2,3,3,8) <_{17} \Delta(2,3,8)$	$[1^2.3^5, 1.2^8, 1.8^2]$	9 realisations
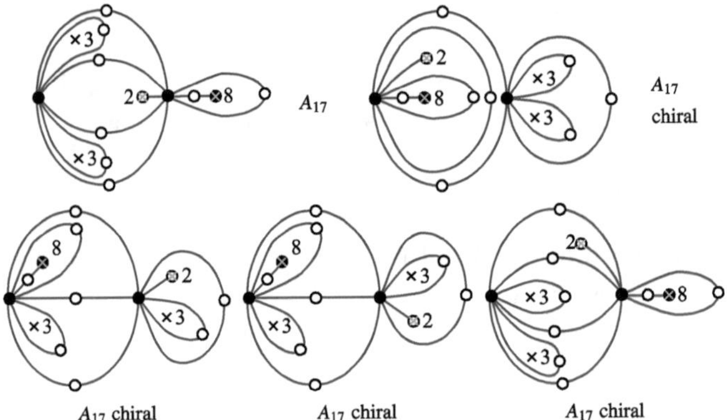			
(S73)	$Q(2,4,5,5) <_{17} \Delta(2,4,5)$	$[1^2.5^3, 1.2^8, 1.4^4]$	8 realisations
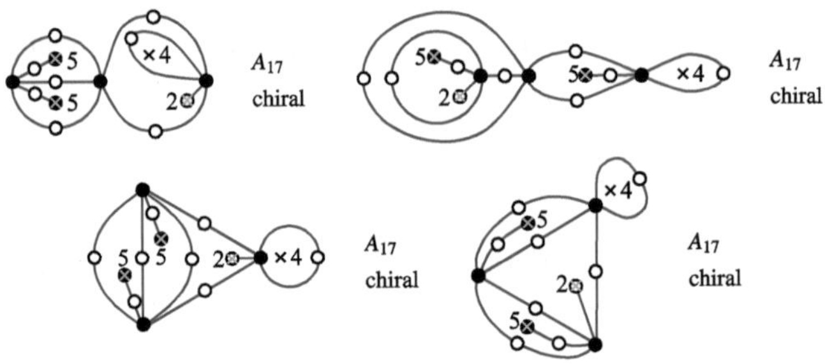			
(S74)	$Q(14,14,14,14) <_{18} \Delta(2,3,14)$	$[1^4.14, 2^9, 3^6]$	1 realisation
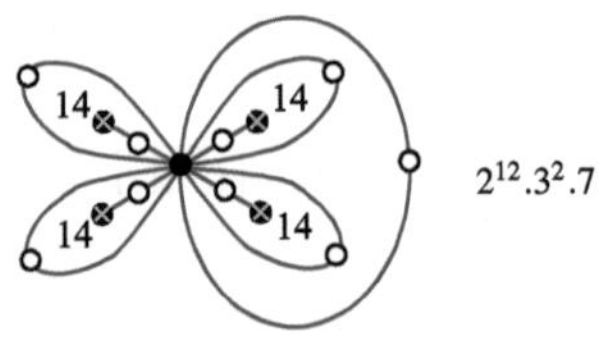			
(S75)	$Q(3,3,3,4) <_{18} \Delta(2,3,8)$	$[2.8^2, 2^9, 1^3.3^5]$	1 realisation
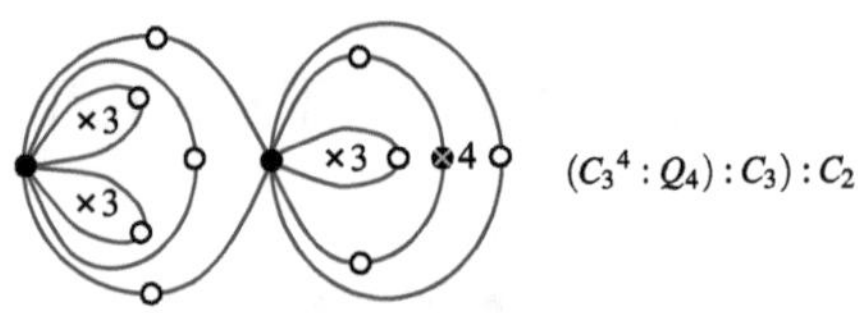			

(continued)

Table 6 (continued)

Label	Inclusion	Passport	Realisation(s)
(S76)	$Q(4,12,12,12) <_{18} \Delta(2,3,12)$	$[1^3.3.12, 2^9, 3^6]$	1 realisation
	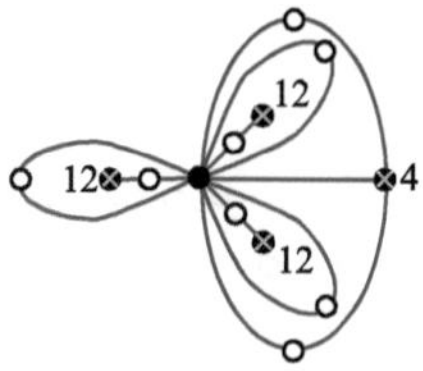	$(((((C_3{}^2 : C_2) \times C_3) : C_2) : C_3) : C_2) \times C_3$	
(S77)	$Q(2,10,10,10) <_{18} \Delta(2,3,10)$	$[1^3.5.10, 2^9, 3^6]$	1 realisation
	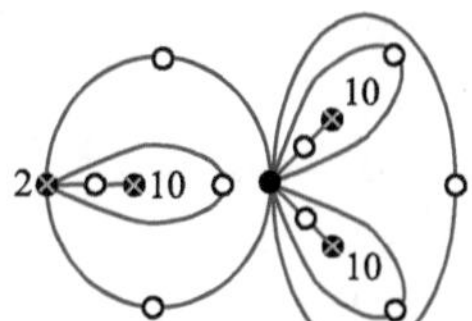	S_{18}	
(S78)	$Q(2,2,8,8) <_{18} \Delta(2,3,8)$	$[1^2.8^2, 1^2.2^8, 3^6]$	8 realisations
	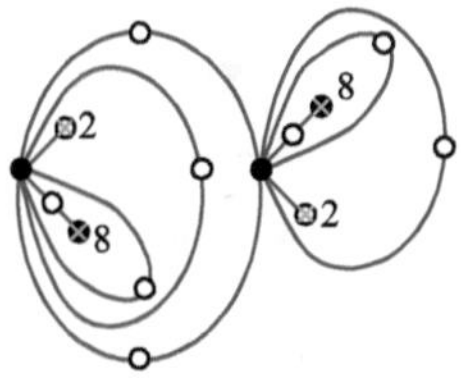 $PSL(2,17)$	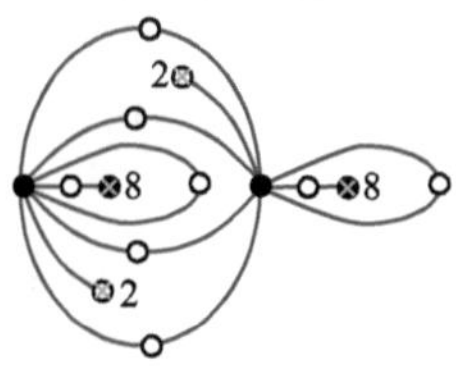 $PSL(2,17)$	
	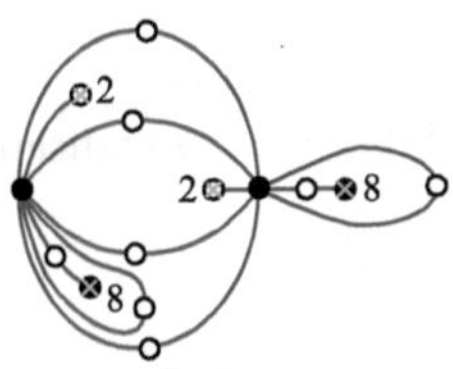 $2^{12}.3^3$ chiral	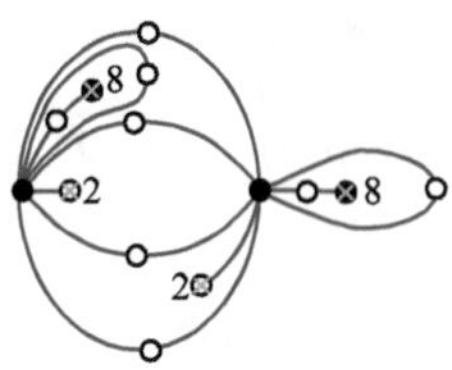 $2^{12}.3^3$ chiral	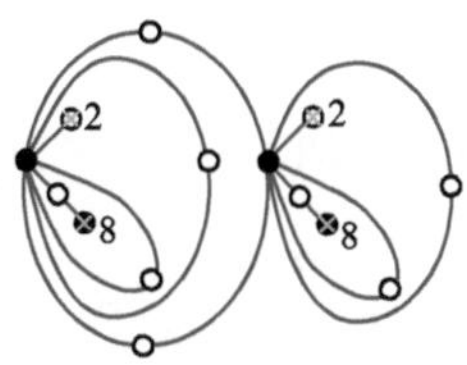 $2^{12}.3^3$ chiral
(S79)	$Q(6,6,12,12) <_{18} \Delta(2,3,12)$	$[1^2.2^2.12, 2^9, 3^6]$	1 realisation
	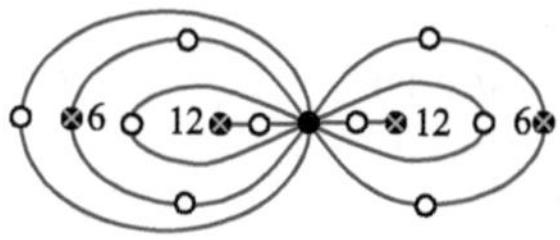	$(C_6{}^2 : C_3) : C_2$	

(continued)

Table 6 (continued)

Label	Inclusion	Passport	Realisation(s)
(S80)	$Q(2,3,4,8) <_{19} \Delta(2,3,8)$	$[1.2.8^2, 1.2^9, 1.3^6]$	10 realisations

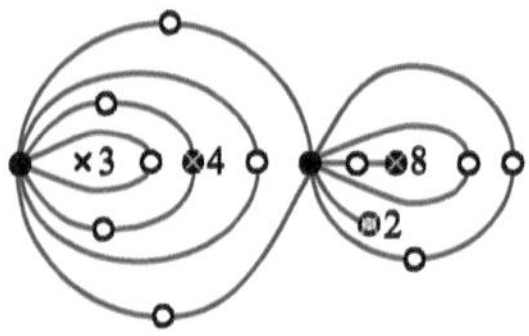

S_{19} chiral

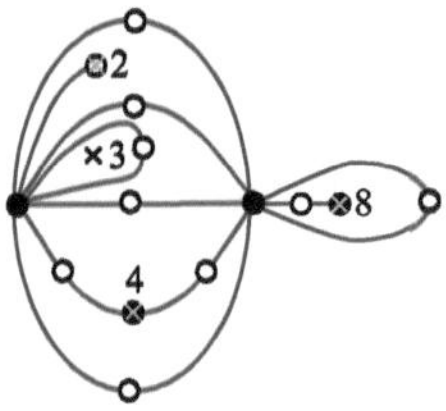

S_{19} chiral

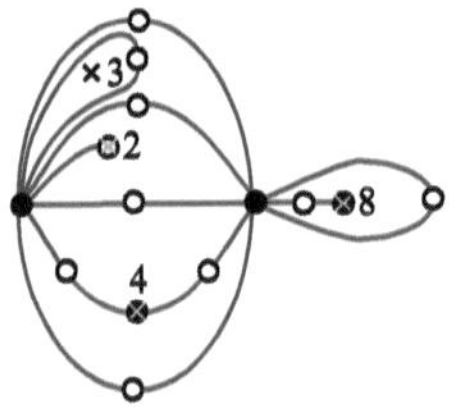

S_{19} chiral

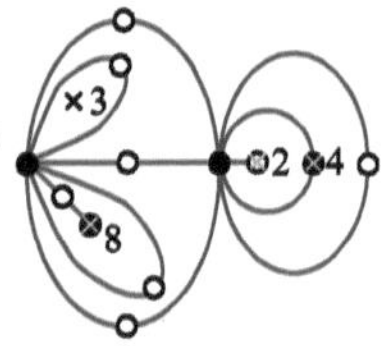

S_{19} chiral

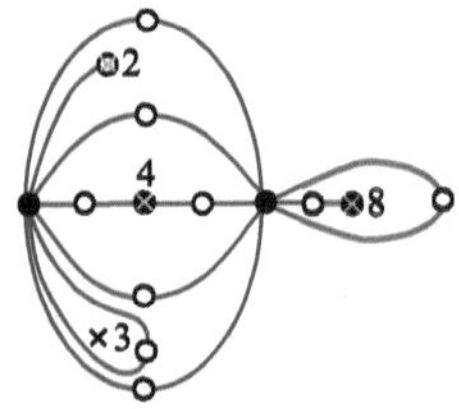

S_{19} chiral

Label	Inclusion	Passport	Realisation(s)
(S81)	$Q(4,4,4,4) <_{20} \Delta(2,4,5)$	$[5^4, 2^{10}, 1^4.4^4]$	3 realisations

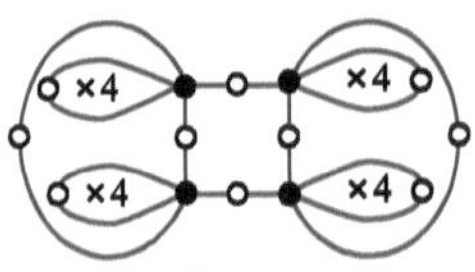

$(C_2{}^4 : C_5) : C_2$

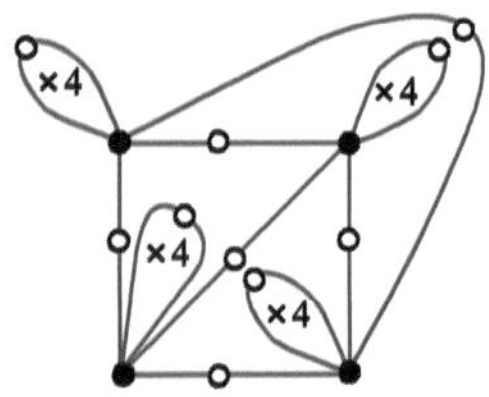

$A_6 \times C_2$

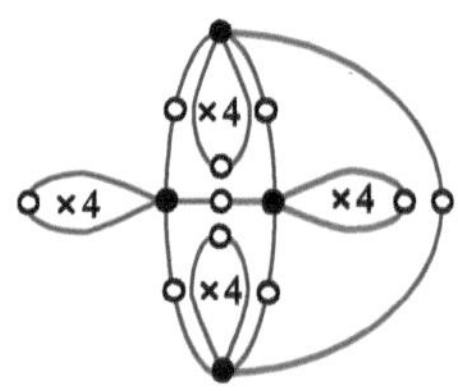

$2^{11}.3.5$

Label	Inclusion	Passport	Realisation(s)
(S82)	$Q(3,3,9,9) <_{20} \Delta(2,3,9)$	$[1^2.9^2, 2^{10}, 1^2.3^6]$	9 realisations

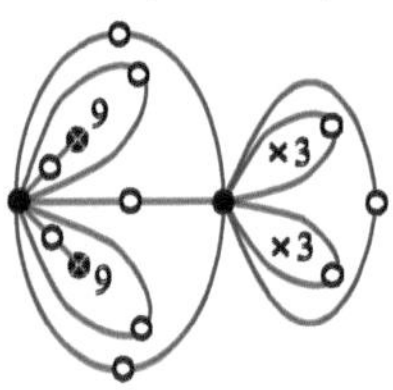

$PSL(2,19)$

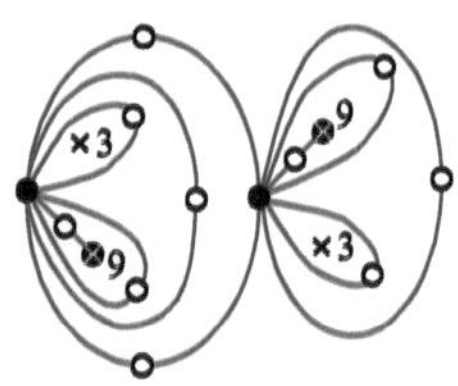

$PSL(2,19)$

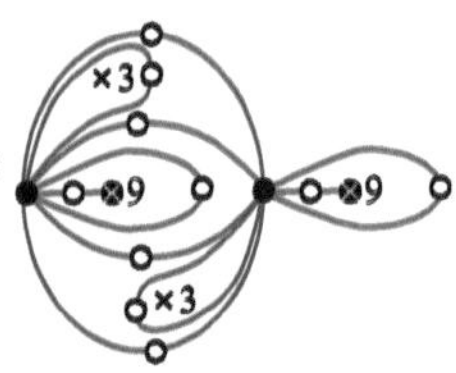

$PSL(2,19)$

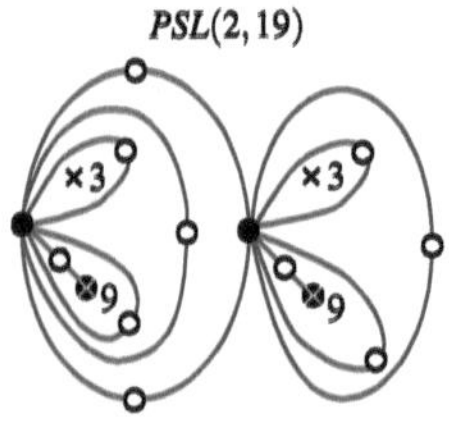

$2^{16}.3^4.5^2.7$ chiral

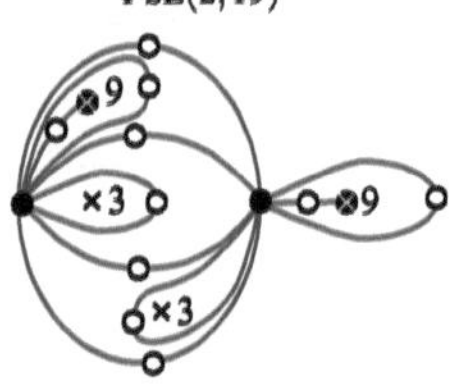

$2^{16}.3^4.5^2.7$ chiral

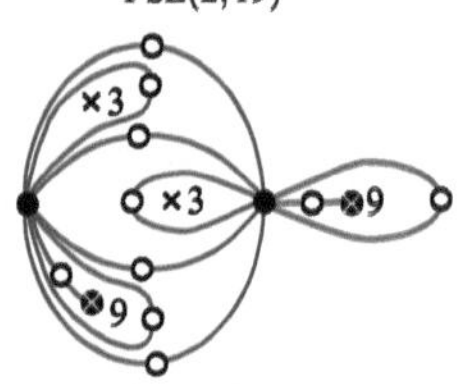

$2^{16}.3^4.5^2.7$ chiral

(continued)

Table 6 (continued)

Label	Inclusion	Passport	Realisation(s)
(S83)	$Q(3,3,4,4) <_{20} \Delta(2,3,8)$	$[2^2.8^2, 2^{10}, 1^2.3^6]$	4 realisations
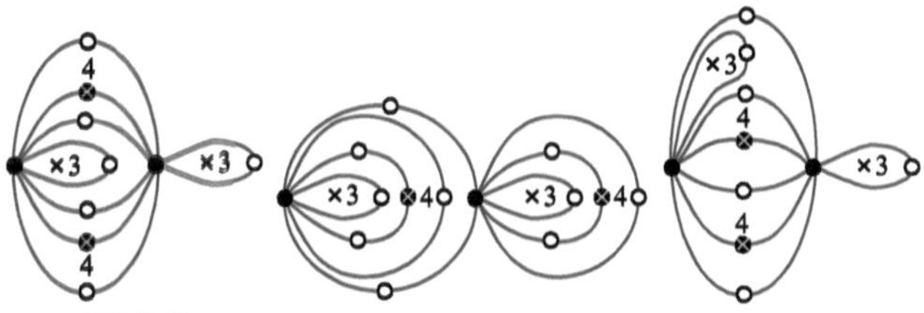 $PGL(2,9)$ $A_{10} \times C_2$ $A_{10} \times C_2$ chiral			
(S84)	$Q(2,3,3,3) <_{21} \Delta(2,3,7)$	$[7^3, 1.2^{10}, 1^3.3^6]$	4 realisations
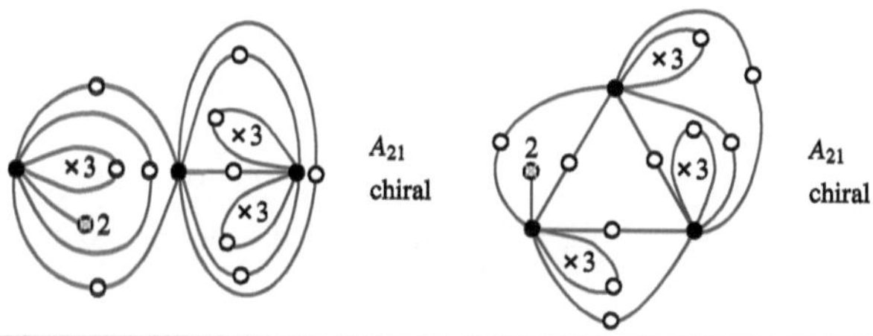			
(S85)	$Q(2,9,9,9) <_{21} \Delta(2,3,9)$	$[1^3.9^2, 1.2^{10}, 3^7]$	2 realisations
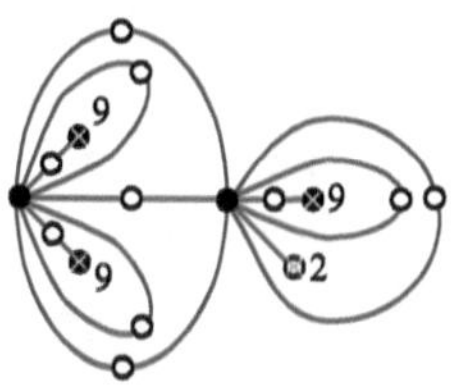 A_{21} chiral			

(continued)

Table 6 (continued)

Label	Inclusion	Passport	Realisation(s)
(S86)	$Q(2,2,3,7) <_{22} \Delta(2,3,7)$	$[1.7^3, 1^2.2^{10}, 1.3^7]$	13 realisations
	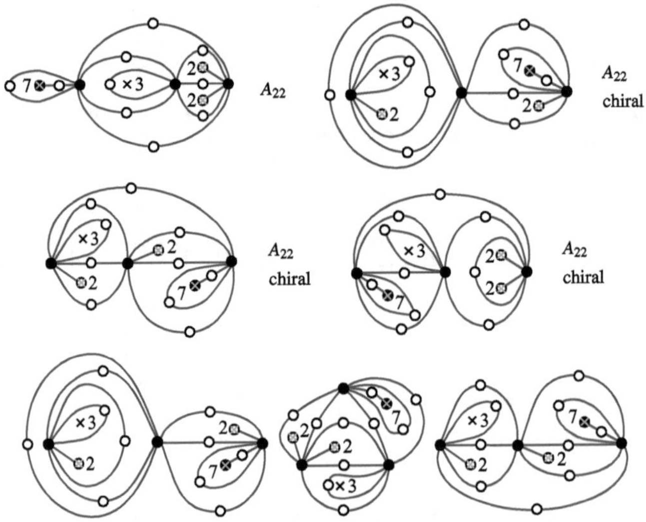		A_{22} A_{22} chiral A_{22} chiral A_{22} chiral A_{22} chiral A_{22} chiral A_{22} chiral
(S87)	$Q(4,4,5,5) <_{22} \Delta(2,4,5)$	$[1^2.5^4, 2^{11}, 1^2.4^5]$	5 realisations
	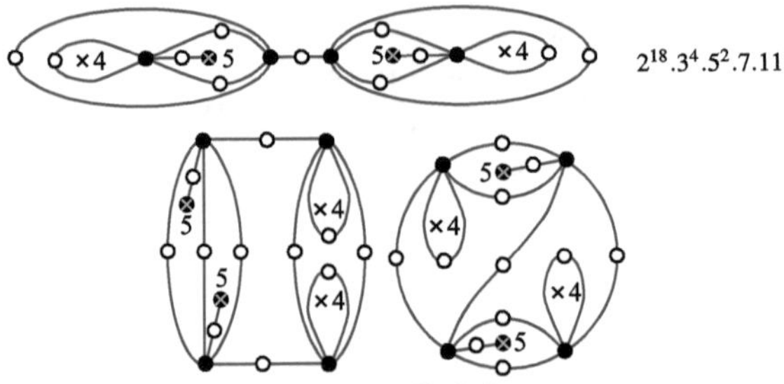		$2^{18}.3^4.5^2.7.11$ $2^{18}.3^4.5^2.7.11$ chiral $2^{18}.3^4.5^2.7.11$ chiral
(S88)	$Q(5,5,5,5) <_{24} \Delta(2,4,5)$	$[1^4.5^4, 2^{12}, 4^6]$	5 realisations
	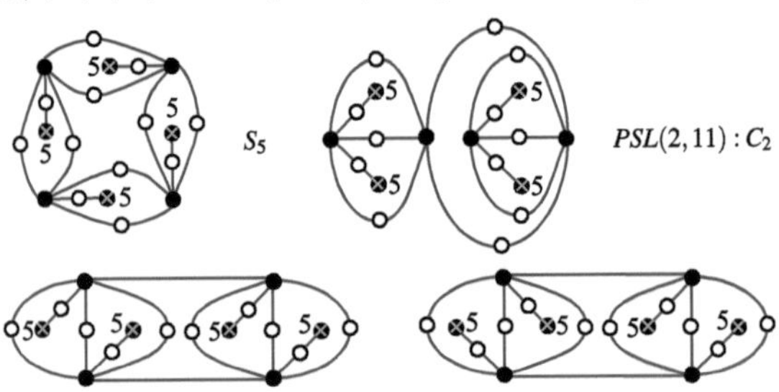		S_5 $PSL(2,11):C_2$ $(C_2{}^5:A_6):C_2$ chiral $PSL(2,11):C_2$

(continued)

Table 6 (continued)

Label	Inclusion	Passport	Realisation(s)
(S89)	$Q(10,10,10,10) <_{24} \Delta(2,3,10)$	$[1^4.10^2, 2^{12}, 3^8]$	3 realisations
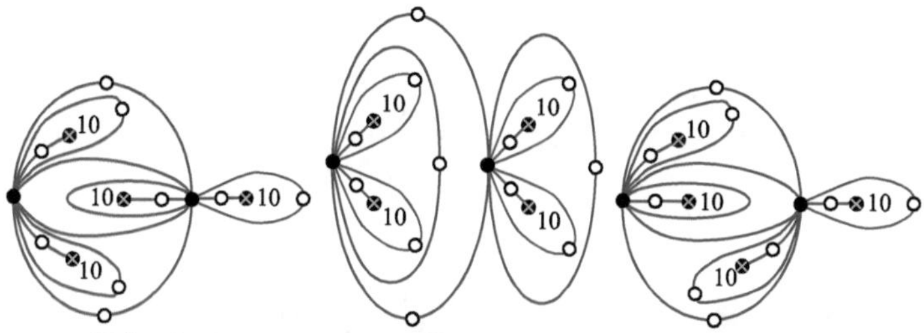	$(C_2{}^5 : A_5) : C_2$	$2^{14}.3.5.11$	$2^{14}.3.5.11$
(S90)	$Q(2,4,8,8) <_{24} \Delta(2,3,8)$	$[1^2.2.4.8^2, 2^{12}, 3^8]$	1 realisation
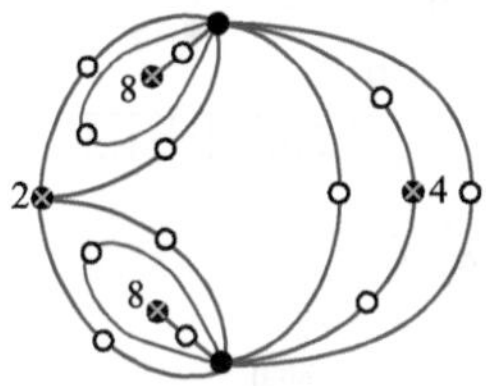	$(((C_4 \times C_2) : C_4) : C_3) : C_2$		
(S91)	$Q(4,4,4,4) <_{24} \Delta(2,3,8)$	$[2^4.8^2, 2^{12}, 3^8]$	1 realisation
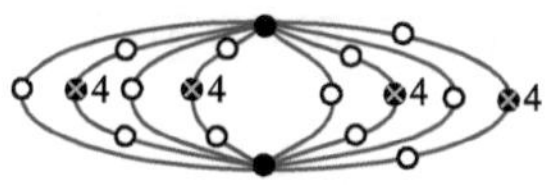	$(C_4{}^2 : C_3) : C_2$		
(S92)	$Q(3,3,8,8) <_{26} \Delta(2,3,8)$	$[1^2.8^3, 2^{13}, 1^2.3^8]$	5 realisations
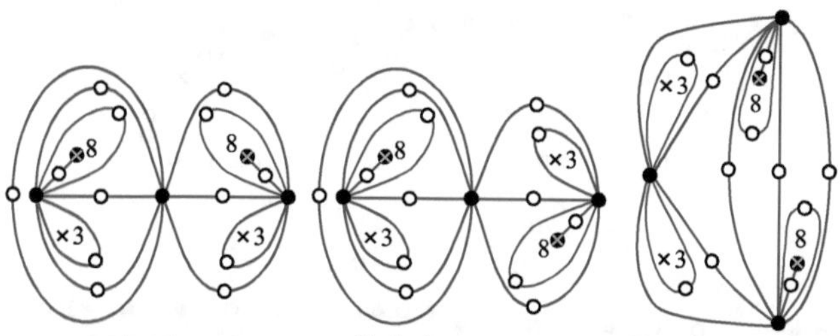	$PSL(2,25) : C_2$	$2^{22}.3^5.5^2.7.11.13$ chiral	$2^{22}.3^5.5^2.7.11.13$ chiral
(S93)	$Q(2,8,8,8) <_{27} \Delta(1,2,3,8)$	$[1^3.8^3, 1.2^{13}, 3^9]$	2 realisations
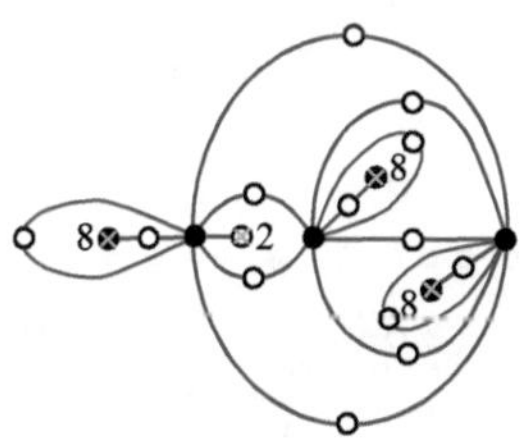	$2^{23}.3^{13}.5^6.7^3.11^2.13^2.17.19.23$ chiral		

(continued)

Table 6 (continued)

Label	Inclusion	Passport	Realisation(s)
(S94)	$Q(3,3,3,3) <_{28} \Delta(2,3,7)$	$[7^4, 2^{14}, 1^4.3^8]$	5 realisations

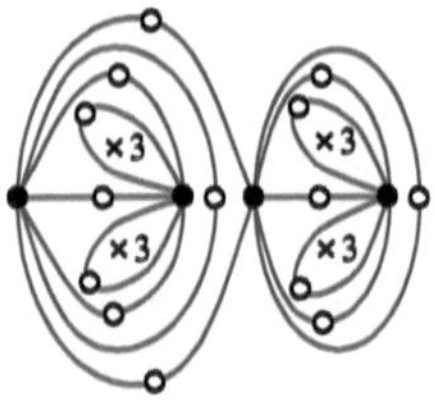

$PSL(2,13)$

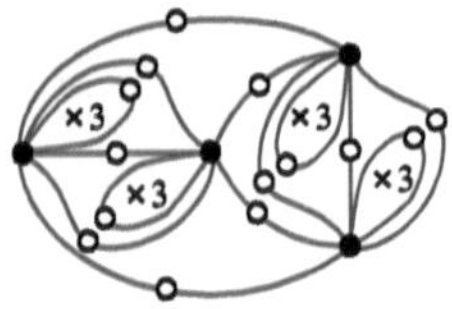

$PSL(2,13)$

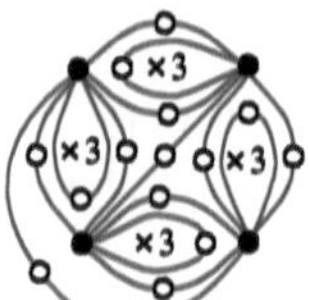

$PSL(2,13)$

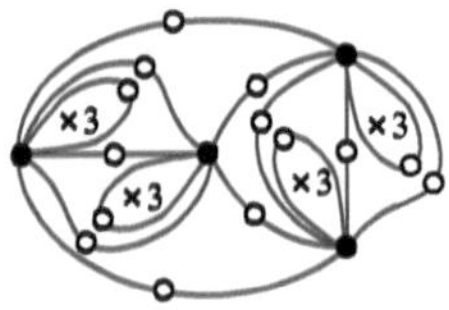

$C_2{}^3 \cdot PSL(3,2)$
chiral

Label	Inclusion	Passport	Realisation(s)
(S95)	$Q(3,4,8,8) <_{28} \Delta(2,3,8)$	$[1^2.2.8^3, 2^{14}, 1.3^9]$	2 realisations

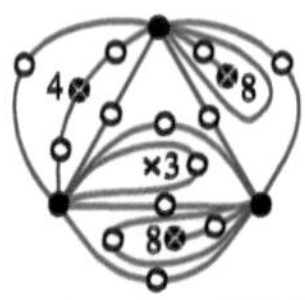

A_{28}
chiral

Label	Inclusion	Passport	Realisation(s)
(S96)	$Q(2,3,3,7) <_{29} \Delta(2,3,7)$	$[1.7^4, 1.2^{14}, 1^2.3^9]$	14 realisations

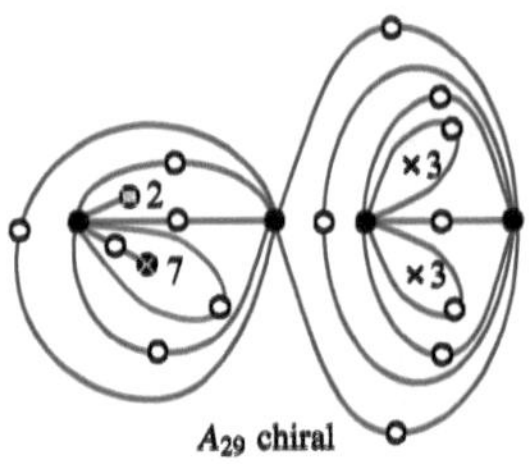

A_{29} chiral

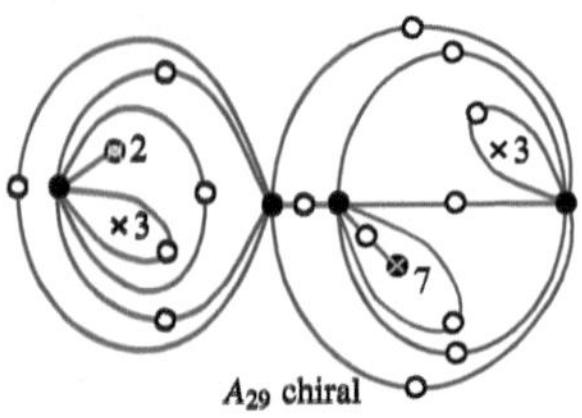

A_{29} chiral

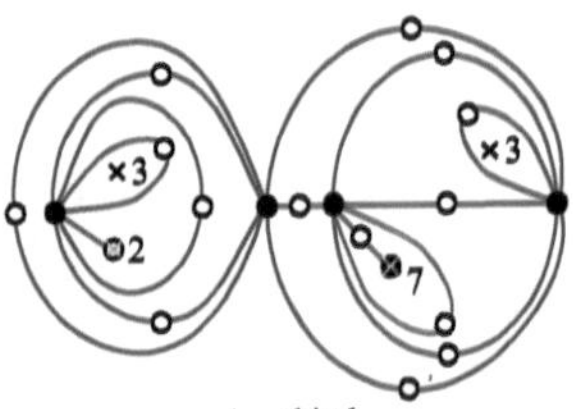

A_{29} chiral

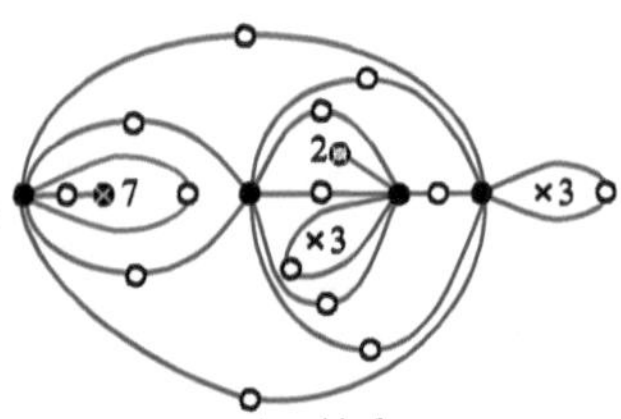

A_{29} chiral

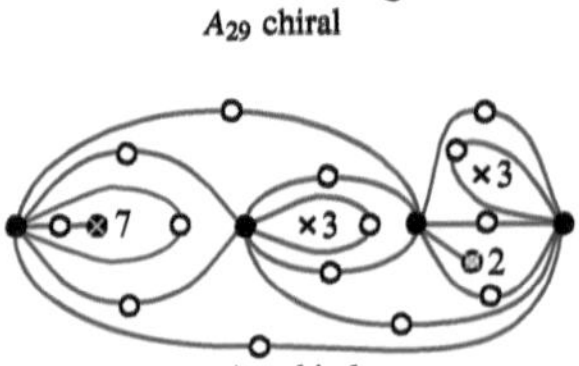

A_{29} chiral

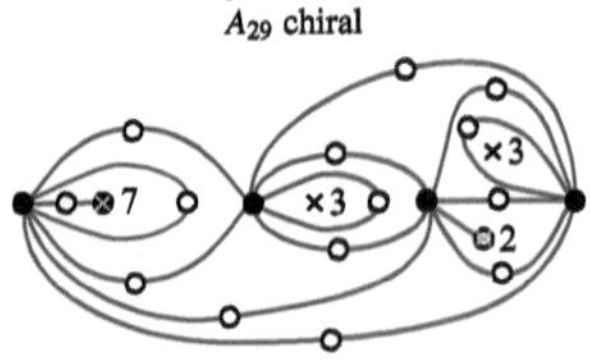

A_{29} chiral

(continued)

Table 6 (continued)

Label	Inclusion	Passport	Realisation(s)

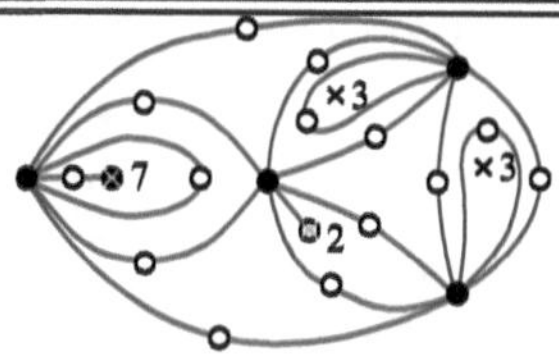

A_{29}
chiral

(S97) $Q(2,2,7,7) <_{30} \Delta(2,3,7)$ $[1^2.7^4, 1^2.2^{14}, 3^{10}]$ 12 realisations

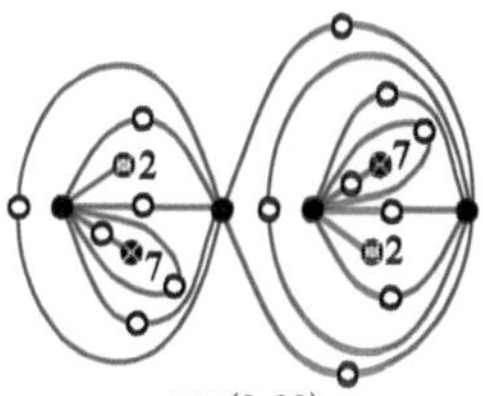

$PSL(2,29)$

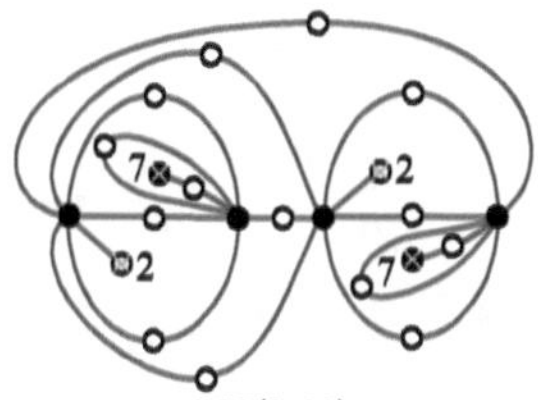

$PSL(2,29)$

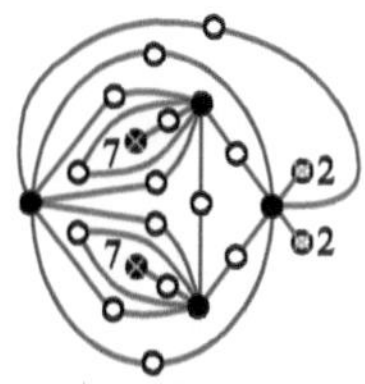

$PSL(2,29)$

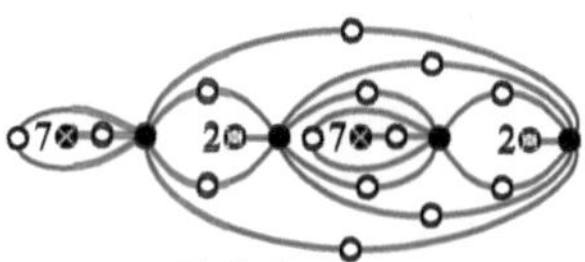

$2^{24}.3^6.5^3.7^2.11.13$

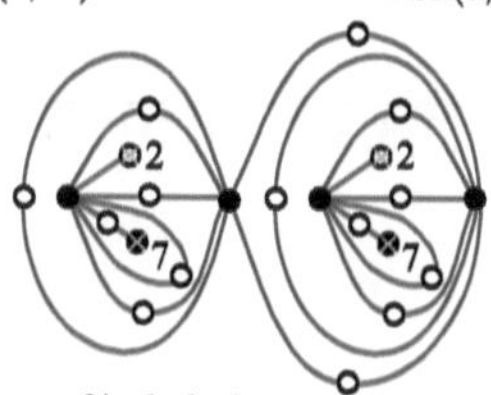

$2^{24}.3^6.5^3.7^2.11.13$ chiral

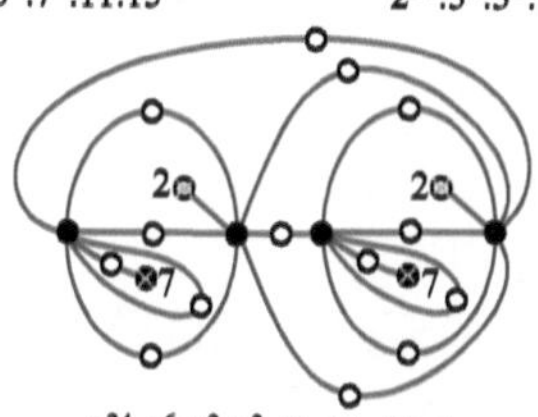

$2^{24}.3^6.5^3.7^2.11.13$ chiral

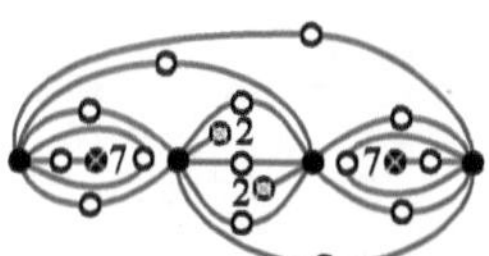

$2^{24}.3^6.5^3.7^2.11.13$ chiral

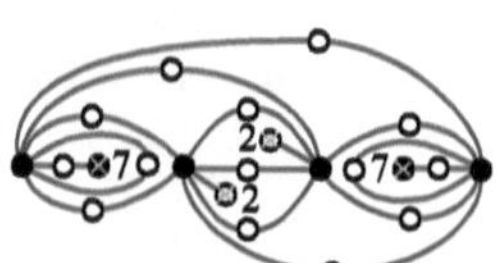

$2^{24}.3^6.5^3.7^2.11.13$ chiral

(S98) $Q(4,4,8,8) <_{30} \Delta(2,3,8)$ $[1^2.2^2.8^3, 2^{15}, 3^{10}]$ 3 realisations

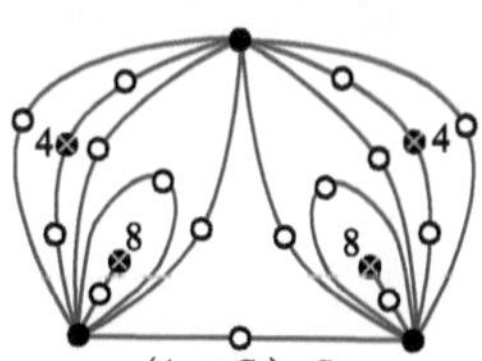

$(A_6 \times C_3) : C_2$

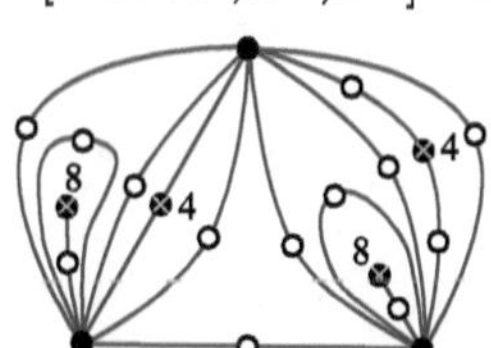

$2^{17}.3.5^3$ chiral

(continued)

Table 6 (continued)

Label	Inclusion	Passport	Realisation(s)
(S99)	$Q(8,8,8,8) <_{36} \Delta(2,3,8)$	$[1^4.8^4, 2^{18}, 3^{12}]$	4 realisations

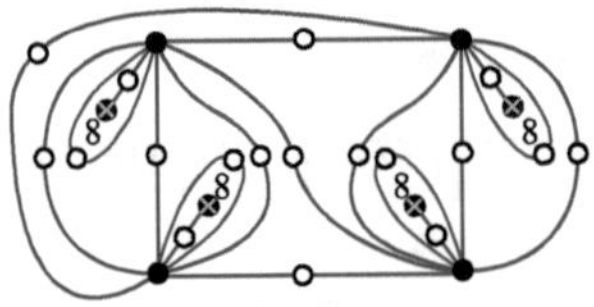

$PSL(2,17) \times C_2$

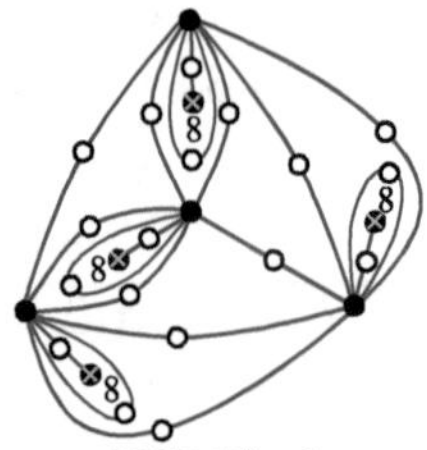

$PSL(2,17) \times C_2$

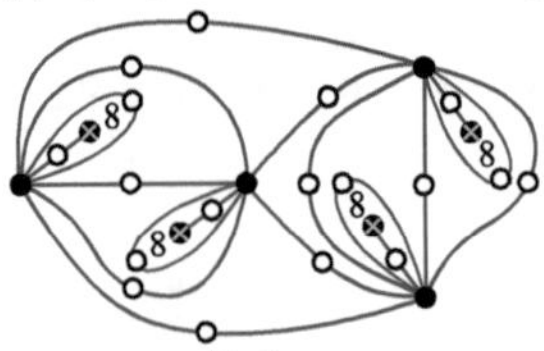

$2^{20}.3^3$ chiral

(S100)	$Q(3,3,3,7) <_{36} \Delta(2,3,7)$	$[1.7^5, 2^{18}, 1^3.3^{11}]$	4 realisations

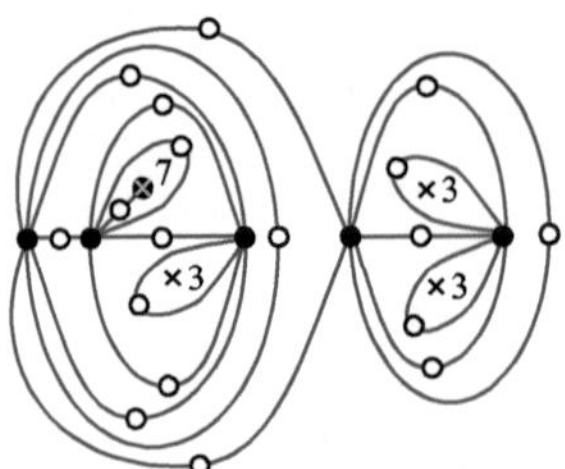

A_{36} chiral

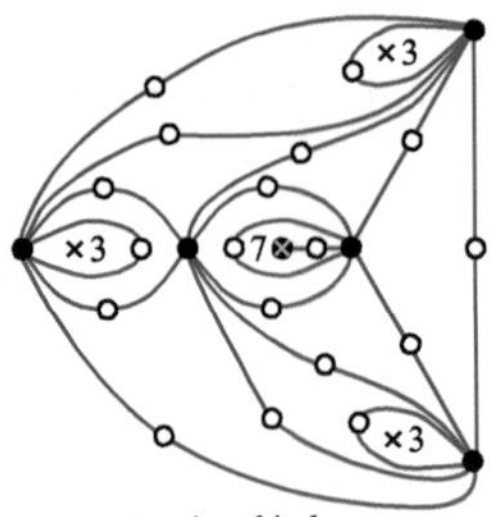

A_{36} chiral

(S101)	$Q(2,3,7,7) <_{37} \Delta(2,3,7)$	$[1^2.7^5, 1.2^{18}, 1.3^{12}]$	15 realisations

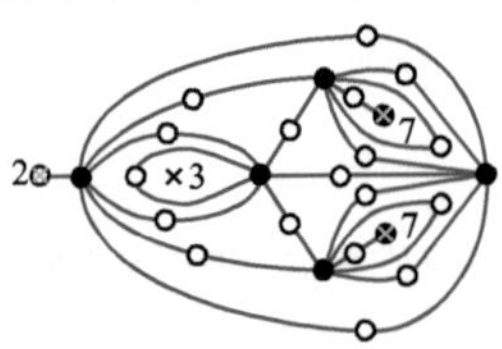

A_{37}

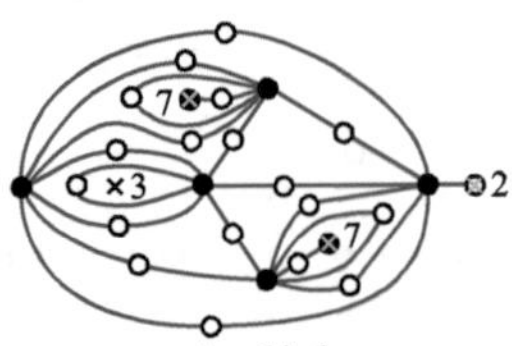

A_{37} chiral

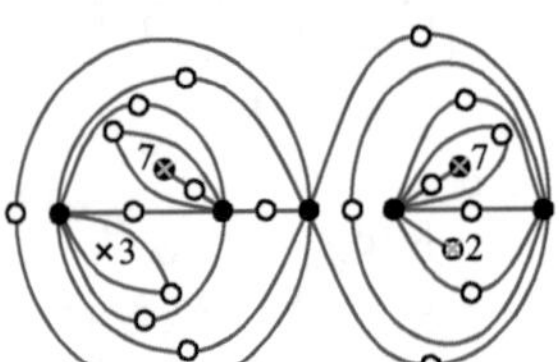

A_{37} chiral

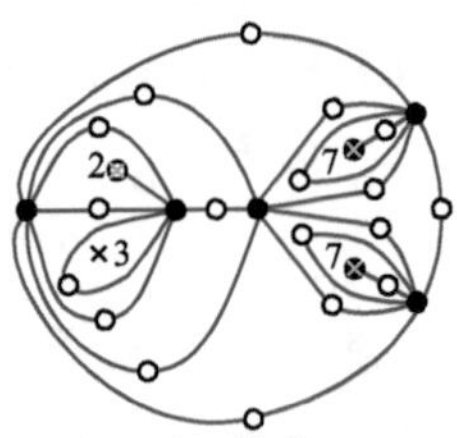

A_{37} chiral

(continued)

Table 6 (continued)

Label	Inclusion	Passport	Realisation(s)

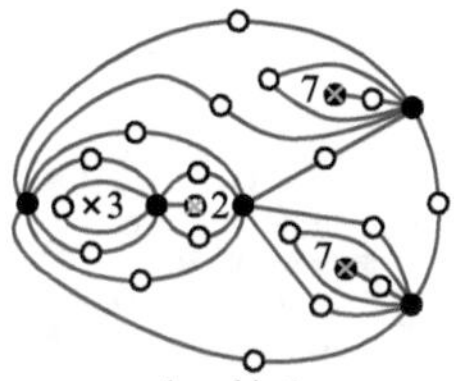

A_{37} chiral

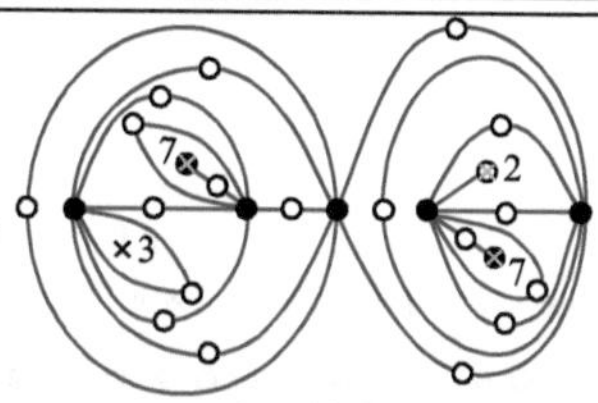

A_{37} chiral

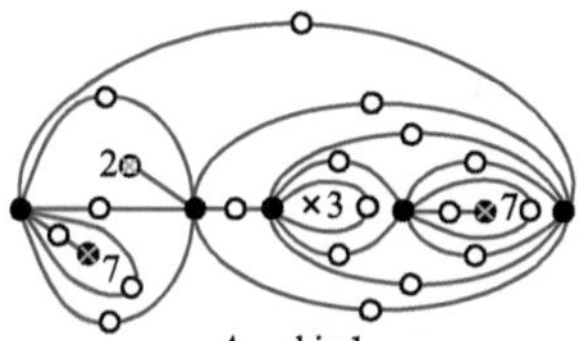

A_{37} chiral

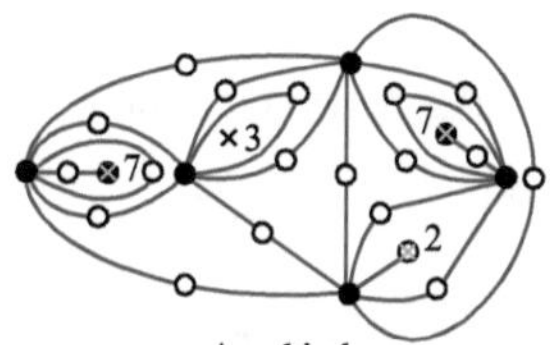

A_{37} chiral

(S102)	$Q(3,3,7,7) <_{44} \Delta(2,3,7)$	$[1^2.7^6, 2^{22}, 1^2.3^{14}]$	16 realisations

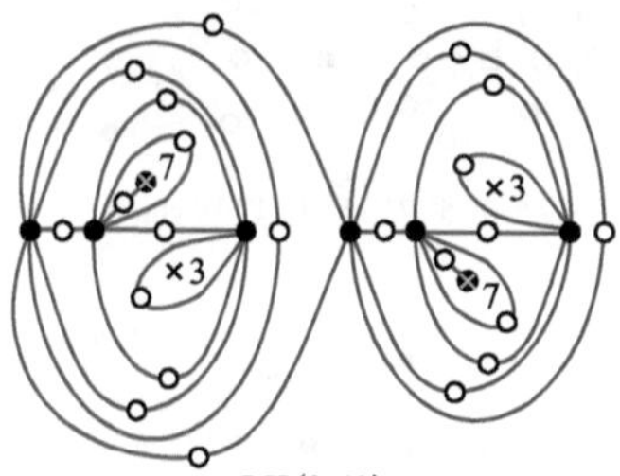

$PSL(2,43)$

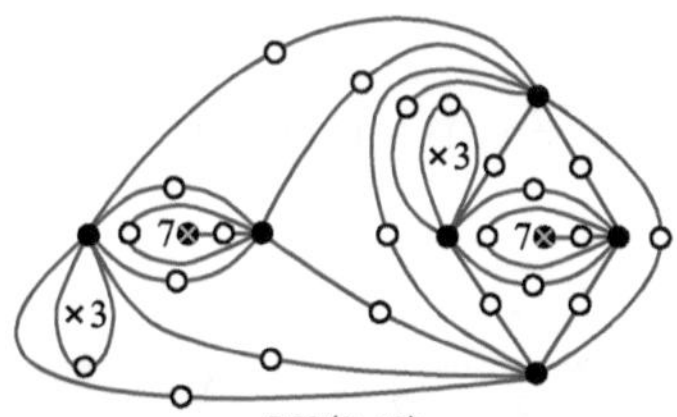

$PSL(2,43)$

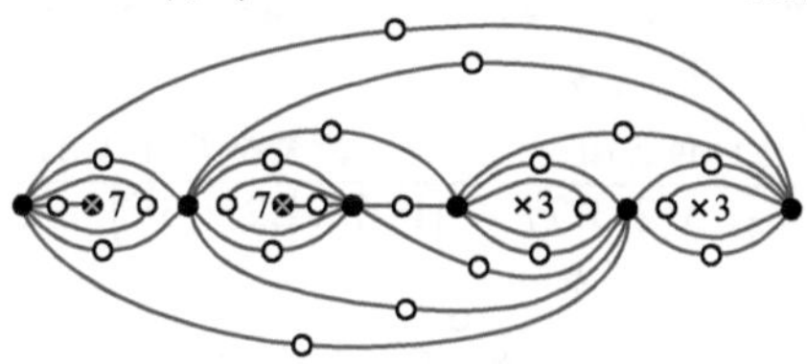

$PSL(2,43)$

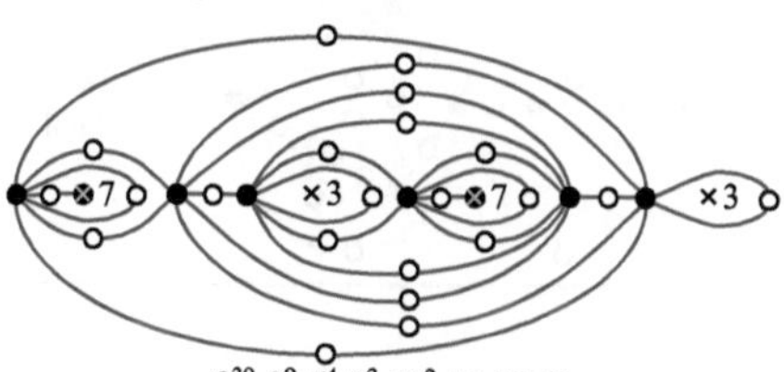

$2^{39}.3^9.5^4.7^3.11^2.13.17.19$

(continued)

Table 6 (continued)

Label	Inclusion	Passport	Realisation(s)

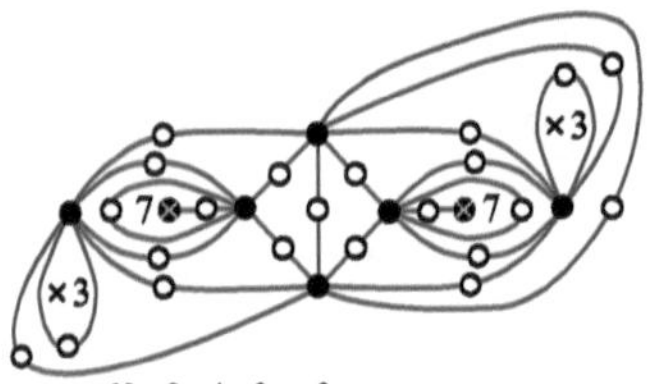

$2^{39}.3^9.5^4.7^3.11^2.13.17.19$ chiral

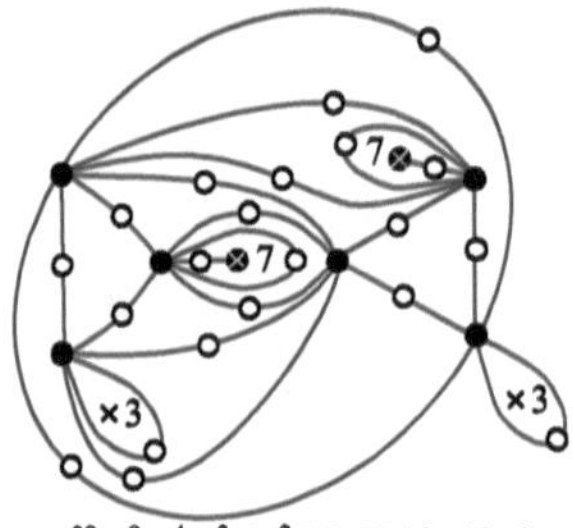

$2^{39}.3^9.5^4.7^3.11^2.13.17.19$ chiral

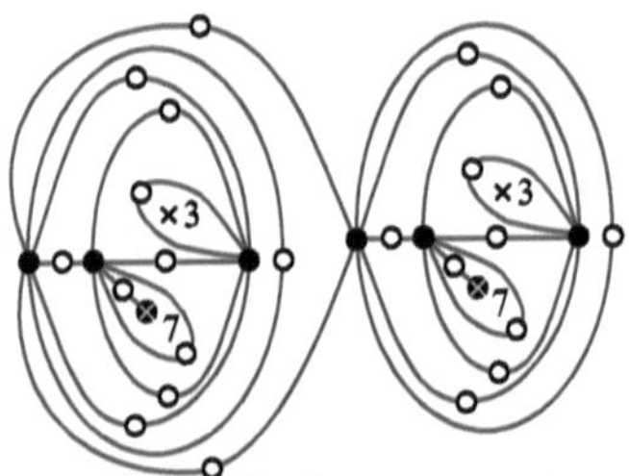

$2^{39}.3^9.5^4.7^3.11^2.13.17.19$ chiral

$2^{39}.3^9.5^4.7^3.11^2.13.17.19$ chiral

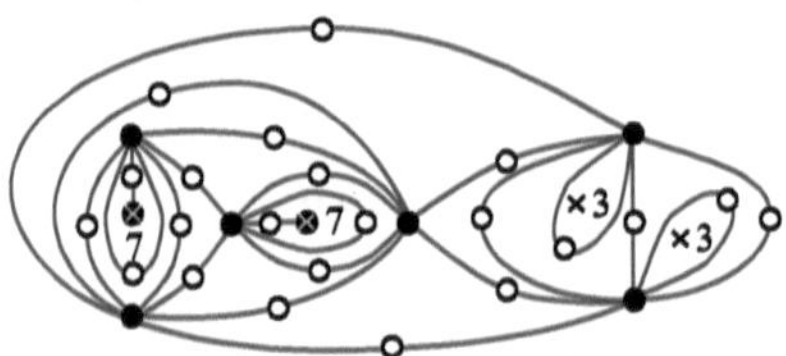

$2^{39}.3^9.5^4.7^3.11^2.13.17.19$ chiral

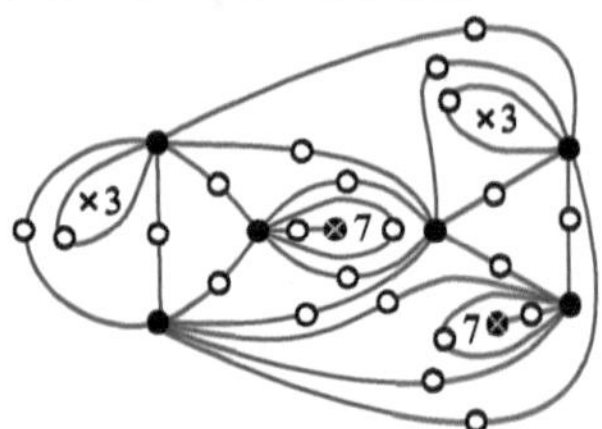

$2^{39}.3^9.5^4.7^3.11^2.13.17.19$ chiral

Label	Inclusion	Passport	Realisation(s)
(S103)	$Q(2,7,7,7) <_{45} \Delta(2,3,7)$	$[1^3.7^6, 1.2^{22}, 3^{15}]$	2 realisations

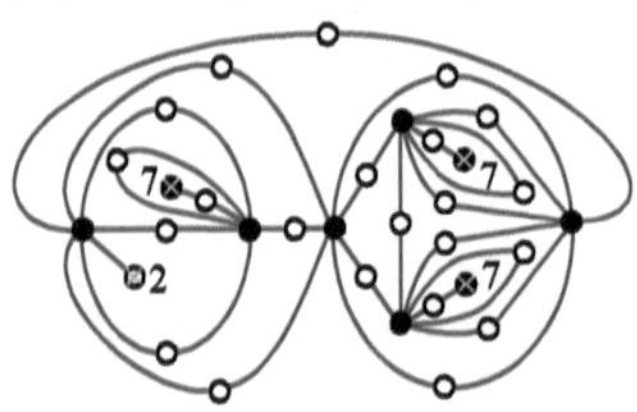

A_{45}
chiral

(continued)

Table 6 (continued)

Label	Inclusion	Passport	Realisation(s)
(S104)	$Q(3,7,7,7) <_{52} \Delta(2,3,7)$	$[1^3.7^7, 2^{26}, 1.3^{17}]$	1 realisation

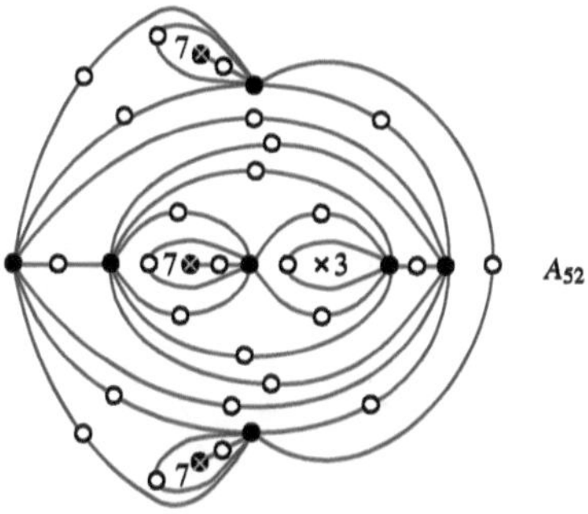

Label	Inclusion	Passport	Realisation(s)
(S105)	$Q(7,7,7,7) <_{60} \Delta(2,3,7)$	$[1^4.7^8, 2^{30}, 3^{20}]$	6 realisations

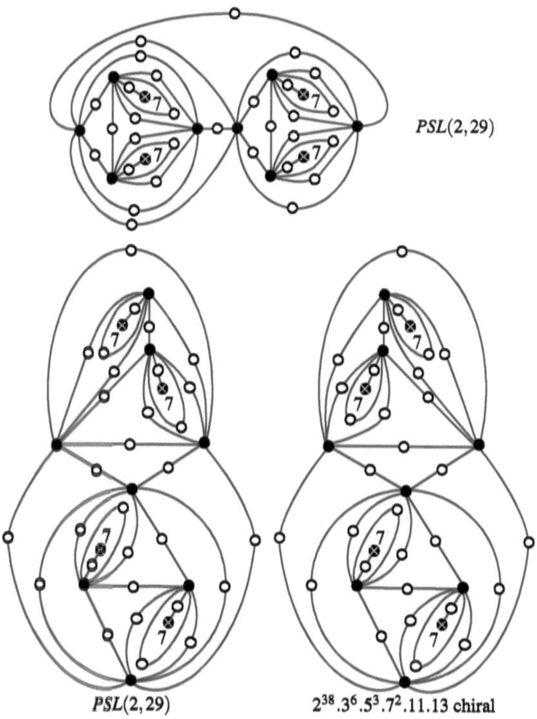

(continued)

Table 6 (continued)

Label	Inclusion	Passport	Realisation(s)

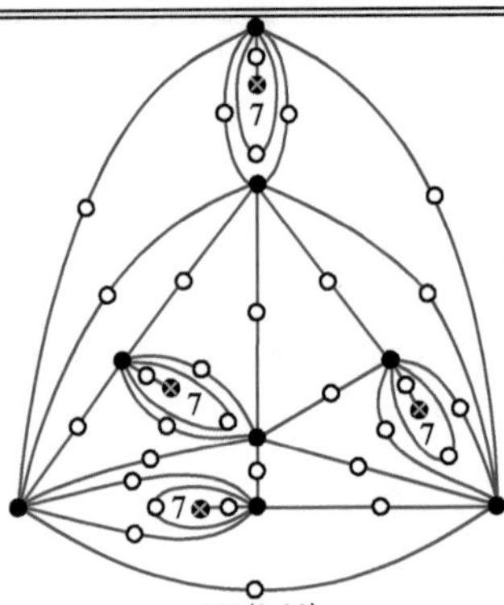

$PSL(2,29)$

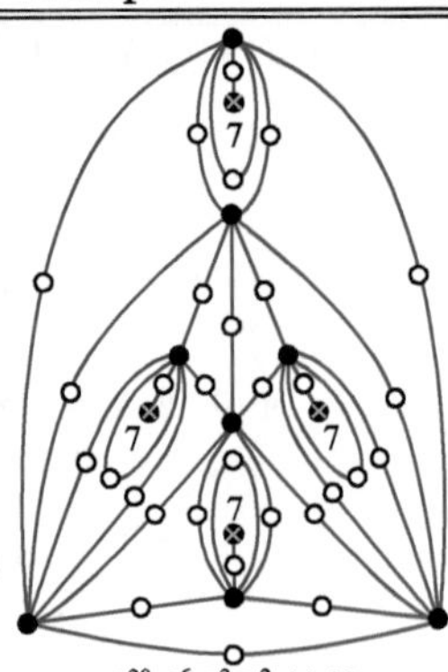

$2^{38}.3^6.5^3.7^2.11.13$

Numerical Solutions with No Corresponding Inclusions

See Table 7.

Table 7 Admissible passports for which there is no inclusion (Forbidden passports)

Numerical solution	Passport	Notes
$(m, 3m, n, n) <_4 (1, 2, 3m, 2n)$	$[1.3, 2^2, 2^2]$	$m = 1: \Delta(3, n, n) \not< \Delta(2, 3, 2n)$
$(n, 3n, 3n, 3n) <_6 (1, 3, 3, 3n)$	$[1^3.3, 3^2, 3^2]$	
$(2, n, 4n, 4n) <_6 (1, 2, 4, 4n)$	$[1^2.4, 2^3, 2.4]$	
$(4, 4, n, 2n) <_6 (1, 2, 4, 4n)$	$[1^2.4, 2^3, 2.4]$	
$(1, 2n, 3n, 6n) <_6 (1, 2, 3, 6n)$	$[1.2.3, 2^3, 3^2]$	
$(5, n, n, n) <_6 (1, 2, 5, 2n)$	$[1.5, 2^3, 2^3]$	
$(2, n, n, n) <_6 (1, 2, 4, 2n)$	$[2^3, 2^3, 2.4]$	
$(n, 5n, 5n, 5n) <_8 (1, 2, 4, 5n)$	$[1^3.5, 2^4, 4^2]$	$n = 1: \Delta(5, 5, 5) \not< \Delta(2, 4, 5)$
$(3, 3, 3n, 5n) <_8 (1, 2, 3, 15n)$	$[3.5, 2^4, 1^2.3^2]$	
$(2n, 3n, 3n, 6n) <_8 (1, 2, 4, 6n)$	$[1.2^2.3, 2^4, 4^2]$	
$(2, n, n, 4n) <_9 (1, 2, 3, 4n)$	$[1.4^2, 1.2^4, 3^3]$	
$(2, 2n, 5n, 5n) <_9 (1, 2, 3, 10n)$	$[2^2.5, 1.2^4, 3^3]$	
$(2, n, n, n) <_9 (1, 2, 3, 3n)$	$[3^3, 1.2^4, 3^3]$	
$(3, n, 2n, 6n) <_{10} (1, 2, 3, 6n)$	$[1.3.6, 2^5, 1.3^3]$	
$(3, n, 3n, 3n) <_{10} (1, 2, 3, 6n)$	$[2^2.6, 2^5, 1.3^3]$	
$(3, n, n, 2n) <_{10} (1, 2, 3, 4n)$	$[2.4^2, 2^5, 1.3^3]$	
$(3, 3n, 4n, 4n) <_{10} (1, 2, 3, 12n)$	$[3^2.4, 2^5, 1.3^3]$	
$(3n, 7n, 21n, 21n) <_{12} (1, 2, 3, 21n)$	$[1^2.3.7, 2^6, 3^4]$	
$(2n, 3n, 12n, 12n) <_{12} (1, 2, 3, 12n)$	$[1^2.4.6, 2^6, 3^4]$	
$(2n, 7n, 7n, 14n) <_{12} (1, 2, 3, 14n)$	$[1.2^2.7, 2^6, 3^4]$	
$(4n, 5n, 10n, 20n) <_{12} (1, 2, 3, 20n)$	$[1.2.4.5, 2^6, 3^4]$	
$(3n, 5n, 5n, 15n) <_{12} (1, 2, 3, 15n)$	$[1.3^2.5, 2^6, 3^4]$	
$(3n, 3n, 4n, 12n) <_{12} (1, 2, 3, 12n)$	$[1.3.4^2, 2^6, 3^4]$	

Table 7 (continued)

Numerical solution	Passport	Notes
$(n, 3n, 3n, 3n) <_{12}$ $(1, 2, 3, 6n)$	$[2^3.6, 2^6, 3^4]$	
$(6n, 10n, 15n, 15n) <_{12}$ $(1, 2, 3, 30n)$	$[2^2.3.5, 2^6, 3^4]$	
$(3n, 4n, 4n, 6n) <_{12}$ $(1, 2, 3, 12n)$	$[2.3^2.4, 2^6, 3^4]$	
$(3, 5, 5, 5) <_8 (1, 2, 5, 6)$	$[2.6, 2^4, 1^3.5]$	
$(2, 2, 6, 6) <_8 (1, 2, 4, 6)$	$[1^2.6, 2^4, 2^2.4]$	
$(5, 5, 5, 5) <_9 (1, 3, 3, 5)$	$[1^4.5, 3^3, 3^3]$	
$(2, 4, 4, 4) <_9 (1, 3, 3, 4)$	$[3^3, 3^3, 1^3.2.4]$	
$(2, 7, 7, 7) <_{10} (1, 2, 4, 7)$	$[2.4^2, 2^5, 1^3.7]$	
$(2, 3, 6, 6) <_{10} (1, 2, 4, 6)$	$[2.4^2, 2^5, 1^2.2.6]$	
$(8, 8, 8, 8) <_{12} (1, 2, 4, 8)$	$[1^4.8, 2^6, 4^3]$	
$(1, 3, 7, 7) <_{16} (1, 2, 3, 7)$	$[1^2.7^2, 2^8, 1.3^5]$	$\Delta(3, 7, 7) \not< \Delta(2, 3, 7)$
$(3, 5, 5, 5) <_{16} (1, 2, 3, 10)$	$[2^3.10, 2^8, 1.3^5]$	
$(2, 3, 4, 4) <_{16} (1, 2, 3, 8)$	$[2^2.4.8, 2^8, 1.3^5]$	
$(2, 5, 5, 5) <_{18} (1, 2, 4, 5)$	$[2.4^4, 2^9, 1^3.5^3]$	
$(2, 2, 8, 8) <_{18} (1, 2, 3, 8)$	$[1^2.4^2.8, 2^9, 3^6]$	
$(5, 5, 5, 5) <_{18} (1, 2, 3, 10)$	$[2^4.10, 2^9, 3^6]$	
$(2, 4, 4, 4) <_{18} (1, 2, 3, 8)$	$[2^3.4.8, 2^9, 3^6]$	
$(2, 4, 4, 8) <_{21} (1, 2, 3, 8)$	$[1.2^2.8^2, 1.2^{10}, 3^7]$	
$(2, 3, 8, 8) <_{22} (1, 2, 3, 8)$	$[1^2.4.8^2, 2^{11}, 1.3^7]$	
$(3, 4, 4, 4) <_{22} (1, 2, 3, 8)$	$[2^3.8^2, 2^{11}, 1.3^7]$	
$(3, 9, 9, 9) <_{24} (1, 2, 3, 9)$	$[1^3.3.9^2, 2^{12}, 3^8]$	

References

1. G. V. Belyĭ: Galois extensions of a maximal cyclotomic field, Izv. Akad. Nauk SSSR Ser. Mat. **43** (1979), no. 2, 267–276, 479.
2. W. Bosma, J. Cannon and C. Playoust: The Magma algebra system. I. The user language, J. Symbolic Comput., **24** (1997), 235–265.
3. A. Breda: A theory of restricted regularity of hypermaps, J. Korean Math. Soc. **43** (2006), No. 5, 991–1018.
4. A. Breda: Riemann surfaces and restrictively-marked hypermaps, Ars Math. Contemp. **3** (2010), 87–98.
5. A. Breda, D. A. Catalano, J. Karabáš and R. Nedela: Maps of Archimedean class and operations on dessins, Discrete Math. **338**, Issue 10 (2015), 1814–1825.
6. A. Breda, D. A. Catalano, J. Karabáš and R. Nedela: Quadrangle group inclusions, to appear.
7. A. Breda, D. A. Catalano, J. Karabáš and R. Nedela: Atlas of quadrangle group inclusions, web page, http://www.savbb.sk/~karabas/science.html#atlas.
8. H. U. Besche, B. Eick, and E. A. O'Brien: The Small Groups library, http://www.icm.tu-bs.de/ag_algebra/software/small/, 2014.

9. The GAP Group: GAP – Groups, Algorithms, and Programming, Version 4.7.6, 2014, http://www.gap-system.org.
10. L. Greenberg: Maximal Fuchsian Groups, Bull. Amer. Math. Soc., **69** (1963), 569–573.
11. G. A. Jones and D. Singerman: Complex Functions, an algebraic and geometric viewpoint, Cambridge University Press, Cambridge 1988.
12. S. Katok: Fuchsian groups, The University of Chicago Press, Chicago 1992.
13. S. K. Lando and A. K. Zvonkin: Graphs on Surfaces and Their Applications, Springer 2004.
14. D. Singerman: Subgroups of Fuchsian groups and finite permutation groups, Bull. London Math. Soc., **2** (1970), 319–323.
15. D. Singerman: Finitely maximal Fuchsian groups, J. London Math. Soc. (2), **6** (1972), 29–38.
16. D. Singerman and R. I. Syddall: The Riemann surface of a uniform dessin, Beiträge Algebra Geom. **44** (2003), no. 2, 413–430.
17. J. Wolfart: The 'Obvious' part of Belyi's Theorem and Riemann Surfaces with many automorphisms, pp. 97–112 in Geometric Galois Actions 1, ed. L. Schneps and P. Lochak, London Math. Soc. Lecture Notes Ser. 242, Cambridge University Press, Cambridge, 1997.

Some Unexpected Consequences of Symmetry Computations

Marston D.E. Conder

Abstract This paper gives some instances of experimental computations involving the action of groups on graphs and maps with a high degree of symmetry, that have led to unexpected theoretical discoveries. These include new presentations for 3-dimensional special linear groups, a closed-form definition for the binary reflected Gray codes, a new theorem on groups expressible as a product of an abelian group and a cyclic group, and some revealing observations about the genus spectrum of particular classes of regular maps on surfaces.

1 Introduction

There is no doubt that the use of high-speed computers and the development of special purpose software have had an enormous impact on mathematics and its applications. They have also stimulated debate on the nature of proof, the reliability of computer hardware, and the acceptability of computer-based arguments. The 1970 s proof by Appel and Haken (and Koch) of the Four Colour Theorem [1, 2] is an interesting example—where some mathematicians were more concerned with the possibility of machine error in analysing 1,936 sub-configurations (or human error in feeding them to the computer) than the possibility of human error in classifying those sub-cases in the first place. The Appel-Haken proof has since been superseded by another by Robertson, Sanders, Seymour and Thomas [22], in which both the generation and analysis of a smaller set of 633 reducible configurations are achieved by computation.

But of course the use of computers goes far beyond helping us prove things. Computer-assisted experimentation or generation of small cases can help us get a clear picture of a mathematical situation, or reveal patterns that might otherwise not easily be seen, and point the way to new discoveries and theorems with computer-free

M.D.E Conder (✉)
Department of Mathematics, University of Auckland, Private Bag 92019,
1142 Auckland, New Zealand
e-mail: m.conder@auckland.ac.nz

J. Širáň and R. Jajcay (eds.), *Symmetries in Graphs, Maps, and Polytopes*,
Springer Proceedings in Mathematics & Statistics 159,
DOI 10.1007/978-3-319-30451-9_3

(and somehow more acceptable) proofs. They can even reveal completely unexpected phenomena. This paper reports on several instances where this has happened, in the context of finitely-presented groups and their actions on graphs and maps.

The consequences include new presentations for 3-dimensional special linear groups (in Sect. 2), a closed-form definition for the binary reflected Gray codes (Sect. 3), a new theorem on groups expressible as a product of an abelian group and a cyclic group, and revealing observations about the genus spectrum of particular classes of regular maps on surfaces (Sect. 4). In each case, it is unlikely the discovery would have been made without the results of the computer-based experimentation or generation of small examples.

These instances underline not only the important role of computing, but also the value of the software that has been developed for enabling such discoveries. In that respect, the author would like to thank (among many others responsible for the wide and helpful range of computational tools now available for handling groups and related structures), both John Cannon for his part in creating MAGMA [5] and its predecessor CAYLEY, and Derek Holt for his development of a fast procedure for finding normal subgroups of small finite index in a finitely-presented group.

2 Arc-Transitive Cubic Graphs and SL(3, $\mathbb{Z}$)

An automorphism of a graph is any bijection of its vertex-set preserving adjacency, and under composition, such bijections form a group known as the *automorphism group* of the graph. A graph is called *arc-transitive* (or *symmetric*) if its automorphism group has a single orbit on ordered pairs (u, v) of adjacent vertices, and *s-arc-transitive* if its automorphism group has a single orbit on directed walks of the form $v_0 - v_1 - v_2 - \cdots - v_{s-1} - v_s$ in which any three consecutive vertices are distinct. Note that any connected arc-transitive graph is necessarily regular.

A remarkable theorem of Tutte (1947) shows that every finite arc-transitive graph of valency 3 is at most 5-arc-transitive; see [23]. In fact there are exactly seven classes of such graphs, which may be classified by the largest value of s for which the graph is s-arc-transitive ($1 \leq s \leq 5$) and the existence or otherwise of an involutory automorphism reversing a given edge; see [18] or [12]. Associated with each class is an amalgamated free product of two small finite groups (corresponding to the stabilizers of a vertex and an edge, with the arc-stabilizer subgroup amalgamated). The resulting seven groups are commonly denoted by G_1, ${G_2}^1$, ${G_2}^2$, G_3, ${G_4}^1$, ${G_4}^2$ and G_5.

Several new examples of arc-transitive 3-valent graphs were found by Biggs and Conway, using a method of inserting additional relations into some of these group presentations, where the new relation corresponds to particular circuits of given length in the graph; see [4]. In particular, associated with a class of 4-arc-transitive graphs containing a circuit of length 12 is the group obtained by inserting an additional relator into the group $G_4{}^1$. This group, denoted by $4^+(a^{12})$ in [4], can be presented as follows:

$$4^+(a^{12}) = \langle\, a, b, \sigma \mid \sigma^2 = (\sigma a)^2 = (\sigma b)^2 = (a^{-1}b)^2 = (a^{-2}b^2)^2 = 1, \\ a^3b^{-3}a^3 = bab,\ a^3b\sigma a^4 = ba^2b,\ a^{12} = 1\,\rangle$$

An open problem described in [4] was to determine whether or not the above group is infinite. Standard computational techniques at the time were not particularly helpful; for example, this group has no subgroup H of small finite index with infinite abelian quotient $H/[H,H]$.

In attempting to prove it is infinite, the author of the current paper found (with some dumb luck) a normal subgroup of index 336 generated by eight conjugates of the element a^6. From this it was easy to construct an 8-dimensional matrix representation of the group, in which the image of a particular element has infinite order, and hence to prove the group is infinite. But the story does not end there.

Using CAYLEY to analyse certain finite images of the matrix group so constructed, the author found that reduction modulo prime p gives an 8-dimensional matrix group of order $2p^3(p^3-1)(p+1)$ for $p = 2, 3, 5, 7$ and 11. This happens to be twice the order of the group $\mathrm{SL}(3, p)$, and that fortunate (and somewhat surprising) observation led to the discovery and computer-free proof of the following:

Theorem 1 *The group* $4^+(a^{12})$ *is isomorphic to* $\mathrm{SL}(3,\mathbb{Z}).C_2$, *the group of all* 3×3 *integer matrices of determinant* 1 *extended by its inverse-transpose automorphism.*

The author's proof (given in [7]) uses the Steinberg presentation for $\mathrm{SL}(3,\mathbb{Z})$.

The underlying reason for the isomorphism has a connection with finite projective planes. As pointed out by Peter Neumann (in a personal communication), a 3-valent graph can be naturally associated with any finite projective plane Π, by taking the quadrangles (sets of four points, no three of which are collinear) and quadrilaterals (sets of four lines, no three of which are copunctual) of Π as its vertices, and joining any quadrangle to each of the three quadrilaterals given by its partitions into two pairs of opposite points. One may take this further, to prove that the existence of a circuit of length 12 corresponds to the Desargues axiom. It would be interesting to study this connection further.

The above isomorphism also leads to the following, as proved in [13]:

Corollary 1

(a) *The group* $\mathrm{SL}(3,\mathbb{Z})$ *has presentation* $\langle\, x, y, z \mid x^3 = y^3 = z^2 = (xz)^3 = (yz)^3 = (x^{-1}zxy)^2 = (y^{-1}zyx)^2 = (xy)^6 = 1\,\rangle$, *and*
(b) *For each odd integer* $k > 1$, *the group* $\mathrm{SL}(3,\mathbb{Z}_k)$ *has presentation*

$$\langle\, x, y \mid x^3 = y^3 = (xy)^6 = (x^{-1}yx^{-1}y^{-1}xy)^2 = (xy^{-1}xyxy^{-1}x^{-1}y^{-1})^k \\ = ((xy^{-1}xyxy^{-1}x^{-1}y^{-1})^{(k-1)/2}xy)^4 = 1\,\rangle\,.$$

Fig. 1 The modified Sierpinski gasket

```
                              x x
                            x x x x
                          x x     x x
                        x x x x x x x x
                      x x             x x
                    x x x x         x x x x
                  x x     x x     x x     x x
                x x x x x x x x x x x x x x x x
              x x                             x x
            x x x x                         x x x x
          x x     x x                     x x     x x
        x x x x x x x x                 x x x x x x x x
      x x             x x             x x             x x
    x x x x         x x x x         x x x x         x x x x
  x x     x x     x x     x x     x x     x x     x x     x x
x x x x x x x x x x x x x x x x x x x x x x x x x x x x x x x x
```

3 Sierpinski's Gasket and Binary Gray Codes

Sierpinski's gasket is another name for Pascal's triangle mod 2. This played a role in the construction of a family $\{X_n\}$ of arc-transitive 4-valent graphs in which the orders of vertex-stabilisers in vertex-transitive subgroups of Aut X_n of smallest possible order form a strictly increasing sequence; see [16].

The author and his PhD student Cameron Walker devised a construction for such a family from a sequence of certain finitely-presented groups, and some small degree permutation representations of these groups, obtained with the help of the `LowIndexSubgroups` facility in MAGMA [5]. In analysing the output of the computations, Cameron Walker observed an interesting pattern in the transpositions of the permutations induced by a number of the involutory generators.

The transpositions that occurred for a particular representation all came from a partition of the degree of the representation into (disjoint) pairs, and the pattern arose from considering which of the transpositions occurred in the permutations induced by each generator. (For example, the permutations $(5,6)$, $(3,4)(5,6)$ and $(1,2)(3,4)$ involve the three transpositions $(1,2)$, $(3,4)$ and $(5,6)$, and can be represented by the strings 001, 011 and 110, respectively.) Then following a helpful discussion with his colleague Joel Schiff who was writing a book on cellular automata, the author found that this pattern can be obtained from Sierpinski's gasket by repeating (or 'doubling') every entry, as illustrated in Fig. 1 (with a pair of 'x's replacing each '1', and a pair of blank spaces replacing each '0').

Observation and identification of this pattern was critical to proving the existence of the family of graphs in [16].

But again the story does not end there.

A *Gray code* of length n is a sequence of n-bit strings (words) on some alphabet, such that each word differs from the next in just one position. A noted family of such codes, called the *Binary reflected Gray codes* G_n, can be defined inductively: G_1 consists of the words 0 and 1 (in that order), and then once G_{n-1} is known, the code G_n can be defined by listing the words of G_{n-1} with each word prefixed by 0, then re-writing them in reverse order with each word prefixed by 1 instead. For

```
 0                              x 1 0
    - - - - - - - - - - - - - - - - - - - - - - - - - - - - - - -
 1                            x x 1 1 0 0
 2                          x x   0 1 1 0
    - - - - - - - - - - - - - - - - - - - - - - - - - - - - - - -
 3                        x x x x 1 1 1 1 0 0 0 0
 4                      x x
 5                    x x x x     0 0 1 1 1 1 0 0
 6                  x x     x x   0 1 1 0 0 1 1 0
    - - - - - - - - - - - - - - - - - - - - - - - - - - - - - - -
 7                x x x x x x x x 1 1 1 1 1 1 1 1 0 0 0 0 0 0 0 0
 8              x x
 9            x x x x
10          x x     x x
11        x x x x x x x x         0 0 0 0 1 1 1 1 1 1 1 1 0 0 0 0
12      x x             x x
13    x x x x         x x x x     0 0 1 1 1 1 0 0 0 0 1 1 1 1 0 0
14  x x     x x     x x     x x   0 1 1 0 0 1 1 0 0 1 1 0 0 1 1 0
    - - - - - - - - - - - - - - - - - - - - - - - - - - - - - - -
```

Fig. 2 Binary Gray codes in the modified Sierpinski gasket

example, $G_3 = (000,\ 001,\ 011,\ 010,\ 110,\ 111,\ 101,\ 100)$. Such codes are used in combinatorics, signal processing, and calculation of correlation coefficients (for a variable subset).

A subsequent chance observation (made by the author after seeing a talk on Gray codes) was that words of the binary reflected Gray codes of small length occur in certain sections of the modified Sierpinski gasket. This is illustrated in Fig. 2. From the top of the figure, take the horizontal bands of depth 1, 2, 4, 8 (demarcated by the dashed lines), and then notice that replacing each 'x' by a '1' and each blank by a '0' reveals copies of the strings from the codes G_1, G_2, G_3 and G_4.

Some further computational experiments and theoretical analysis showed that this holds more generally, via deletion of specified rows of the modified gasket (namely those obtained by taking horizontal bands of depth 2^k for $k = 0, 1, 2, 3, \ldots$, from the top, and then in the band of depth 2^k, retaining only the rows in positions $1, 2, 4, \ldots, 2^k$, measured from the bottom of the band).

In turn this led to the following, first ever closed-form definition for the words of the binary reflected Gray codes (see [8]):

Theorem 2 *For $1 \leq j \leq n$ and $1 \leq k \leq 2^n$, the jth letter of the kth word of the binary Gray code of length n is the parity (modulo 2) of the binomial coefficient $C(2^n - 2^{n-j} - 1, [2^n - 2^{n-j-1} - k/2])$, where $[\cdot]$ is the integer floor function.*

A proof is easy by induction: show the given binomial coefficients satisfy the required reflective property of the code G_n, and use induction on the length n.

The significant point here is that the theorem might not have been found at all without the use of group-theoretic computation to investigate a seemingly unrelated class of finitely-presented groups. But also it would be interesting to know if there is a deeper connection between binomial coefficients and the Gray codes (or related objects such as Hamilton paths in hypercubes).

4 Regular Maps

Regular maps are generalisations of the Platonic solids (viewed as tessellations of the sphere) to surfaces of higher genus. Their formal study was initiated by Brahana in the 1920s [6] and continued by Coxeter (see [17]) and others much later. Regular maps on the sphere and the torus and other orientable surfaces of small genus are now well understood, but until recently, the situation for surfaces of higher genus was something of a mystery. In particular, some long-standing questions have remained open about the genera of orientable surfaces that carry no rotary map without reflectional symmetry or that carry no regular map without multiple edges.

A *map* is an embedding of a connected graph or multigraph into a surface, such that each component (or *face*) of the complement is simply-connected. The *genus* and the *Euler characteristic* of the map M are defined as the genus and the Euler characteristic of the supporting surface. The topological *dual* of M (denoted by M^*) is obtained from M by interchanging the roles of vertices and faces in the usual way.

An automorphism of a map is an adjacency- and incidence-preserving bijection from the map to itself, taking vertices to vertices, edges to edges, and faces to faces. By connectedness, every automorphism of the map is uniquely determined by its effect on any *flag* (an incident vertex-edge pair (v, e) taken together with a chosen side along the edge e).

A map M is called *regular* if its automorphism group $\mathrm{Aut}(M)$ acts regularly on the set of all flags, and an orientable map M is called *orientably-regular* (or sometimes *rotary*) if the group $G = \mathrm{Aut}^o(M)$ of all its orientation-preserving automorphisms acts regularly on the set of oriented edges (or *arcs*) of M. An orientably-regular map that admits an orientation-reversing automorphism is called *reflexible* (and is then regular), while otherwise it is said to be *chiral*.

In every regular or orientably-regular map M, the faces have constant size k (say) and the vertices have constant valency m (say), and the map M is then said to have *type* $\{k, m\}$. Moreover, for any incident vertex-edge-face triple (v, e, f), there is a k-fold rotation X about the face f and an m-fold rotation Y about the vertex v, such that XY is an involutory rotation about the edge e. By connectedness, X and Y generate either $\mathrm{Aut}^o(M)$ in the orientable case, or $\mathrm{Aut}(M)$ in the non-orientable case. Thus either $\mathrm{Aut}^o(M)$ or $\mathrm{Aut}(M)$ is a quotient of the ordinary $(k, m, 2)$ *triangle group* $\Delta^o(k, m, 2) = \langle\, x, y, z \mid x^k = y^m = z^2 = xyz = 1\,\rangle$, under an epimorphism taking x to X and y to Y. The dual M^* is also regular, with the roles of X and Y interchanged, and the map M (or its dual M^*) is reflexible if and only if the group generated by X and Y admits an automorphism of order 2 taking X to X^{-1} and Y to Y^{-1}.

Conversely, given any epimorphism θ from $\Delta^o(k, m, 2)$ to a finite group G with torsion-free kernel, a map M can be constructed using (right) cosets of the images of $\langle x\rangle$, $\langle y\rangle$ and $\langle z\rangle$ as vertices, faces and edges, with incidence given by non-empty intersection, and then G acts regularly on the ordered edges of M by (right) multiplication. From this point of view the study of regular maps is simply the study of smooth quotients of triangle groups—with 'smooth' here meaning that the orders of the generators x, y and z are preserved.

Deep connections exist between maps and other branches of mathematics, however, which go far beyond group theory, and include hyperbolic geometry, Riemann surfaces and, rather surprisingly, number fields and Galois theory, based on observations made by Belyĭ and Grothendieck; see [21] for example.

But here we concentrate on just three recent developments.

One is in the study of *regular Cayley maps*. A *Cayley graph* $C(A, S)$ for a group A with respect to some generating subset S is the graph with vertex-set A and edge-set $\{\{u, ux\} : u \in A, x \in S\}$, and a regular Cayley map is an embedding of a Cayley graph $C(A, S)$ as an orientably-regular map in such a way that the Cayley group A preserves the embedding, or equivalently, an orientably-regular map whose automorphism group has a subgroup acting regularly on vertices. (An easy example is the spherical embedding of the 3-cube, which is a Cayley graph for the dihedral group of order 8.) Any such embedding is determined by an ordering of the elements of the generating set $S \cup S^{-1}$ for the group A, and may be associated with a *skew morphism* of A; see [11] for further details.

If M is a regular Cayley map for an abelian group A, then the Cayley group A acts regularly on the vertex-set of M, and so the group $\mathrm{Aut}^o(M)$ is expressible in the form $\mathrm{Aut}^o(M) = A\langle Y\rangle$, with $A \cap \langle Y\rangle = \{1\}$. In a computational investigation of examples of such maps (carried out with the help of MAGMA), the author discovered that in all small cases, the derived subgroup of $\mathrm{Aut}^o(M)$ is isomorphic to a subgroup of A. A theorem of Itô [20] on the product of two abelian groups ensures that $\mathrm{Aut}^o(M)$ is metabelian, but this is a much stronger conclusion. There being no obvious reason for this phenomenon, the author sought advice from group-theoretic colleagues on the special case where A is cyclic, and a helpful answer from Marty Isaacs and subsequent joint work with him led to the following extension of Itô's theorem (in which the 'rank' is the minimum cardinality of a generating set):

Theorem 3 *If the group G is a product AB of two abelian subgroups A and B, such that at least one of A and B is finite, and at least one of A and B is cyclic, then* $\mathrm{rank}(G'/G' \cap A) \le \mathrm{rank}(B)$ *and* $\mathrm{rank}(G'/G' \cap B) \le \mathrm{rank}(A)$.

Also if G is finite then $G'/(G' \cap A)$ is isomorphic to a subgroup of $B/(A \cap B)$, and $G'/(G' \cap B)$ is isomorphic to a subgroup of $A/(A \cap B)$. See [10] for a proof of the above theorem and further details.

The second development is quite significant. Recently Derek Holt and his student David Firth developed a computational procedure for finding normal subgroups of small index in finitely-presented groups [19], and this has been implemented in MAGMA. In 2006 the author used the new procedure to find all normal subgroups of appropriately small finite index in triangle groups, in order to extend the census of all regular maps of small Euler characteristic $\chi(M)$ to the range $-200 \le \chi(M) < 0$; see [9]. This not only improved the previous range ($-28 \le \chi(M) < 0$) by a considerable margin, but also led to many new discoveries and revealed patterns in the genus spectrum of various kinds of regular maps never seen before.

For example, this work showed for the first time that an orientable surface could carry no (reflexible) regular map without multiple edges in its underlying graph. In fact this happens for genus 20, 32, 38, 44, 62, 68, 74, 80 and 98, and for many higher genera as well, of the form $p+1$ where p is a prime congruent to 1 mod 6. Similarly, it revealed a sequence of genera (also of the form $p+1$ for various primes) such that every orientably-regular map on a surface of one of those genera is reflexible, and thereby exhibiting a sequence of gaps in the genus spectrum of orientably-regular maps that are chiral. These observations (and others by the author in joint work with Jozef Širáň and Tom Tucker) led to a complete, computer-free classification of all orientably-regular maps M of genus $g \geq 0$ for which $g-1$ and $|\mathrm{Aut}^o(M)|$ are relatively prime, and as a four-part corollary, the following:

Theorem 4 (a) *If p is a prime such that $p-1$ is not divisible by* 3, 5 *or* 8, *then every orientably-regular map of genus $g = p+1$ is reflexible*;

(b) *If M is an orientably-regular but chiral map of genus $g = p+1$, where p is prime, and $p-1$ is not divisible by* 5 *or* 8, *then either M or its topological dual M^* has multiple edges*;

(c) *If M is a reflexible orientably-regular map of genus $g = p+1$, where p is prime and $p > 13$, then either M or M^* has multiple edges, and if $p \equiv 1$ mod* 6, *then both M and M^* have multiple edges*; *and*

(d) *There exists no non-orientable regular map of genus $p+2$ where p is a prime congruent to* 1 *mod* 12, *except when $p = 13$.*

Part (d) was known previously (see [3]). The full proof (which is remarkably short, and includes a new proof of part (d)) is given in [14].

Finally, the consequence of Theorem 3 about the structure of the quotients $G'/(G' \cap A)$ and $G'/(G' \cap B)$ was instrumental in the very recent complete classification of all regular Cayley maps for cyclic groups; see [15].

Acknowledgments The author is grateful for support for this work from the Marsden Fund and a James Cook Fellowship of the Royal Society of New Zealand, and acknowledges the considerable help of MAGMA [5] in showing what was possible.

References

1. K. Appel and W. Haken, Every planar map is four colorable. Part I. Discharging, *Illinois J. Math.* **21** (1977), 429–490.
2. K. Appel, W. Haken and J. Koch, Every planar map is four colorable. Part II. Reducibility, *Illinois J. Math.* **21** (1977), 491–567.
3. A.B. D'Azevedo, R. Nedela and J. Širáň, Classification of regular maps of negative prime Euler characteristic, *Trans. Amer. Math. Soc.* **357** (2005), 4175–4190.
4. N. Biggs, Presentations for cubic graphs, in *Computational Group Theory* (ed. M. Atkinson), Academic Press, 1984, pp. 57–63.
5. W. Bosma, J. Cannon and C. Playoust, The MAGMA Algebra System I: The User Language, *J. Symbolic Computation* **24** (1997), 235–265.
6. H.R. Brahana, Regular maps and their groups, *Amer. J. Math.* **49** (1927), 268–284.

7. M.D.E. Conder, A surprising isomorphism, *J.Algebra* **129** (1990) 494–501.
8. M.D.E. Conder, Explicit definition of the binary reflected Gray codes, *Discrete Math.* **195** (1999), 245–249.
9. M.D.E. Conder, Regular maps and hypermaps of Euler characteristic 1 to 200, *J. Combinatorial Theory, Ser. B* **99** (2009), 455–459.
10. M.D.E. Conder and I.M. Isaacs, Derived subgroups of products of an abelian and a cyclic subgroup, *J. London Math. Society* **69** (2004), 333–348.
11. M.D.E. Conder, R. Jajcay and T.W. Tucker, Regular Cayley maps for finite abelian groups, *J. Algebraic Combinatorics* **25** (2007), 259–283.
12. M.D.E. Conder and P.J. Lorimer, Automorphism groups of symmetric graphs of valency 3, *J. Combinatorial Theory, Ser. B* **47** (1989), 60–72.
13. M.D.E. Conder, E.F. Robertson and P.R. Williams, Presentations for 3-dimensional special linear groups over integer rings, *Proc. Amer. Math. Soc.* **115** (1992), 19–26.
14. M.D.E. Conder, J. Širáň and T.W. Tucker, The genera, reflexibility and simplicity of regular maps, *J. European Math. Soc.* **12** (2010), 343–364.
15. M.D.E. Conder and T.W. Tucker, Regular Cayley maps for cyclic groups, *Trans. Amer. Math. Soc.* **366** (2014), 3585–3609.
16. M.D.E. Conder and C.G. Walker, Vertex-transitive graphs with arbitrarily large vertexstabilizers, *J. Algebraic Combinatorics* **8** (1998), 29–38.
17. H.S.M. Coxeter and W.O.J. Moser, *Generators and Relations for Discrete Groups*, 4th ed., Springer (1980).
18. D. Ž. Djoković and G.L. Miller, Regular groups of automorphisms of cubic graphs, *J. Combinatorial Theory Ser. B* **29** (1980), 195–230.
19. D. Firth, *An algorithm to find normal subgroups of a finitely presented group up to a given index*, PhD Thesis, University of Warwick, 2005.
20. N. Itô, Über das Produkt von zwei abelschen Gruppen. *Math. Zeitschrift* **62** (1955), 400–401.
21. G.A. Jones and D. Singerman, Belyĭ functions, hypermaps, and Galois groups, *Bull. London Math. Soc.* **28** (1996), 561–590.
22. N. Robertson, D. Sanders, P. Seymour and R. Thomas, The four-colour theorem, *J. Combin. Theory Ser. B* **70** (1997), 2–44.
23. W.T. Tutte, On the symmetry of cubic graphs, *Canad. J. Math.* **11** (1959), 621–624.

A 3D Spinorial View of 4D Exceptional Phenomena

Pierre-Philippe Dechant

Abstract We discuss a Clifford algebra framework for discrete symmetry groups (such as reflection, Coxeter, conformal and modular groups), leading to a surprising number of new results. Clifford algebras allow for a particularly simple description of reflections via 'sandwiching'. This extends to a description of orthogonal transformations in general by means of 'sandwiching' with Clifford algebra multivectors, since all orthogonal transformations can be written as products of reflections by the Cartan-Dieudonné theorem. We begin by viewing the largest non-crystallographic reflection/Coxeter group H_4 as a group of rotations in two different ways—firstly via a folding from the largest exceptional group E_8, and secondly by induction from the icosahedral group H_3 via Clifford spinors. We then generalise the second way by presenting a construction of a 4D root system from any given 3D one. This affords a new, spinorial, perspective on 4D phenomena, in particular as the induced root systems are precisely the exceptional ones in 4D, and their unusual automorphism groups are easily explained in the spinorial picture; we discuss the wider context of Platonic solids, Arnold's trinities and the McKay correspondence. The multivector groups can be used to perform concrete group-theoretic calculations, e.g. those for H_3 and E_8, and we discuss how various representations can also be constructed in this Clifford framework; in particular, representations of quaternionic type arise very naturally.

1 Introduction

Reflections are the building blocks for a large class of discrete symmetries that are of interest in both pure and applied mathematics. Coxeter groups, root systems and polytopes are intimately related to Lie groups and algebras, as well as to the geometry of various dimensions. The geometry of three dimensions has manifold obvious implications for physics, chemistry and biology; in particular, in our recent work

P.-P. Dechant (✉)
Department of Mathematics, University of York, Heslington, York YO10 5GG, UK
e-mail: ppd22@cantab.net

J. Širáň and R. Jajcay (eds.), *Symmetries in Graphs, Maps, and Polytopes*, Springer Proceedings in Mathematics & Statistics 159,
DOI 10.1007/978-3-319-30451-9_4

we were interested in the role of icosahedral symmetry to virus structure, fullerenes and quasicrystals [13–15]. Lie groups and algebras, as well as their root lattices and Coxeter/Weyl groups are also ubiquitous in high energy physics [4, 20, 21, 27]. As we shall see later, even conformal and modular groups fall into this category of discrete groups; the relevant areas in mathematical physics include conformal field theory [30] and Moonshine phenomena [17–19, 32].

In this reflection framework one always has an inner product on the n-dimensional vector space in question; one can thus in fact always construct the corresponding 2^n-dimensional Clifford algebra, which contains the original n-dimensional vector space as a subspace. This is particularly useful, as Clifford algebra allows a dramatic simplification when it comes to handling reflections. We have explored such a Clifford algebra framework in [6, 7, 9, 10] from the pure mathematics perspective; this has led to a number of conceptual and computational simplifications, as well as some very profound results on the nature of four-dimensional (4D) geometry and its interplay with the geometry of three dimensions (3D), in particular that of rotations. Here we present an account of this work tailored to finite group theorists, exploring various pure connections, as well as presenting new work on group and representation theory.

This paper is organised as follows. Section 2 introduces root systems, reflection and Coxeter groups, and their graphical representations. In Sect. 3 we present some Clifford algebra background, in particular the unique reflection prescription and the resulting versor formalism via the Cartan-Dieudonné theorem. We discuss two ways of viewing the reflection group H_4 as a group of rotations—in Sect. 4 we discuss H_4 as a subgroup of E_8; in Sect. 5 we consider the group of 120 multivectors generated by the simple root vectors of H_3 via multiplication in the Clifford algebra. This is the binary icosahedral group $2I$, and its multivector components are exactly the roots of H_4. We then generalise this observation: this yields a remarkable theorem which induces a 4D root system from every 3D root system in a constructive way (Sect. 6). We discuss this construction—which uses Clifford spinors—and its relation to the 4D Platonic solids along with their peculiar symmetries, as well as the wider context of Arnold's trinities and the McKay correspondence, which this construction puts into a wider framework. In Sect. 7 we revisit some of the earlier group-theoretic calculations and show how Clifford multivectors can be used to construct more conventional matrix representations; in particular, certain representations of quaternionic type of a number of specific groups arise uniformly in this construction. We conclude with a summary and possible further work in Sect. 8.

2 Root Systems and Reflection Groups

In this section, we introduce reflection/Coxeter groups as generated by their root systems. Let V be an n-dimensional Euclidean vector space endowed with a positive definite bilinear form $(\cdot|\cdot)$. A *root system* is a collection Φ of non-zero vectors (called root vectors) satisfying the following two axioms:

1. Φ only contains a root α and its negative, but no other scalar multiples: $\Phi \cap \mathbb{R}\alpha = \{-\alpha, \alpha\}$ for every $\alpha \in \Phi$.
2. Φ is invariant under all reflections corresponding to root vectors. That is, if s_α is the reflection of V in the hyperplane with normal α, we require that $s_\alpha \Phi = \Phi$ for every $\alpha \in \Phi$.

For a crystallographic root system, a subset Δ of Φ, called *simple roots* $\alpha_1, \ldots, \alpha_n$, is sufficient to express every element of Φ via $\mathbb{Z}$-linear combinations with coefficients of the same sign. Φ is therefore completely characterised by this basis of simple roots. In the case of the non-crystallographic root systems H_2, H_3 and H_4, the same holds for the extended integer ring $\mathbb{Z}[\tau] = \{a + \tau b | a, b \in \mathbb{Z}\}$, where τ is the golden ratio $\tau = \frac{1}{2}(1+\sqrt{5}) = 2\cos\frac{\pi}{5}$, and σ is its Galois conjugate $\sigma = \frac{1}{2}(1-\sqrt{5})$ (the two solutions to the quadratic equation $x^2 = x + 1$). For the crystallographic root systems, the classification in terms of Dynkin diagrams essentially follows the one familiar from Lie groups and Lie algebras, as their Weyl groups are the crystallographic Coxeter groups. A mild generalisation to so-called Coxeter-Dynkin diagrams is necessary for the non-crystallographic root systems, where nodes correspond to simple roots, orthogonal roots are not connected, roots at $\frac{\pi}{3}$ have a simple link, and other angles $\frac{\pi}{m}$ have a link with a label m. The *Cartan matrix* of a set of simple roots $\alpha_i \in \Delta$ is defined as the matrix $A_{ij} = 2(\alpha_i|\alpha_j)/(\alpha_j|\alpha_j)$. For instance, the root system of the icosahedral group H_3 has one link labelled by 5 (via the above relation $\tau = 2\cos\frac{\pi}{5}$), as does its four-dimensional analogue H_4. A plethora of examples of diagrams is presented later, in Table 2 in Sect. 5.

The reflections in the second axiom of the root system generate a reflection group. A Coxeter group is a mathematical abstraction of the concept of a reflection group via involutory generators (i.e. their square is the identity, which captures the idea of a reflection), subject to mixed relations that represent m-fold rotations (since two successive reflections generate a rotation in the plane spanned by the two roots). A *Coxeter group* is a group generated by a set of involutory generators $s_i, s_j \in S$ subject to relations of the form $(s_i s_j)^{m_{ij}} = 1$ with $m_{ij} = m_{ji} \geq 2$ for $i \neq j$. The finite Coxeter groups have a geometric representation where the involutions are realised as reflections at hyperplanes through the origin in a Euclidean vector space V, i.e. they are essentially just the classical reflection groups. In particular, then the abstract generator s_i corresponds to the simple reflection $s_i : \lambda \rightarrow s_i(\lambda) = \lambda - 2\frac{(\lambda|\alpha_i)}{(\alpha_i|\alpha_i)}\alpha_i$ in the hyperplane perpendicular to the simple root α_i. The action of the Coxeter group is to permute these root vectors, and its structure is thus encoded in the collection $\Phi \in V$ of all such roots, which in turn form a root system.

We now move onto the Clifford algebra framework, which affords a uniquely simple prescription for performing reflections (and thus any orthogonal transformation) in spaces of any dimension and signature. For any root system, the quadratic form mentioned in the definition always allows one to enlarge the n-dimensional vector space V to the corresponding 2^n-dimensional Clifford algebra. The Clifford algebra is therefore a very natural object to consider in this context, as its unified structure simplifies many problems both conceptually and computationally (as we shall see in the next section), rather than applying the linear structure of the space and the inner product separately.

3 Clifford Versor Framework

Clifford algebra can be viewed as a deformation of the (perhaps more familiar) exterior algebra by a quadratic form—though we do not necessarily advocate this point of view; they are isomorphic as vector spaces, but not as algebras, and Clifford algebra is in fact much richer due to the invertibility of the algebra product, as we shall see. The *geometric product* of Geometric/Clifford Algebra is defined by $xy = x \cdot y + x \wedge y$—where the scalar product (given by the symmetric bilinear form) is the symmetric part $x \cdot y = (x|y) = \frac{1}{2}(xy + yx)$ and the exterior product the antisymmetric part $x \wedge y = \frac{1}{2}(xy - yx)$ [16, 22–24]. It provides a very compact and efficient way of handling reflections in any number of dimensions, and thus by the *Cartan-Dieudonné theorem* in fact of any orthogonal transformation. For a unit vector α, the two terms in the above formula for a reflection of a vector v in the hyperplane orthogonal to α simplify to the double-sided ('sandwiching') action of α via the geometric product

$$v \rightarrow s_\alpha v = v' = v - 2(v|\alpha)\alpha = v - 2\frac{1}{2}(v\alpha + \alpha v)\alpha = v - v\alpha^2 - \alpha v\alpha = -\alpha v\alpha. \quad (1)$$

This prescription for reflecting vectors in hyperplanes is remarkably compact (note that α and $-\alpha$ encode the same reflection and thus provide a double cover). Via the Cartan-Dieudonné theorem, any orthogonal transformation can be written as the product of reflections, and thus by performing consecutive reflections each given via 'sandwiching', one is led to define a versor as a Clifford multivector $A = a_1 a_2 \cdots a_k$, that is the product of k unit vectors a_i [23]. Versors form a multiplicative group called the versor/pinor group Pin under the single-sided multiplication with the geometric product, with inverses given by $\tilde{A}A = A\tilde{A} = \pm 1$, where the tilde denotes the reversal of the order of the constituent vectors $\tilde{A} = a_k \cdots a_2 a_1$, and the $\pm$-sign defines its parity. Every orthogonal transformation $\underline{\mathrm{A}}$ of a vector v can thus be expressed by means of unit versors/pinors via

$$\underline{\mathrm{A}} : v \rightarrow v' = \underline{\mathrm{A}}(v) = \pm \tilde{A} v A. \quad (2)$$

Unit versors are double-covers of the respective orthogonal transformation, as A and $-A$ encode the same transformation. Even versors R, that is, products of an even number of vectors, are called spinors or rotors. They form a subgroup of the Pin group and constitute a double cover of the special orthogonal group, called the Spin group. Clifford algebra therefore provides a particularly natural and simple construction of the Spin groups. Thus the remarkably simple construction of the binary polyhedral groups in Sect. 6 is not at all surprising from a Clifford point of view, but appears to be unknown in the Coxeter community, and ultimately leads to the novel result of the spinor induction theorem of (exceptional) 4D root systems in Sect. 6.

The versor realisation of the orthogonal group is much simpler than conventional matrix approaches. Table 1 summarises the various action mechanisms of

Table 1 Versor framework for a unified treatment of the chiral, full, binary and pinor polyhedral groups

Continuous group	Discrete subgroup	Multivector action
$SO(n)$	Rotational/chiral	$x \rightarrow \tilde{R}xR$
$O(n)$	Reflection/full	$x \rightarrow \pm\tilde{A}xA$
Spin(n)	Binary	Spinors R under $(R_1, R_2) \rightarrow R_1R_2$
Pin(n)	Pinor	Pinors A under $(A_1, A_2) \rightarrow A_1A_2$

multivectors: a rotation (e.g. the continuous group $SO(3)$ or the discrete subgroup, the chiral icosahedral group $I = A_5$) is given by double-sided action of a spinor R, whilst these spinors themselves form a group under single-sided action/multiplication (e.g. the continuous group Spin(3) $\sim SU(2)$ or the discrete subgroup, the binary icosahedral group $2I$). Likewise, a reflection (continuous $O(3)$ or the discrete subgroup, the full icosahedral group the Coxeter group H_3) corresponds to sandwiching with the versor A, whilst the versors single-sidedly form a multiplicative group (the Pin(3) group or the discrete analogue, the double cover of H_3, which we denote Pin(H_3)). In the conformal geometric algebra setup one uses the fact that the conformal group $C(p, q)$ is homomorphic to $SO(p+1, q+1)$ to treat translations as well as rotations in a unified versor framework [5, 8, 9, 16, 24]. [8, 9] also discuss reflections, inversions, translations and modular transformations in this way.

Example 1 The Clifford/Geometric algebra of three dimensions Cl(3) is spanned by three orthogonal—and thus anticommuting—unit vectors e_1, e_2 and e_3. It also contains the three bivectors e_1e_2, e_2e_3 and e_3e_1 that all square to -1, as well as the highest grade object $e_1e_2e_3$ (trivector and pseudoscalar), which also squares to -1. Therefore, in Clifford algebra various geometric objects arise that provide imaginary structures; however, there can be different ones and they can have non-trivial commutation relations with the rest of the algebra.

$$\underbrace{\{1\}}_{\text{1 scalar}} \quad \underbrace{\{e_1, e_2, e_3\}}_{\text{3 vectors}} \quad \underbrace{\{e_1e_2 = Ie_3, e_2e_3 = Ie_1, e_3e_1 = Ie_2\}}_{\text{3 bivectors}} \quad \underbrace{\{I \equiv e_1e_2e_3\}}_{\text{1 trivector}}. \tag{3}$$

4 H_4 as a Rotation Rather Than Reflection Group I: From E_8

The largest exceptional Coxeter group E_8 and the largest non-crystallographic Coxeter group H_4 are closely related. This connection between E_8 and H_4 can be shown via Coxeter-Dynkin diagram foldings on the level of Coxeter groups [31] or as a projection relating the root systems [14, 29]. On the level of the root system this is due to the existence of a projection which maps the 240 roots of E_8 onto the 120 roots of H_4 and their τ-multiples. We now consider the Dynkin diagram folding picture in more detail.

Fig. 1 Coxeter-Dynkin diagram folding from E_8 to H_4. Note that deleting nodes α_1 and α_7 yields corresponding results for $D_6 \rightarrow H_3$, and likewise for $A_4 \rightarrow H_2$ by further removing α_2 and α_6

We take the simple roots α_1 to α_8 of E_8 as shown in Fig. 1, and consider the Clifford algebra in 8D with the usual Euclidean metric. The simple reflections corresponding to the simple roots are thus just given via sandwiching $s_\alpha v = -\alpha v \alpha$ as in Eq. (1). The Coxeter element w is defined as the product of all these eight simple reflections, and in Clifford algebra it is therefore simply given by the corresponding (s)pinor $W = \alpha_1 \cdots \alpha_8$ acting via sandwiching. Its order, the Coxeter number h (that is, the smallest h such that $W^h = \pm 1$), is 30 for E_8.

As illustrated in Fig. 1, one can define certain combinations of pairs of reflections (corresponding to roots on top of each other in the Dynkin diagram folding), e.g. $s_{a_1} = s_{\alpha_1} s_{\alpha_7}$ etc., and in a Clifford algebra sandwiching way these are given by the products of root vectors $a_1 = \alpha_1\alpha_7$, $a_2 = \alpha_2\alpha_6$, $a_3 = \alpha_3\alpha_5$ and $a_4 = \alpha_4\alpha_8$ (which is essentially a partial folding of the usual alternating folding used in the construction of the Coxeter plane with symmetry group $I_2(h)$). It is easy to show that the subgroup with the generators s_{a_i} in fact satisfies the relations of an H_4 Coxeter group [3, 31]: because of the Coxeter relations for E_8 and the orthogonality of the combined pair the combinations s_a are easily seen to be involutions, and the 3-fold relations are similarly obvious from the Coxeter relations; only for the 5-fold relation does one have to perform an explicit calculation in terms of the reflections with respect to the root vectors. This is thus particularly easy by multiplying together vectors in the Clifford algebra, rather than by concatenating two reflection formulas of the usual type—despite containing just two terms, concatenation gets convoluted quickly, which is not the case in the multiplication of multivectors.

Since the combinations s_a are pairs of reflections, they are obviously rotations in the eight-dimensional space, so this H_4 group acts as rotations in the full space, but as a reflection group in a 4D subspace. The H_4 Coxeter element is given by multiplying together the four combinations a_i—its Coxeter versor is therefore trivially seen to be the same as that of E_8 (up to sign, since orthogonal vectors anticommute) and the Coxeter number of H_4 is thus the same as that of E_8, 30. The projection of the E_8 root system onto the Coxeter plane consists of two copies of the projection of H_4 into the Coxeter plane, with a relative factor of τ.

5 H_4 as a Rotation Rather Than Reflection Group II: From H_3

In the previous section we have considered certain generators and multivectors in the algebra. We now consider the whole (s)pinor group generated by the simple reflections of H_3. The simple roots are taken as $\alpha_1 = e_2$, $\alpha_2 = -\frac{1}{2}((\tau - 1)e_1 +$

$e_2 + \tau e_3)$, and $\alpha_3 = e_3$. Under free multiplication, these generate a group with 240 elements (pinors), and the even subgroup consists of 120 elements (spinors), for instance of the form $\alpha_1\alpha_2 = -\frac{1}{2}(1 - (\tau - 1)e_1e_2 + \tau e_2e_3)$ and $\alpha_2\alpha_3 = -\frac{1}{2}(\tau - (\tau - 1)e_3e_1 + e_2e_3)$. These are the double covers of $I = A_5$ and $H_3 = A_5 \times \mathbb{Z}_2$, respectively. With these groups of multivectors one can perform standard group theory calculations, such as finding inverses and conjugacy classes. The spinors have four components $(1, e_1e_2, e_2e_3, e_3e_1)$; by taking the components of these 120 spinors as a set of vectors in 4D one obtains the 120 roots in the H_4 root system. This is very surprising from a Coxeter perspective, as one usually thinks of H_3 as a subgroup of H_4, and therefore of H_4 as more 'fundamental'; however, one now sees that H_4 does not in fact contain any structure that is not already contained in H_3, and can therefore think of H_3 as more fundamental. We will present a general and uniform construction explaining and systematising this fact in the next section.

From a Clifford perspective it is not surprising to find this group of 120 spinors, which is the binary icosahedral group, since as we have seen Clifford algebra provides a simple construction of the Spin groups. This spinor group, the binary icosahedral group $2I$, has 120 elements and 9 conjugacy classes, and calculations in the Clifford algebra are very straightforward; standard approaches would have to somehow move from $SO(3)$ rotation matrices to $SU(2)$ matrices for the binary group—here both are treated in the same framework. The fact that the rotational icosahedral group I (given by double-sided action of spinors R as $\tilde{R}xR$) has five conjugacy classes and it being of order 60 imply that this group has five irreducible representations of dimensions $1, 3, \bar{3}, 4$ and 5 (since the sum of the dimensions squared gives the order of the group $\sum d_i^2 = |G|$). The nine conjugacy classes of the binary icosahedral group $2I$ of order 120 (given by the spinors R under algebra multiplication) imply that this acquires a further four irreducible spinorial representations 2_s, $2'_s$, 4_s and 6_s.

The binary icosahedral group has a curious connection with the affine Lie algebra E_8^+ (which also applies to the other binary polyhedral groups and the affine Lie algebras of ADE-type) via the so-called McKay correspondence [28], which is twofold. First, we may define a graph by assigning a node to each irreducible representation of the binary icosahedral group with the following rule for connecting nodes: each node corresponding to a certain irreducible representation is connected to the nodes corresponding to those irreducible representations that are contained in its tensor product with the irrep 2_s. For instance, tensoring the trivial representation 1 with 2_s trivially gives 2_s and thus the only link 1 has is with 2_s; $2_s \otimes 2_s = 1 + 3$, such that 2_s is connected to 1 and 3, etc. The graph that is built up in this way is precisely the Dynkin diagram of affine E_8, as shown in Fig. 2. The second connection is the

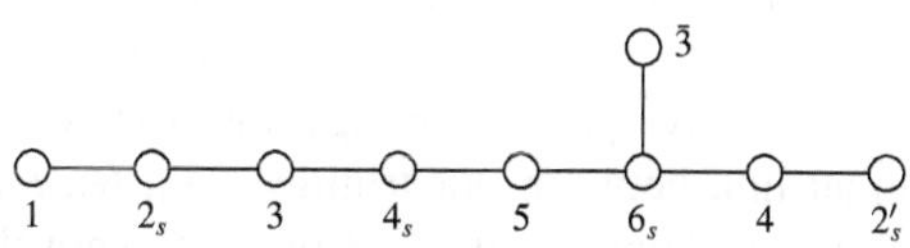

Fig. 2 The graph depicting the tensor product structure of the binary icosahedral group $2I$ is the same as the Dynkin diagram for the affine extension of E_8, E_8^+

following observation: the Coxeter element is the product of all the simple reflections $\alpha_1 \cdots \alpha_8$ and its order, the Coxeter number h, is 30 for E_8. This also happens to be the sum of the dimensions of the irreducible representations of $2I$. This extends to all other cases of polyhedral groups and ADE-type affine Lie algebras, as shown in the second and third columns in Table 2 and in Fig. 3.

6 The General Construction: Spinor Induction and the 4D Platonic Solids, Trinities and McKay Correspondence

In this section we systematise the above observation. Starting with any 3D root system, we present a construction that yields a 4D root system; the intermediate steps involve Clifford spinor techniques. We begin by an auxiliary result.

Proposition 1 **($O(4)$-structure of spinors)** *The space of* Cl(3)*-spinors* $R = a_0 + a_1Ie_1 + a_2Ie_2 + a_3Ie_3$ *can be endowed with a* 4D Euclidean norm $|R|^2 = R\tilde{R} = a_0^2+a_1^2+a_2^2+a_3^2$ *induced by the* inner product $(R_1, R_2) = \frac{1}{2}(R_1\tilde{R}_2+R_2\tilde{R}_1)$ *between two spinors* R_1 *and* R_2.

This allows one to reinterpret the group of 3D spinors generated from a 3D root system as a set of 4D vectors, which in fact can be shown to satisfy the axioms of a root system as given above.

Theorem 1 **(Induction Theorem)** *Any 3D root system gives rise to a spinor group G which induces a root system in 4D.*

Proof It is sufficient to check the two axioms for the root system Φ consisting of the set of 4D vectors given by the 3D spinor group:

1. By construction, Φ contains the negative of a root R since spinors provide a double cover of rotations, i.e. if R is in a spinor group G, then so is $-R$, but no other scalar multiples (normalisation to unity).
2. Φ is invariant under all reflections with respect to the inner product (R_1, R_2) in Proposition 1 since $R_2' = R_2 - 2(R_1, R_2)/(R_1, R_1)R_1 = -R_1\tilde{R}_2R_1 \in G$ for $R_1, R_2 \in G$ by the closure property of the group G (in particular $-R$ and $\tilde{R}$ are in G if R is).

Since the number of irreducible 3D root systems is limited to (A_3, B_3, H_3), this yields a definite list of induced root systems in 4D; this turns out to be (D_4, F_4, H_4), which are exactly the exceptional root systems in 4D. In fact, the two triples may be regarded as trinities in Arnold's sense, originally applying to the trinity $(\mathbb{R}, \mathbb{C}, \mathbb{H})$ and later extended to projective spaces, Lie algebras, spheres, Hopf fibrations etc. [1, 2]. Arnold's original link between our trinities (A_3, B_3, H_3) and (D_4, F_4, H_4) was extremely convoluted, and our construction presents a novel direct link between the two.

Table 2 Clifford spinor construction and McKay correspondence: connections between 3D, 4D and ADE-type root systems. $\|\Phi\|$, $\sum d_i$ and h are 6, 12, 18 and 30, respectively

3D root system		4D root system/binary polyhedral group		Affine Lie algebra	
A_1^3		A_1^4		D_4^+	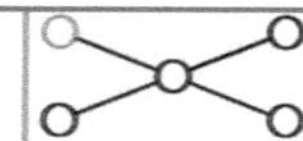
A_3	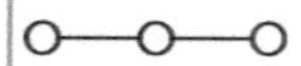	D_4	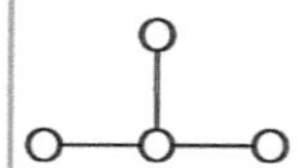	E_6^+	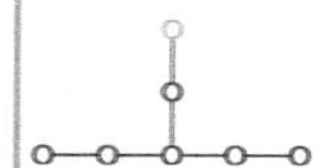
B_3	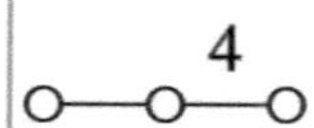	F_4	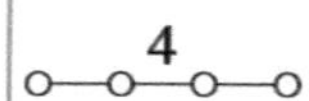	E_7^+	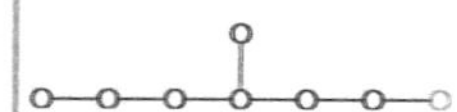
H_3	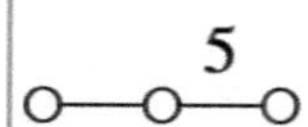	H_4	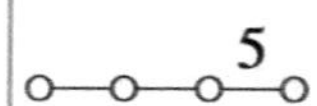	E_8^+	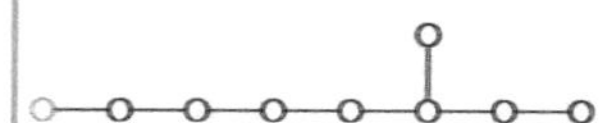

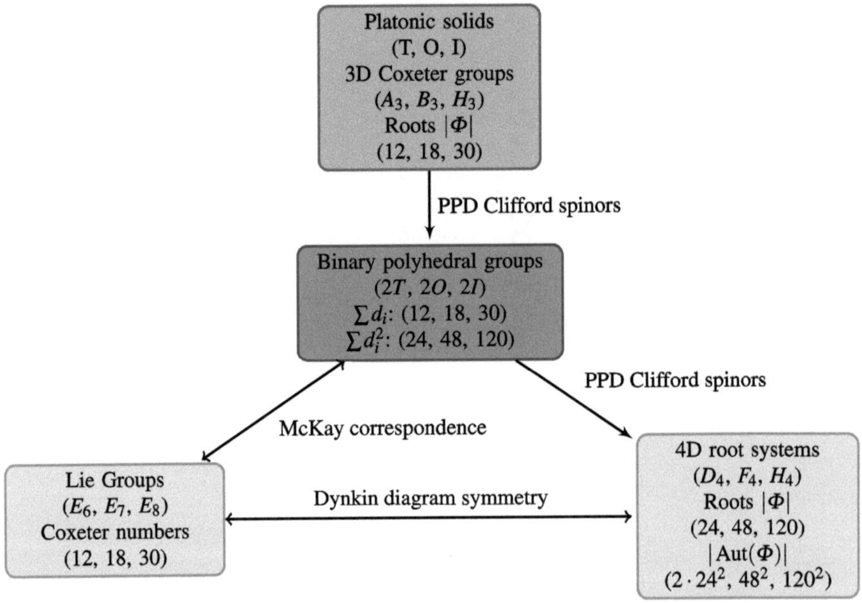

Fig. 3 Web of connections putting the original McKay correspondence and trinities into a wider context. The connection between the sum of the dimensions of the irreducible representations d_i of the binary polyhedral groups and the Coxeter number of the Lie algebras actually goes all the way back to the number of roots in the 3D root systems (12, 18, 30)—these then induce the binary polyhedral groups (linked to McKay) and the 4D root systems via the Clifford spinor construction

These root systems are intimately linked to the Platonic solids—there are 5 in three dimensions and 6 in four dimensions: A_3 is the root system relevant to the tetrahedron, B_3 generates the symmetries of the cube and octahedron (which are dual under the exchange of faces and vertices), and H_3 describes the symmetries of the dual pair icosahedron and dodecahedron (the rotational subgroup is denoted by $I = A_5$).

Likewise, the 4D Coxeter groups describe the symmetries of the 4D Platonic solids, but this time the connection is more immediate since the root systems are actually Platonic solids themselves: D_4 is the 24-cell (self-dual), an analogue of the tetrahedron, which is also related to the F_4 root system, and the H_4 root system is the Platonic solid the 600-cell. Its dual, the 120-cell obviously has the same symmetry. The root system A_1^3 generates the root system A_1^4, which constitutes the vertices of the Platonic solid 16-cell, with the 8-cell as its dual. There is thus an abundance of root systems in 4D that are related to the Platonic solids, and in fact the only one not equal or dual to a root system is the 5-cell with symmetry group A_4—which of course could not be a root system, as it has an odd number (5) of vertices. This abundance of root systems in 4D can in some sense be thought of as due to the accidentalness of this construction. In particular, the induced root systems are precisely the exceptional (i.e. they do not have counterparts in other dimensions) root systems in 4D: D_4 has the triality symmetry (permutation symmetry of the three legs in the diagram) that is exceptional in 4D, F_4 is the only F-type root system, and H_4 is the largest non-crystallographic root system. In contrast, in arbitrary dimensions there are only A_n (n-simplex), and B_n (n-hypercube and n-hyperoctahedron).

Not only is there an abundance of root systems related to the Platonic solids as well as their exceptional nature, but they also have unusual automorphism groups, in that the order of the groups grows as the square of the number of roots. This is also explained via the above spinor construction by means of the following result (which is simply guaranteed by closure of the spinor group under group multiplication, reversal and multiplication by -1):

Theorem 2 (Spinorial symmetries) *A root system induced via the Clifford spinor construction of a binary polyhedral spinor group G has an automorphism group that trivially contains two factors of the respective spinor group G acting from the left and from the right.*

This systematises many case-by-case observations on the structure of the automorphism groups [25, 26]. For instance, the automorphism group of the H_4 root system is $2I \times 2I$; in the spinor picture, it is not surprising that $2I$ yields both the root system and the two factors in the automorphism group.

We noted earlier that the binary polyhedral spinor groups and the ADE-type affine Lie algebras are connected via the McKay correspondence [28], for instance the binary polyhedral groups $(2T, 2O, 2I)$ and the Lie algebras (E_6, E_7, E_8)—for these (12, 18, 30) is both the Coxeter number of the respective Lie algebra and the sum of the dimensions of the irreducible representation of the polyhedral group.

However, the connection between (A_3, B_3, H_3) and (E_6, E_7, E_8) via Clifford spinors does not seem to be known. In particular, we note that (12, 18, 30) is exactly the number of roots Φ in the 3D root systems (A_3, B_3, H_3), which feeds through to the binary polyhedral groups and via the McKay correspondence all the way to the affine Lie algebras. Our construction therefore makes deep connections between trinities and puts the McKay correspondence into a wider framework, as shown in Table 2 and Fig. 3. It is also striking that the affine Lie algebras and the 4D root systems trinities have identical Dynkin diagram symmetries: D_4 and E_6^+ have triality S_3, F_4 and E_7^+ have an S_2 symmetry and H_4 and E_8^+ only have S_1, but are intimately related as explained in Sect. 4.

7 Group and Representation Theory with Clifford Multivectors

The usual picture of orthogonal transformations on an n-dimensional vector space is via $n \times n$ matrices acting on vectors, immediately establishing connections with representations. The above spinor techniques are somewhat unusual; however, it is easy to construct representations in this picture. Orthogonal transformations in the

2^n-dimensional Clifford algebra leave the original n-dimensional vector space invariant; one can therefore consider various representation matrices acting on different subspaces of the Clifford algebra such as—but not limited to—the original vector space.

The scalar subspace of the Clifford algebra is one-dimensional. Double-sided action of spinors R gives the trivial representation, since $\tilde{R}1R = \tilde{R}R = 1$, and likewise pinors A give the parity.

The double-sided action of spinors R on a vector x in the n-dimensional vector space gives an $n \times n$-dimensional representation, which is just the usual $SO(n)$ representation in terms of rotation matrices; similar applies to pinors and $O(n)$. For instance, for the spinor examples considered above, $\alpha_1\alpha_2$ and $\alpha_2\alpha_3$, the corresponding rotation matrices with the spinors acting as $\tilde{R}xR$ are

$$\frac{1}{2}\begin{pmatrix} \tau & \tau-1 & -1 \\ 1-\tau & -1 & -\tau \\ -1 & \tau & 1-\tau \end{pmatrix} \text{ and } \frac{1}{2}\begin{pmatrix} \tau & 1-\tau & -1 \\ 1-\tau & 1 & -\tau \\ 1 & \tau & \tau-1 \end{pmatrix}.$$

The characters $\chi(g)$ are obviously 0 and τ in these cases, and correspond to two different conjugacy classes of the icosahedral group, as shown in Table 3. For a general spinor $R = a_0 + a_1Ie_1 + a_2Ie_2 + a_3Ie_3$ one has

$$\frac{1}{2}\begin{pmatrix} a_0^2+a_1^2-a_2^2-a_3^2 & -2a_0a_3+2a_1a_2 & 2a_0a_2+2a_1a_3 \\ 2a_0a_3+2a_1a_2 & a_0^2-a_1^2+a_2^2-a_3^2 & -2a_0a_1+2a_2a_3 \\ -2a_0a_2+2a_1a_3 & 2a_0a_1+2a_2a_3 & a_0^2-a_1^2-a_2^2+a_3^2 \end{pmatrix} \text{ and } 3a_0^2-a_1^2-a_2^2-a_3^2,$$

so one can read off the character directly from the spinor components. If the spinors were acting as $Rx\tilde{R}$ (or alternatively one considers $\alpha_2\alpha_1$ and $\alpha_3\alpha_2$), then the rotation matrices would be given by

$$\frac{1}{2}\begin{pmatrix} \tau & 1-\tau & -1 \\ \tau-1 & -1 & \tau \\ -1 & -\tau & 1-\tau \end{pmatrix} \text{ and } \frac{1}{2}\begin{pmatrix} \tau & 1-\tau & 1 \\ 1-\tau & 1 & \tau \\ -1 & -\tau & \tau-1 \end{pmatrix},$$

with the same characters as before. One sees that the first example are 3-fold rotations and the second are 5-fold rotations; swapping the action of the spinor changes to the contragredient representation: if R is in $12C_5$ then $\tilde{R}$ is in $12C_5^2$, and they both have the same character τ—i.e. one exchanges the 3 and the $\bar{3}$ by this operation.

Table 3 Character table for the icosahedral group I.

I	1	$20C_3$	$15C_2$	$12C_5$	$12C_5^2$
1	1	1	1	1	1
3	3	0	-1	τ	σ
$\bar{3}$	3	0	-1	σ	τ
4	4	1	0	-1	-1
5	5	-1	1	0	0

However, rather than restricting oneself to the n-dimensional vector space, one can also define representations by $2^n \times 2^n$-matrices acting on the whole Clifford algebra. Likewise, one can define $2^{(n-1)} \times 2^{(n-1)}$-dimensional representations as acting on the even subalgebra. For instance, for the spinors considered above which have components in $(1, e_1e_2, e_2e_3, e_3e_1)$, multiplication with another spinor will reshuffle these components $(1, e_1e_2, e_2e_3, e_3e_1)$; this reshuffling can therefore be described by a 4×4-matrix. For the examples used above, for the two specific spinors $\alpha_1\alpha_2$ and $\alpha_2\alpha_3$ multiplying a generic spinor $R = a_4 + a_1Ie_1 + a_2Ie_2 + a_3Ie_3$ from the left reshuffles the components (a_1, a_2, a_3, a_4) with the matrices given as

$$\frac{1}{2}\begin{pmatrix} -1 & \tau-1 & 0 & -\tau \\ 1-\tau & -1 & -\tau & 0 \\ 0 & \tau & -1 & \tau-1 \\ \tau & 0 & 1-\tau & -1 \end{pmatrix} \text{ and } \frac{1}{2}\begin{pmatrix} -\tau & 0 & 1-\tau & -1 \\ 0 & -\tau & -1 & \tau-1 \\ \tau-1 & 1 & -\tau & 0 \\ 1 & 1-\tau & 0 & -\tau \end{pmatrix},$$

with characters -2 and -2τ. Of course there is a corresponding set of matrices where the spinor acts by right multiplication.

These matrices are part of a representation of the icosahedral group of the so-called quaternionic type. Other polyhedral groups also have representations of quaternionic type, which seems to be regarded as deeply significant yet appears to be poorly understood. Since the 3D unit spinors $(1, e_1e_2, e_2e_3, e_3e_1)$ are isomorphic to the quaternion algebra, the appearance of quaternionic representations is not very surprising from a Clifford algebra point of view. In fact, the above construction constructs the representations of quaternionic type in a uniform way, for any of the polyhedral groups (though irreducibility is a separate issue). The existence of these representations is therefore linked to the existence of the Clifford algebras and the structure of the Spin groups. These representations are therefore also much clearer in the Clifford algebra framework.

One can easily verify the quaternionic nature of the above representation and the corresponding cases for the other polyhedral groups. Representations of quaternionic type χ are characterised by $||\chi||^2 = \frac{1}{|G|}\sum_{g\in G}|\chi(g)|^2 = 4$; with real and complex type representations having 1 and 2 on the right-hand side, respectively. It is straightforward to calculate the corresponding 120 4×4 matrices and confirm that indeed $||\chi||^2 = 480/120 = 4$, in analogy with the two computational examples above. In fact it is easily shown that the representation matrix that belongs to a general spinor $R = b_4 + b_1Ie_1 + b_2Ie_2 + b_3Ie_3$ is given by

$$\begin{pmatrix} b_4 & b_3 & -b_2 & b_1 \\ -b_3 & b_4 & b_1 & b_2 \\ b_2 & -b_1 & b_4 & b_3 \\ -b_1 & -b_2 & -b_3 & b_4 \end{pmatrix} \text{ and } \chi = 4b_4,$$

such that the character is just given by four times the scalar component of the spinor.

The more general ways of constructing representations outlined above hold in any dimension, and because of the characterisation of Clifford algebras as matrix algebras over ($\mathbb{R}$, $\mathbb{C}$, $\mathbb{H}$), one expects these to yield a mixture of representations of real, complex and quaternionic type.

8 Conclusion

In this paper, we have discussed a Clifford algebra framework for certain discrete groups, based on the simple prescription for performing reflections in Clifford algebra. In fact, the Clifford algebra framework is more natural, as the existence of a quadratic form on the vector space considered in the context of root systems means that the corresponding Clifford algebra is always implicit. The reflection symmetries are the building blocks for many discrete symmetries that are interesting for applications also in mathematical physics and biology. However, the framework itself has led to new insights in pure mathematics, as regards the interplay of the geometry of dimensions three and four, Trinities and the McKay correspondence, as well as group and representation theory. It remains to explore the implications of these results in pure and applied mathematics; this has been begun in pure mathematics in [6, 7], but there could also be interesting consequences in high energy physics, where 4D root systems are ubiquitous in String Theory, M-Theory and Grand Unified Theories. In particular, recently I was also able to derive the E_8 root system consisting of 240 roots via an analogous Clifford construction of a double cover of the full icosahedral group H_3 of order 120 [11, 12] and using a certain reduced inner product due to Wilson [33].

Acknowledgments I would like to thank Reidun Twarock, Anne Taormina, David Hestenes, Anthony Lasenby, John Stillwell, Jozef Siran, Robert Wilson and Ben Fairbairn.

References

1. Vladimir Igorevich Arnold. Symplectization, complexification and mathematical trinities. *The Arnoldfest*, pages 23-37, 1999.
2. Vladimir Igorevich Arnold. *Mathematics: Frontiers and perspectives*. Amer Mathematical Society, 2000.
3. Nicolas Bourbaki. *Groupes et algèbres de Lie*, chapitres 4, 5 et 6. Masson, Paris, 1981.
4. T. Damour, M. Henneaux, and H. Nicolai. E_{10} and a 'small tension expansion' of M-Theory. *Physical Review Letters*, 89:221601, 2002.
5. Pierre-Philippe Dechant. *Models of the Early Universe*. PhD thesis, University of Cambridge, UK, 2011.
6. Pierre-Philippe Dechant. Clifford algebra unveils a surprising geometric significance of quaternionic root systems of Coxeter groups. *Advances in Applied Clifford Algebras*, 23(2):301-321, 2013, doi:10.1007/s00006-012-0371-3.
7. Pierre-Philippe Dechant. Platonic solids generate their four-dimensional analogues. *Acta Crystallographica Section A: Foundations of Crystallography*, 69(6):592-602, 2013.

8. Pierre-Philippe Dechant. A Clifford algebraic framework for Coxeter group theoretic computations. *Advances in Applied Clifford Algebras*, 24(1):89-108, 2014.
9. Pierre-Philippe Dechant. Clifford algebra is the natural framework for root systems and Coxeter groups. group theory: Coxeter, conformal and modular groups. *Advances in Applied Clifford Algebras*, 2015, doi:10.1007/s00006-015-0584-3.
10. Pierre-Philippe Dechant. Rank-3 root systems induce root systems of rank 4 via a new Clifford spinor construction. *Journal of Physics: Conference Series*, 597(1):012027, 2015.
11. Pierre-Philippe Dechant. The birth of E_8 out of the (s)pinors of the icosahedron *submitted to Proceedings of the Royal Society A 20150504*, 2016, doi:10.1098/rspa.2015.0504.
12. Pierre-Philippe Dechant. The E_8 geometry from a Clifford perspective, *Advances in Applied Clifford Algebras*, 2016.
13. Pierre-Philippe Dechant, Céline Boehm, and Reidun Twarock. Novel Kac-Moody-type affine extensions of non-crystallographic Coxeter groups. *Journal of Physics A: Mathematical and Theoretical*, 45(28):285202, 2012.
14. Pierre-Philippe Dechant, Céline Boehm, and Reidun Twarock. Affine extensions of noncrystallographic Coxeter groups induced by projection. *Journal of Mathematical Physics*, 54(9), 2013.
15. Pierre-Philippe Dechant, Jess Wardman, Tom Keef, and Reidun Twarock. Viruses and fullerenes—symmetry as a common thread? *Acta Crystallographica Section A*, 70(2):162–167, Mar 2014.
16. Chris Doran and Anthony N. Lasenby. *Geometric Algebra for Physicists*. Cambridge University Press, Cambridge, 2003.
17. Tohru Eguchi, Hirosi Ooguri, and Yuji Tachikawa. Notes on the K3 surface and the Mathieu group M_{24}. *Experimental Mathematics*, 20(1):91-96, 2011.
18. Tohru Eguchi, Yuji Sugawara, and Anne Taormina. Liouville field, modular forms and elliptic genera. *Journal of high energy physics*, 2007(03):119, 2007.
19. Terry Gannon. *Moonshine beyond the Monster: The bridge connecting algebra, modular forms and physics*. Cambridge University Press, 2006.
20. David J. Gross, Jeffrey A. Harvey, Emil J. Martinec, and Ryan Rohm. Heterotic String Theory. 1. The Free Heterotic String. *Nucl.Phys.*, B256:253, 1985.
21. M. Henneaux, D. Persson, and P. Spindel. Spacelike Singularities and Hidden Symmetries of Gravity. *Living Reviews in Relativity*, 11:1-+, April 2008.
22. David Hestenes. *Space-Time Algebra*. Gordon and Breach, New York, 1966.
23. David Hestenes. *New foundations for classical mechanics*; 2nd ed. Fundamental theories of physics. Kluwer, Dordrecht, 1999.
24. David Hestenes and Garret Sobczyk. *Clifford algebra to geometric calculus: a unified language for mathematics and physics*. Fundamental theories of physics. Reidel, Dordrecht, 1984.
25. M. Koca, M. Al-Barwani, and R. Koç. Quaternionic root systems and subgroups of the Aut(F_4). *Journal of Mathematical Physics*, 47(4):043507-+, April 2006.
26. M. Koca, R. Koç, and M. Al-Barwani. Quaternionic roots of SO(8), SO(9), F_4 and the related Weyl groups. *Journal of Mathematical Physics*, 44:3123-3140, July 2003.
27. Mehmet Koca, Ramazan Koc, and Muataz Al-Barwani. Noncrystallographic Coxeter group H_4 in E_8. *Journal of Physics A: Mathematical and General*, 34(50):11201, 2001.
28. John McKay. Graphs, singularities, and finite groups. In *Proc. Symp. Pure Math*, volume 37, pages 183-186, 1980.
29. R. V. Moody and J. Patera. Quasicrystals and icosians. *Journal of Physics A: Mathematical and General*, 26(12):2829, 1993.
30. A. N. Schellekens. Introduction to Conformal Field Theory. *Fortschritte der Physik*, 44:605–705, 1996.
31. O. P. Shcherbak. Wavefronts and reflection groups. *Russian Mathematical Surveys*, 43(3):149, 1988.
32. Anne Taormina and Katrin Wendland. A twist in the M_{24} moonshine story. *arXiv preprint*arXiv:1303.3221, 2013.
33. R. A. Wilson. *Geometriae Dedicata*, 20:157, 1986.

Möbius Inversion in Suzuki Groups and Enumeration of Regular Objects

Martin Downs and Gareth A. Jones

Abstract We compute the Möbius function for the subgroup lattice of the simple Suzuki group $Sz(q)$, and use it to enumerate regular objects such as maps, hypermaps, dessins d'enfants and surface coverings with automorphism groups isomorphic to $Sz(q)$.

1 Introduction

Hall's theory of Möbius inversion in groups [13] allows one to enumerate various objects associated with a given finite group G. In particular, it shows that the (necessarily finite) number $n_\Gamma(G)$ of normal subgroups N of a finitely generated group Γ with $\Gamma/N \cong G$ is given by

$$n_\Gamma(G) = \frac{1}{|\mathrm{Aut}\, G|} \sum_{H \le G} \mu_G(H)|\mathrm{Hom}(\Gamma, H)|, \tag{1}$$

where μ_G is the Möbius function on the lattice of subgroups of G, defined recursively by

$$\sum_{K \ge H} \mu_G(K) = \delta_{H,G}. \tag{2}$$

(Here $\delta_{H,G}$ is the Kronecker delta function, equal to 1 or 0 as $H = G$ or $H < G$.) This equation arises from noting that

$$n_\Gamma(G) = |\mathrm{Epi}(\Gamma, G)/\mathrm{Aut}\, G| = |\mathrm{Epi}(\Gamma, G)|/|\mathrm{Aut}\, G|,$$

G.A. Jones announces with deep sadness that Martin Downs died on 16 May 2015.

M. Downs and G.A. Jones (✉)
School of Mathematics, University of Southampton, Southampton SO17 1BJ, UK
e-mail: G.A.Jones@maths.soton.ac.uk

J. Širáň and R. Jajcay (eds.), *Symmetries in Graphs, Maps, and Polytopes*,
Springer Proceedings in Mathematics & Statistics 159,
DOI 10.1007/978-3-319-30451-9_5

where $\operatorname{Aut} G$ acts semi-regularly by composition on the set $\operatorname{Epi}(\Gamma, G)$ of epimorphisms $\Gamma \to G$, and then applying Möbius inversion to the equation

$$|\operatorname{Hom}(\Gamma, G)| = \sum_{H \leq G} |\operatorname{Epi}(\Gamma, H)|. \tag{3}$$

For example, if $\Gamma = F_k$, the free group of finite rank k, then (1) becomes

$$n_\Gamma(G) = \frac{1}{|\operatorname{Aut} G|} \sum_{H \leq G} \mu_G(H)|H|^k, \tag{4}$$

giving the number $d_k(G)$ of orbits of $\operatorname{Aut} G$ on generating k-tuples for G. A similar principle applies to the enumeration of torsion-free normal subgroups of Γ with quotient G: one simply counts the smooth homomorphisms and epimorphisms $\Gamma \to H$, that is, those preserving the orders of torsion elements.

In certain categories $\mathfrak{C}$, the objects $\mathscr{O}$ can be identified with the permutation representations of a particular group Γ, and the regular objects (the connected objects with maximum symmetry) correspond to the representations of Γ as a regular permutation group $G \cong \operatorname{Aut} \mathscr{O}$, or equivalently to the normal subgroups N of Γ with $\Gamma/N \cong G$. For instance, in the case of coverings of a suitably 'nice' topological space X we take $\Gamma = \pi_1 X$, the fundamental group of X. For maps or hypermaps, we take Γ to be a free product $V_4 * C_2$ or $C_2 * C_2 * C_2$, and for oriented maps or hypermaps its even subgroup $C_\infty * C_2$ or F_2. For example $d_2(G)$, as given by Eq. (4), is

- the number of isomorphism classes of orientably regular hypermaps with automorphism group G;
- the number of regular unbranched coverings of the sphere minus three points (or of the torus minus one point) with covering group G;
- the number of regular dessins (in Grothendieck's terminology [11]) with automorphism group G.

Implementing Eq. (1) for a specific group G depends on knowing the value of $\mu_G(H)$ for each subgroup $H \leq G$. Sometimes, though, $n_\Gamma(G)$ can be calculated without directly using the Möbius function of G. For instance, in the case where G is a Suzuki group $Sz(2^e)$, $n_\Gamma(G)$ has been calculated for orientably regular maps of type $\{4, 5\}$ by Silver and the second author [19], and for regular maps and polytopes by Hubard and Leemans [15] and by Kiefer and Leemans [22]. These results were achieved without the knowledge of the complete Möbius function; essentially, in each case the authors used a restricted form of Möbius inversion, concentrating mainly on subgroups $H \cong Sz(2^f)$ where f divides e. The aim of this paper is to use the case $G = Sz(2^e)$ to illustrate the advantage of invoking Hall's theory in full; the above results are recovered in a more unified manner; more importantly, the theory allows the enumerations of a significantly broader range of regular objects with automorphism group G, compared with those previously known. In doing this, we first have to determine μ_G, and a full derivation of this will be presented in this paper.

The Suzuki groups $G = Sz(q) = {}^2B_2(q)$, where $q = 2^e$ for some odd $e > 1$, form a family of finite simple groups. These groups, discovered in 1960 by Suzuki [29, 30], are important for several reasons: their low dimension, as subgroups of $GL_4(q)$, and their doubly transitive action on Tits ovoids, make them objects of great interest in finite geometry [25] and in the theory of finite permutation groups [16, Sect. XI.3]; moreover, as the only non-abelian finite simple groups of order coprime to 3 they often need to be treated as exceptional cases when proving theorems by inspection, as in the work of Breuillard, Green and Tao [1] on expanders.

After explaining the connections between certain categories $\mathfrak{C}$ and groups Γ in Sect. 2, and discussing various techniques for evaluating $|\mathrm{Hom}\,(\Gamma, H)|$ in Sect. 3, we describe the Suzuki groups $G = Sz(q)$ and their subgroups H in Sect. 4. We give the values of $\mu_G(H)$ in Table 1 in Sect. 4.4, with a proof in Sects. 5–7. Specifically, this table gives the values of $\mu_G(H)$ and $|N_G(H)|$ for a set $\mathcal{T}$ of representatives H of the conjugacy classes of subgroups of G on which μ_G can take non-zero values. This information is sufficient for applications of Eq. (1): since $|\mathrm{Aut}\,G| = e|G|$, this now takes the form

$$n_\Gamma(G) = \frac{1}{e}\sum_{H\in\mathcal{T}}\frac{\mu_G(H)|\mathrm{Hom}(\Gamma, H)|}{|N_G(H)|}. \qquad (5)$$

The second aim of this paper is to apply this equation to enumerate regular objects in various categories with automorphism groups isomorphic to $Sz(q)$. For example, it follows that for the Suzuki groups G we have

$$d_2(G) = \frac{1}{e}\sum_{f|e}\mu\left(\frac{e}{f}\right)2^f(2^{4f} - 2^{3f} - 9),$$

where μ is the classical Möbius function on $\mathbb{N}$, given by $\mu(n) = (-1)^k$ if n is a product of k distinct primes, and $\mu(n) = 0$ otherwise. The value of this formula therefore gives the number of regular objects with automorphism group G itemised earlier. Similar formulae, enumerating regular maps and hypermaps of various types, normal subgroups of Hecke groups, and regular coverings of orientable and non-orientable surfaces, are given in Sect. 8, and in more detail for the smallest simple Suzuki group $Sz(8)$ in Sect. 9.

2 Categories and Groups

We will consider some categories $\mathfrak{C}$ for which there is a group $\Gamma = \Gamma_{\mathfrak{C}}$ such that the set $\mathcal{R}(G) = \mathcal{R}_{\mathfrak{C}}(G)$ of regular objects in $\mathfrak{C}$ with automorphism group G is in bijective correspondence with the set $\mathcal{N}(G) = \mathcal{N}_\Gamma(G)$ of normal subgroups of Γ with quotient group G (see [18] for further details). We will call Γ the *parent group* of $\mathfrak{C}$. In particular, if Γ is finitely generated and G is finite then these two sets have the same finite cardinality

$$r(G) = r_{\mathfrak{C}}(G) := |\mathscr{R}_{\mathfrak{C}}(G)| = n(G) = n_{\Gamma}(G) := |\mathscr{N}_{\Gamma}(G)|, \tag{6}$$

so that Eq. (1) gives

$$r_{\mathfrak{C}}(G) = \frac{1}{|\mathrm{Aut}\, G|} \sum_{H \leq G} \mu_G(H)|\mathrm{Hom}(\Gamma, H)|. \tag{7}$$

2.1 Maps, Hypermaps and Groups

A map $\mathscr{M}$ is regular (in the category $\mathfrak{M}$ of all maps) if its automorphism group $G = \mathrm{Aut}\,\mathscr{M} = \mathrm{Aut}_{\mathfrak{M}}\mathscr{M}$ acts transitively on vertex-edge-face flags, or equivalently on the faces of the barycentric subdivision $\mathscr{B}$ of $\mathscr{M}$. In this case G is generated by automorphisms r_i $(i = 0, 1, 2)$ which change (in the only possible way) the i-dimensional component of a particular flag, while preserving its j-dimensional components for each $j \neq i$. If $\mathscr{M}$ has type $\{m, n\}$ in the notation of [3, Chap. 8], meaning that its faces are all m-gons and its vertices all have valency n, these generators satisfy

$$r_i^2 = (r_0r_1)^m = (r_0r_2)^2 = (r_1r_2)^n = 1.$$

It follows that there is an epimorphism $\theta : \Gamma \to G$, $R_i \mapsto r_i$, where

$$\Gamma = \Gamma_{\mathfrak{M}} = \langle R_0, R_1, R_2 \mid R_i^2 = (R_0R_2)^2 = 1\rangle, \tag{8}$$

so $\mathscr{M}$ determines a normal subgroup $N = \ker\theta$ of Γ with $\Gamma/N \cong G$. Conversely, each such normal subgroup determines a regular map $\mathscr{M}$ with $\mathrm{Aut}\,\mathscr{M} \cong G$. Two such maps are isomorphic if and only if they correspond to the same normal subgroup, so the set $\mathscr{R}(G) = \mathscr{R}_{\mathfrak{M}}(G)$ of regular maps with automorphism group G is in bijective correspondence with the set $\mathscr{N}(G) = \mathscr{N}_{\Gamma}(G)$. If G is finite then the preceding argument gives

$$r_{\mathfrak{M}}(G) = \frac{1}{|\mathrm{Aut}\, G|} \sum_{H \leq G} \mu_G(H)|\mathrm{Hom}(\Gamma, H)|. \tag{9}$$

This group $\Gamma_{\mathfrak{M}}$, a free product of its subgroups $\langle R_0, R_2\rangle \cong V_4$ and $\langle R_1\rangle \cong C_2$, can be regarded as the extended triangle group $\Delta[\infty, 2, \infty]$, generated by reflections in the sides of a hyperbolic triangle with angles $0, \pi/2, 0$. Other triangle groups play a similar role for related categories. For example, the extended triangle group

$$\Gamma = \Delta[n, 2, m] = \langle R_0, R_1, R_2 \mid R_i^2 = (R_0R_1)^m = (R_0R_2)^2 = (R_1R_2)^n = 1\rangle$$

is the parent group for maps of all types $\{m', n'\}$ dividing $\{m, n\}$ (meaning that m' divides m and n' divides n). For maps of type $\{m, n\}$ one must restrict attention to normal subgroups N of Γ such that R_0R_1 and R_1R_2 have images of orders m and n in Γ/N.

For the category $\mathfrak{H}$ of hypermaps, where hyperedges may be incident with any number of hypervertices and hyperfaces, we delete the relation $(R_0R_2)^2 = 1$ from the presentation (8), giving the group

$$\Gamma_{\mathfrak{H}} = \Delta[\infty, \infty, \infty] \cong C_2 * C_2 * C_2.$$

For hypermaps of types dividing (p, q, r), we use the extended triangle group

$$\Delta[p, q, r] = \langle R_0, R_1, R_2 \mid R_i^2 = (R_0R_1)^r = (R_0R_2)^q = (R_1R_2)^p = 1\rangle.$$

The parent groups for the categories $\mathfrak{M}^+$ and $\mathfrak{H}^+$ of oriented maps and hypermaps are the orientation-preserving subgroups of index 2 in $\Gamma_{\mathfrak{M}}$ and $\Gamma_{\mathfrak{H}}$, generated by the elements $X = R_1R_0$, $Y = R_0R_2$ and $Z = R_2R_1$ satisfying $XYZ = 1$. These are the triangle groups

$$\Gamma_{\mathfrak{M}^+} = \Delta(\infty, 2, \infty) = \langle X, Y, Z \mid Y^2 = XYZ = 1\rangle \cong C_\infty * C_2$$

and

$$\Gamma_{\mathfrak{H}^+} = \Delta(\infty, \infty, \infty) = \langle X, Y, Z \mid XYZ = 1\rangle \cong C_\infty * C_\infty \cong F_2.$$

For oriented maps of types dividing $\{m, n\}$, or oriented hypermaps of types dividing (p, q, r), we use the triangle groups $\Delta(n, 2, m)$ and $\Delta(p, q, r)$, restricting attention to torsion-free normal subgroups for maps and hypermaps of these exact types. Similarly, for the categories $\mathfrak{M}_k$ and $\mathfrak{M}_k^+$ of maps and of oriented maps in which all vertices have valency dividing k we use the parent groups $\Gamma_{\mathfrak{M}_k} = \Delta[k, 2, \infty]$ and $\Gamma_{\mathfrak{M}_k^+} = \Delta(k, 2, \infty) \cong C_k * C_2$. (See [20] for further background.)

2.2 *Reflexibility*

The regular objects in the categories $\mathfrak{M}^+$ and $\mathfrak{H}^+$ are often referred to as orientably regular, since they need not be regular as objects in the larger categories $\mathfrak{M}$ and $\mathfrak{H}$. Let $\mathscr{H}$ be an orientably regular hypermap of type (p, q, r), corresponding to a normal subgroup N of $\Gamma_{\mathfrak{H}^+} = F_2$ with $F_2/N \cong \mathrm{Aut}_{\mathfrak{H}^+}\mathscr{H} \cong G$ for some group G. Then the following are equivalent:

- $\mathscr{H}$ is regular in the category $\mathfrak{H}$ of all hypermaps;
- $\mathscr{H}$ has an orientation-reversing automorphism;
- N is normal in $\Gamma_{\mathfrak{H}} = C_2 * C_2 * C_2$;

- some (and hence each) pair of generators from the canonical generating triple x, y, z for G is inverted by an automorphism of G.

If these conditions hold we say that $\mathscr{H}$ is reflexible; otherwise it is chiral, and $\mathscr{H}$ and its mirror image $\overline{\mathscr{H}}$ form a chiral pair, isomorphic in $\mathfrak{H}$ but not in $\mathfrak{H}^+$.

If $\mathscr{H}$ is reflexible then $\tilde{G} := \mathrm{Aut}_{\mathfrak{H}}\mathscr{H} \cong \Gamma_{\mathfrak{H}}/N$ is a semidirect product of G by a complement C_2 generated by the image $r_i \in \tilde{G}$ of any R_i $(i = 0, 1, 2)$. The elements of $\tilde{G} \setminus G$ act by conjugation on the normal subgroup G; if one of them induces an inner automorphism then they all do, and we say that $\mathscr{H}$ is inner reflexible. In this case, r_i induces conjugation by some $g \in G$, so $c := r_i g$ centralises G and hence c^2 is in the centre $Z(G)$ of G. If $Z(G)$ is trivial then $\tilde{G} = G \times C$ where $C = \langle c \rangle \cong C_2$, and there is a non-orientable regular hypermap $\tilde{\mathscr{H}} = \mathscr{H}/C \in \mathscr{R}_{\mathfrak{H}}(G)$ with orientable double cover $\mathscr{H}$; this gives a monomorphism $\mathscr{H} \mapsto \tilde{\mathscr{H}}$, from the inner reflexible maps in $\mathscr{R}_{\mathfrak{H}^+}(G)$ to $\mathscr{R}_{\mathfrak{H}}(G)$. If, in addition, G has no subgroup of index 2, then each hypermap in $\mathscr{R}_{\mathfrak{H}}(G)$ is non-orientable, with an inner reflexible orientable double cover in $\mathscr{R}_{\mathfrak{H}^+}(G)$, so this monomorphism is a bijection. This proves the first part of the following result; the second part is obvious:

Proposition 1 *(a) For any finite group G with trivial centre and no subgroup of index 2, the inner reflexible hypermaps in $\mathscr{R}_{\mathfrak{H}^+}(G)$ are the orientable double covers of the hypermaps in $\mathscr{R}_{\mathfrak{H}}(G)$; there are $r_{\mathfrak{H}}(G)$ of them.*

(b) If, in addition, Out *G has odd order, then every reflexible hypermap in $\mathscr{R}_{\mathfrak{H}^+}(G)$ is inner reflexible, and there are $r_{\mathfrak{H}}(G)$ of them.*

Every non-abelian finite simple group G satisfies (a), and the Suzuki groups also satisfy (b) (see Sect. 4.2(2)). The function $\mathscr{H} \mapsto \tilde{\mathscr{H}}$ preserves types of hypermaps, so the above proposition also applies to maps.

2.3 Covering Spaces

Under suitable conditions (namely, that X is path connected, locally path connected, and semilocally simply connected [27, Chap. 13]), the equivalence classes of unbranched coverings $Y \to X$ of a topological space X form a category $\mathfrak{C}$ in which the connected objects correspond to the conjugacy classes of subgroups of the fundamental group $\Gamma = \pi_1 X$; among these, the regular coverings correspond to the normal subgroups N of Γ, with covering group isomorphic to Γ/N. If, in addition, X is a compact Hausdorff space, then Γ is finitely generated [27, p. 500], so one can use the methods described earlier to count regular coverings of X with a given finite covering group. In particular, this applies if X is a compact manifold or orbifold. Indeed, the categories of maps and hypermaps described above can be regarded as obtained in this way from suitable orbifolds X, such as a triangle with angles $\pi/p, \pi/q, \pi/r$ for hypermaps of type dividing (p, q, r), or a sphere with three cone-points of orders p, q, r in the oriented case. Similarly, Grothendieck's dessins d'enfants [10, 11] are the finite coverings of a sphere minus three points, so their parent group is its fundamental group F_2.

3 Counting Homomorphisms

In order to apply this method to a specific pair of groups Γ and G, one needs to be able to count homomorphisms $\Gamma \to H$ for certain subgroups $H \leq G$. Given a presentation for Γ with generators X_i and defining relations $R_j(X_i) = 1$, this amounts to counting the solutions (x_i) in H of the equations $R_j(x_i) = 1$. For certain groups Γ, such as free products of cyclic groups, this calculation is straightforward; in some other cases, the character table of H gives this information, as illustrated by the following three theorems, the first of which is due to Frobenius [8] (see [28, Theorem 7.2.1] for a proof of a generalisation of this).

Theorem 2 *Let $\mathscr{C}_i$ $(i = 1, 2, 3)$ be conjugacy classes in a finite group H. Then the number of solutions of the equation $x_1x_2x_3 = 1$ in H, with $x_i \in \mathscr{C}_i$ for $i = 1, 2, 3$, is given by the formula*

$$\frac{|\mathscr{C}_1||\mathscr{C}_2||\mathscr{C}_3|}{|H|} \sum_{\chi} \frac{\chi(x_1)\chi(x_2)\chi(x_3)}{\chi(1)} \tag{10}$$

where $x_i \in \mathscr{C}_i$ and χ ranges over the irreducible complex characters of H.

If Γ is the triangle group

$$\Delta(m_1, m_2, m_3) = \langle X_1, X_2, X_3 \mid X_1^{m_1} = X_2^{m_2} = X_3^{m_3} = X_1X_2X_3 = 1\rangle$$

of type (m_1, m_2, m_3) for some integers m_i, then $|\mathrm{Hom}(\Gamma, H)|$ can be found by summing (10) over all choices of triples of conjugacy classes $\mathscr{C}_i$ of elements of orders dividing m_i. Similarly, the number of smooth homomorphisms $\Gamma \to H$ can be found by restricting the summation to triples of classes of elements of order equal to m_i.

When Γ is an orientable surface group, that is, the fundamental group

$$\Pi_g = \pi_1\mathscr{S}_g = \langle A_i, B_i \ (i = 1, \ldots, g) \mid \prod_{i=1}^{g}[A_i, B_i] = 1\rangle$$

of a compact orientable surface $\mathscr{S}_g$ of genus $g \geq 1$, with $[a, b]$ denoting the commutator $a^{-1}b^{-1}ab$, the following theorem of Frobenius [8] and Mednykh [26] is useful (see [17] for applications):

Theorem 3 *In any finite group H, the number of solutions (a_i, b_i) of the equation $\prod_{i=1}^{g}[a_i, b_i] = 1$ is given by the formula*

$$|H|^{2g-1}\sum_{\chi}\chi(1)^{2-2g} \tag{11}$$

where χ ranges over the irreducible complex characters of H.

The formula in (11) gives $|\mathrm{Hom}(\Pi_g, H)|$ in terms of the degrees $\chi(1)$ of the irreducible characters of H. When Γ is a non-orientable surface group

$$\Pi_g^- = \langle A_i\ (i = 1, \ldots, g) \mid \prod_{i=1}^{g} A_i^2 = 1\rangle$$

of genus $g \geq 1$, the corresponding result, due to Frobenius and Schur [9], is as follows:

Theorem 4 *In any finite group H, the number of solutions (a_i) of the equation $\prod_{i=1}^{g} a_i^2 = 1$ is given by the formula*

$$|H|^{g-1}\sum_{\chi} c_\chi^g\,\chi(1)^{2-g} \tag{12}$$

where χ ranges over the irreducible complex characters of H.

Here $c_\chi = |H|^{-1}\sum_{h\in H}\chi(h^2)$ is the Frobenius-Schur indicator of χ, equal to $1, -1$ or 0 as χ is respectively the character of a real representation, the real character of a non-real representation, or a non-real character.

4 The Suzuki Groups and Their Subgroups

For the rest of this paper, G will denote a Suzuki group $Sz(q)$. This section is based mainly on Suzuki's description of these groups in [30]; see also [16, Sect. XI.3] and [31, Sect. 4.2] for further information. We have largely followed Suzuki's notation for elements and subgroups, except that we use the symbol F for the subgroup denoted in [30] by H (a Frobenius group of order $q^2(q-1)$), while we use H for an arbitrary subgroup of G. Also, our notation for certain cyclic subgroups, namely A_1 and A_2 in Sect. 4.1, differs from that used by Suzuki; our usage better respects the subgroup structure.

The method used for calculating the values of μ_G is as follows. Hall [13, Theorem 2.3] showed that, in any finite group G, a subgroup H satisfies $\mu_G(H) = 0$ unless H is an intersection of maximal subgroups of G. In our case, with $G = Sz(q)$, instead of directly determining the set $\mathscr{I}$ of such intersections, we first describe, in Sect. 4.3.3, a more convenient set $\mathscr{S}$ of subgroups of G such that every subgroup in $\mathscr{I}$ is conjugate to a subgroup in $\mathscr{S}$ (see Theorem 6). Since μ_G is invariant under conjugation, it sufficient to find its values on $\mathscr{S}$; then the set $\mathscr{T}$ appearing in Eq. (5) is simply the subset of $\mathscr{S}$ on which μ_G can take non-zero values.

For each pair of subgroups $H, K \in \mathscr{S}$, we determine in Table 2 (Sect. 7) the number $N(H;K)$ of conjugates of K containing H. Since $\mu_G(K) = 0$ for all $K \notin \mathscr{S}$, Eq. (2) gives

$$\mu_G(H) = -\sum_{H<K\in\mathscr{S}} N(H;K)\mu_G(K) \tag{13}$$

where $H \in \mathscr{S} \setminus \{G\}$. This allows $\mu_G(H)$ to be calculated recursively, starting with $\mu_G(G) = 1$ and then using the values of $\mu_G(K)$ for the subgroups $K \in \mathscr{S}$ properly containing H.

4.1 The Definition of the Suzuki Groups

Let $\mathbb{F} = \mathbb{F}(e)$ be the finite field $\mathbb{F}_q$ of $q = 2^e$ elements for some odd $e \geq 1$, and let θ be the automorphism $\alpha \mapsto \alpha^r$ of $\mathbb{F}$ where $r = \sqrt{2q} = 2^{(e+1)/2}$, so that θ^2 is the Frobenius automorphism $\alpha \mapsto \alpha^2$.

For any $\alpha, \beta \in \mathbb{F}$ let (α, β) denote the 4×4 matrix

$$(\alpha,\beta) = \begin{pmatrix} 1 & & & \\ \alpha & 1 & & \\ \alpha^{\theta+1}+\beta & \alpha^\theta & 1 & \\ \alpha^{\theta+2}+\alpha\beta+\beta^\theta & \beta & \alpha & 1 \end{pmatrix}.$$

Since $(\alpha, \beta)(\gamma, \delta) = (\alpha + \gamma, \alpha\gamma^\theta + \beta + \delta)$, these matrices (α, β) form a group $Q = Q(e)$ of order q^2, with identity element $(0, 0)$.

For each $\kappa \in \mathbb{F}^* := \mathbb{F} \setminus \{0\}$ let a_κ denote the 4×4 diagonal matrix with diagonal entries ζ_i where $\zeta_1^\theta = \kappa^{1+\theta}$, $\zeta_2^\theta = \kappa$, $\zeta_3 = \zeta_2^{-1}$ and $\zeta_4 = \zeta_1^{-1}$. These matrices form a cyclic group $A_0 = A_0(e) \cong \mathbb{F}^*$ of order $q - 1$. Since

$$a_\kappa^{-1}(\alpha, \beta)a_\kappa = (\alpha\kappa, \beta\kappa^{1+\theta}),$$

the group $F = F(e)$ generated by Q and A_0 in $GL_4(q)$ is a semidirect product of a normal subgroup Q by a complement A_0, so it has order $q^2(q-1)$.

We define $G = G(e)$ to be the subgroup of $GL_4(q)$ generated by F and the 4×4 matrix with entries 1 on the minor diagonal and 0 elsewhere (denoted by τ in [30]). This is the Suzuki group associated with $\mathbb{F}_q$, usually denoted by $Sz(q)$ or ${}^2B_2(q)$. It is, in fact, the subgroup of the symplectic group $Sp_4(q) = B_2(q)$ fixed by a certain automorphism of order 2.

In its natural action $g : [v] \mapsto [vg]$ on the projective space $\mathbb{P}^3(\mathbb{F})$, G acts as a doubly transitive permutation group of degree $q^2 + 1$ on the ovoid

$$\Omega = \Omega(e) = \{[\alpha^{\theta+2} + \alpha\beta + \beta^\theta,\ \beta,\ \alpha,\ 1] \mid \alpha, \beta \in \mathbb{F}\} \cup \{\infty\} \subset \mathbb{P}^3(\mathbb{F}),$$

where

$$\infty := [1,0,0,0] \in \mathbb{P}^3(\mathbb{F}).$$

The subgroup G_∞ of G fixing ∞ is F. This acts as a Frobenius group on $\Omega \setminus \{\infty\}$: its Frobenius kernel is Q, acting regularly on $\Omega \setminus \{\infty\}$, and A_0 is a Frobenius complement $G_{\infty,\omega}$, fixing a second point

$$\omega := [0,0,0,1] \in \Omega$$

and acting semiregularly on $\Omega \setminus \{\infty, \omega\}$. Thus the stabiliser of any three points in Ω is the identity subgroup I, so G acts on Ω as a Zassenhaus group.

There are cyclic subgroups of G of mutually coprime odd orders

$$2^e \pm 2^{(e+1)/2} + 1 = q \pm r + 1,$$

contained in Singer subgroups of $GL_4(q)$: note that since $r = \sqrt{2q}$, it follows that $q+r+1$ and $q-r+1$ are coprime and

$$(q+r+1)(q-r+1) = q^2+1,$$

which divides q^4-1. Take any subgroup of order $a_1(e) := q+r+1$ or $q-r+1$, depending on which is divisible by 5, and denote it by $A_1 = A_1(e)$; similarly take any subgroup of order $a_2(e) = q+r+1$ or $q-r+1$ not divisible by 5, and denote it by $A_2 = A_2(e)$. (This rule for indexing A_1 and A_2 differs from that used in [16, 30], where the rule is that $|A_1(e)| > |A_2(e)|$ for all e.)

4.2 Basic Properties of Suzuki Groups

Here we record some basic properties of G; see [16, 30] for proofs.

1. G has order $q^2(q^2+1)(q-1)$, and is simple if $e > 1$. (The group $G(1)$ is isomorphic to $AGL_1(5)$, of order 20.)
2. $\mathrm{Aut}\, G$ is a semidirect product of $\mathrm{Inn}\, G \cong G$ by a cyclic group of order e acting as the Galois group $\mathrm{Gal}\,\mathbb{F}$ on matrix entries, so $|\mathrm{Aut}\, G| = e|G|$.
3. Any two subgroups of G conjugate to Q intersect trivially, and any two subgroups conjugate to F have their intersection conjugate to A_0.
4. Q is a Sylow 2-subgroup of G of order q^2 and of exponent 4. The centre Z of Q consists of the identity and the involutions of Q (the matrices $(0,\beta)$ for $\beta \in \mathbb{F}$), with Z and Q/Z elementary abelian of order q.
5. $ZA_0 \cong F/Z \cong AGL_1(q)$, with A_0 acting regularly by conjugation on the non-identity elements of Z and of Q/Z.

6. The involutions of G are all conjugate, as are the cyclic subgroups of order 4; however an element of order 4 is not conjugate to its inverse.
7. All elements of G except those in a conjugate of Q have odd order. Each maximal cyclic subgroup of G of odd order is conjugate to A_0, A_1 or A_2; the intersection of any two of them is trivial.
8. A non-identity element of G has two fixed points on Ω, one fixed point, or none as it is conjugate to an element of A_0, of Q or of A_i for $i = 1, 2$, or, equivalently, as it has order dividing $q - 1$, 4 or $q^2 + 1$.

4.3 Some Particular Subgroups

Here we list some particular subgroups of G, in the anticipation that any subgroup H not conjugate to a member of the list satisfies $\mu_G(H) = 0$, and can therefore be ignored in the enumerations mentioned in Sect. 1.

4.3.1 Subgroups Associated with Subfields

If f divides e then restricting matrix entries to the subfield $\mathbb{F}(f)$ of $\mathbb{F}$ of order 2^f yields a subgroup $G(f) = Sz(2^f)$ of G. This acts doubly transitively, with degree $2^{2f} + 1$, on the subset $\Omega(f)$ of Ω defined over $\mathbb{F}(f)$. Since the point ∞ is defined over the prime field $\mathbb{F}(1)$, its stabiliser in $G(f)$ is $F(f) := F \cap G(f)$, which acts faithfully on $\Omega(f) \setminus \{\infty\}$ as a Frobenius group with kernel $Q(f) := Q \cap G(f) = \{(\alpha, \beta) \mid \alpha, \beta \in \mathbb{F}(f)\}$ and complement $A_0(f) := A_0 \cap G(f) = \{a_\kappa \mid \kappa \in \mathbb{F}(f)^*\}$. (Note that since r and $r + 1$ are coprime to $q - 1$, we have $a_\kappa \in G(f)$ if and only if $\kappa \in \mathbb{F}(f)^*$.) Let $Z(f)$ denote the centre of $Q(f)$, an elementary abelian group of order 2^f. If f and h are divisors of e, and f divides h, then

$$G(f) \leq G(h),\ \ F(f) \leq F(h),\ \ Q(f) \leq Q(h),\ \ Z(f) \leq Z(h),\ \ A_0(f) \leq A_0(h).$$

For each odd integer $f \geq 1$, let $a_1(f)$ and $a_2(f)$ denote the unique member of the pair $2^f \pm 2^{(f+1)/2} + 1$ which respectively is and is not divisible by 5. It is easy to show that

$$a_i(f) = 2^f - (-1)^i \chi(f) 2^{(f+1)/2} + 1$$

for $i = 1, 2$, where $\chi(f) := 1$ or -1 as $f \equiv \pm 1$ or $\pm 3 \pmod 8$. If e is an odd multiple of f, a simple argument using modular arithmetic [7, Sect. 2.3.1] shows that $a_i(f)$ divides $a_i(e)$ for $i = 1, 2$. This implies that for $i = 1, 2$, the cyclic group $A_i(e)$ of G has a subgroup $A_i(f)$ of order $a_i(f)$ for each f dividing e; furthermore, if $f|h|e$ then $A_i(f) \leq A_i(h)$. Note, however, that $A_i(f)$ is not necessarily a subgroup of $G(f)$, though it is conjugate to such a subgroup.

We have now identified seven families of subgroups of G, each forming a lattice isomorphic to the lattice $\Lambda(e)$ of divisors of e, a fact that is useful in evaluating μ_G. In the next subsection we will identify three others.

4.3.2 The Normalisers of Some Subgroups

The normaliser B_0 of A_0 in G is a dihedral group of order $2(q-1)$; it is the subgroup $G_{\{\infty,\omega\}}$ of G preserving the subset $\{\infty, \omega\}$ of Ω, with its subgroup A_0 fixing these two elements and its involutions transposing them. Let us choose a particular involution $c \in B_0$ and, for each f dividing e, define

$$B_0(f) := \langle A_0(f), c\rangle \leq B_0,$$

a dihedral group of order $2(2^f - 1)$ (so $B_0(1) \cong C_2$). If $f > 1$, then $B_0(f)$ is self-normalising whereas the normaliser of $A_0(f)$ is B_0.

For $i = 1, 2$ the normaliser B_i of A_i in G is a semidirect product of A_i and a cyclic group of order 4 generated by an element c_i satisfying $c_i^{-1}ac_i = a^{2^e}$ for all $a \in A_i$. For each f dividing e let

$$B_i(f) := \langle A_i(f), c_i\rangle \leq B_i,$$

so $|B_i(f)| = 4a_i(f)$, with $B_2(1) \cong C_4$. If $i = 1$ or $f > 1$ then $B_i(f)$ is self-normalising, whereas the normaliser of $A_i(f)$ is B_i.

By their construction, these groups $B_i(f)$ are (abstract) Frobenius groups of degree $a_i(f)$, and they satisfy $B_i(f) \leq B_i(h)$ for $i = 0, 1$ and 2 whenever $f \mid h \mid e$.

4.3.3 An Important Set of Subgroups

For each f dividing e, we have defined the following subgroups of G, with the symbols (f) usually omitted when $f = e$:

$$G(f),\ F(f),\ Q(f),\ Z(f),\ B_i(f),\ A_i(f)\quad (i = 0, 1, 2). \tag{14}$$

Let $\mathscr{S}$ denote the set consisting of the subgroups in (14) for all f dividing e. The conjugacy class in G of any of these groups will be denoted by changing the appropriate italic capital letter to the corresponding script capital; thus $\mathscr{G}(f), \mathscr{F}, \ldots$ denote the conjugacy classes containing $G(f)$, F, and so on.

We note the following coincidences, conjugacies (denoted by $\sim$) and isomorphisms:

$$G(1) \sim B_1(1) \cong AGL_1(5),\quad F(1) = Q(1) \sim B_2(1) \cong C_4,$$
$$B_0(1) \sim Z(1) \cong C_2,\quad A_2(1) = A_0(1) = I.$$

In addition, if 3 divides e then

$$B_1(1) = B_1(3), \quad A_1(1) = A_1(3) \cong C_5.$$

Apart from these, any two distinct terms in (14) represent non-conjugate subgroups of G. In view of their special role in the following calculations, we will denote the class $\mathscr{A}_2(1) = \mathscr{A}_0(1)$ by $\mathscr{C}_1$, the class $\mathscr{B}_0(1) = \mathscr{Z}(1)$ by $\mathscr{C}_2$, and the class $\mathscr{F}(1) = \mathscr{Q}(1) = \mathscr{B}_2(1)$ by $\mathscr{C}_4$, since these consist of the cyclic subgroups of G of orders 1, 2 and 4.

4.4 *The Möbius Function of a Suzuki Group*

We can now present our main result in the form of Table 1, which gives the non-zero values of $\mu_G(H)$ for the subgroups H of G; any subgroups H not appearing in Table 1 (such as $Q(f)$ and $Z(f)$ for $f > 1$) satisfy $\mu_G(H) = 0$, and can therefore be ignored in applying equations such as (1).

Because of the conjugacies listed in Sect. 4.3.3, some conjugacy classes appear 'under an alias': for instance $\mathscr{F}(1)$ appears as $\mathscr{B}_2(1)$, and if 3 divides e then $\mathscr{G}(1)$ and $\mathscr{B}_1(1)$ appear as $\mathscr{B}_1(3)$. In the second column, $a_i(f) = 2^f - (-1)^i\chi(f)2^{(f+1)/2} + 1$ for $i = 1, 2$ (see Sect. 4.3.1). In the third column, the values of $|N_G(H)|$ are given for applications of Eq. (5). In the final column, μ is the classical Möbius function on $\mathbb{N}$, defined by

$$\sum_{m|n} \mu(m) = \delta_{n,1}$$

Table 1 Information about the subgroups H with non-zero values for $\mu_G(H)$

Conjugacy class of H	$\|H\|$	$\|N_G(H)\|$	$\mu_G(H)$
$\mathscr{G}(f),\ 1 < f \mid e$	$2^{2f}(2^{2f}+1)(2^f-1)$	$\|H\|$	$\mu(e/f)$
$\mathscr{F}(f),\ 1 < f \mid e$	$2^{2f}(2^f-1)$	$\|H\|$	$-\mu(e/f)$
$\mathscr{B}_0(f),\ 1 < f \mid e$	$2(2^f-1)$	$\|H\|$	$-\mu(e/f)$
$\mathscr{A}_0(f),\ 1 < f \mid e$	2^f-1	$2(q-1)$	$2\frac{(2^e-1)}{(2^f-1)}\mu(e/f)$
$\mathscr{B}_1(f),\ 1 < f \mid e$	$4a_1(f)$	$\|H\|$	$-\mu(e/f)$
$\mathscr{B}_2(f),\ 1 < f \mid e$	$4a_2(f)$	$\|H\|$	$-\mu(e/f)$
$\mathscr{B}_2(1) = \mathscr{C}_4$	4	$2q$	$-2^e\mu(e)$
$\mathscr{B}_0(1) = \mathscr{C}_2$	2	q^2	$-2^{2e-1}\mu(e)$
$\mathscr{A}_0(1) = \mathscr{C}_1$	1	$\|G\|$	$\|G\|\mu(e)$

for all $n \in \mathbb{N}$, with the consequence that $\mu(n) = (-1)^k$ or 0 as n is or is not a product of k distinct primes for some $k \geq 0$. Our aim is to show that the final column of this table is correct, by proving the following theorem:

Theorem 5 *Let G be a Suzuki group $Sz(2^e)$ for some odd $e > 1$, and let H be a subgroup of G. If $\mu_G(H) \neq 0$ then $\mu_G(H)$ is as given by Table 1.*

5 Subgroups H with $\mu_G(H) \neq 0$

As a first step towards proving Theorem 5, in this section we find some necessary conditions for a subgroup H of G to satisfy $\mu_G(H) \neq 0$.

5.1 *Maxint Subgroups*

If G is any finite group, we shall say that a subgroup H of G is *maxint* if it is the intersection of a set of maximal subgroups of G (when $H = G$ this set is empty). The set of maxint subgroups of G will be denoted by $\mathscr{I}$. Hall proved in [13, Theorem 2.3] that if $H \notin \mathscr{I}$ then $\mu_G(H) = 0$, so in determining μ_G one may restrict attention to the subgroups $H \in \mathscr{I}$. Since μ_G is preserved under conjugacy, it is sufficient to consider a set of representatives of the conjugacy classes of subgroups in $\mathscr{I}$. The main step in the proof of Theorem 5 is to show that if G is a Suzuki group $G(e)$ then the set $\mathscr{S}$ defined in Sect. 4.3.3 contains such a set of representatives:

Theorem 6 *If $H \in \mathscr{I}$ then H is conjugate in G to a subgroup in $\mathscr{S}$, that is, $\mathscr{I}$ is contained in the union of the conjugacy classes*

$$\mathscr{G}(f),\ \mathscr{F}(f),\ \mathscr{Q}(f),\ \mathscr{L}(f),\ \mathscr{B}_i(f),\ \mathscr{A}_i(f)$$

of subgroups of G, where f divides e and $i = 0, 1$ or 2.

The rest of this section is devoted to a proof of this theorem. We will use the following criterion for a subgroup H of G to be in $\mathscr{I}$. Let $\mathscr{M}$ denote the set of all maximal subgroups of G, and let $\mathscr{M}(H)$ denote the set of those containing a particular subgroup H of G. Then the following, valid for any finite group G, is evident:

Lemma 1 *Let $H \leq G$. Then*

$$H \leq \bigcap_{M \in \mathscr{M}(H)} M,$$

with equality if and only if $H \in \mathscr{I}$.

5.2 Maximal Subgroups

We will systematically apply Lemma 1 to the various subgroups H of G, using the following result:

Proposition 7 *The set $\mathscr{M}$ of maximal subgroups of G is given by*

$$\mathscr{M} = \bigcup_{e/f \text{ prime}} \mathscr{G}(f) \cup \mathscr{F} \cup \mathscr{B}_0 \cup \mathscr{B}_1 \cup \mathscr{B}_2.$$

This result is an immediate consequence of the following classification, due to Suzuki [30, Theorems 9 and 10]:

Proposition 8 *If $H \leq G$ then either $H \in \mathscr{G}(f)$ for some f dividing e, or H is a subgroup of a group in $\mathscr{F}$ or in $\mathscr{B}_i$ for some $i = 0, 1$ or 2.*

In the first case H is either simple or isomorphic to $G(1) \cong AGL_1(5)$, and in the second case H is solvable. Finite solvable groups H all satisfy Hall's theorems [12] on the existence and conjugacy of Hall π-subgroups for any set π of primes, generalising Sylow's theorems for single primes. We will use this fact several times, mainly with π the set $2'$ of odd primes.

Since $G(f) \in \mathscr{S}$ for each f dividing e, it follows from Proposition 8 that, in proving Theorem 6, it is sufficient to assume that H is a subgroup of a group in $\mathscr{F}$ or $\mathscr{B}_i$ for $i = 0, 1$ or 2. We will deal with these cases in turn.

In preparation for applying Lemma 1 in the first case, we will consider how the various maximal subgroups of G intersect F.

5.3 Point-Stabilisers in Maximal Subgroups

Recall that F is the stabiliser in G of the point $\infty \in \Omega$. If $H \leq F$ then

$$\bigcap_{M \in \mathscr{M}(H)} M = \bigcap_{M \in \mathscr{M}(H)} (M \cap F),$$

so in applying Lemma 1 to H one can restrict attention to the stabilisers $M_\infty = M \cap F$ of ∞ for the various maximal subgroups M of G. The following result describes the possibilities for these stabilisers.

Lemma 2 *Let M be a maximal subgroup of G.*

1. *If $M = F^g \in \mathscr{F}$, then $M \cap F = F$ or $M \cap F = G_{\infty,\infty g} \in \mathscr{A}_0$ as $g \in F$ or not.*
2. *If $M = G(f)^g \in \mathscr{G}(f)$ for some $f|e$, then $M \cap F = F(f)^g \in \mathscr{F}(f)$ or $M \cap F = I \in \mathscr{C}_1$ as $g \in G(f)F$ or not.*
3. *If $M \in \mathscr{B}_0$ then $M \cap F \in \mathscr{A}_0 \cup \mathscr{C}_1 \cup \mathscr{C}_2$.*
4. *If $M \in \mathscr{B}_i$ for $i = 1, 2$ then $M \cap F \in \mathscr{C}_1 \cup \mathscr{C}_2 \cup \mathscr{C}_4$.*

In order to prove the second part, we need the following lemma:

Lemma 3 *If f divides e then $G(f)$ acts semi-regularly on $\Omega \setminus \Omega(f)$.*

Proof By Sect. 4.2(8), a non-identity element of G fixes 2, 1 or 0 elements of Ω as it has order dividing $q-1, 4$ or q^2+1. Similarly, a non-identity element of $G(f)$ fixes 2, 1 or 0 elements of $\Omega(f)$ as it has order dividing $2^f-1, 4$ or $2^{2f}+1$ respectively. Since 2^f-1 divides $q-1$, and $2^{2f}+1$ divides q^2+1, a non-identity element of $G(f)$ can have no further fixed points in $\Omega \setminus \Omega(f)$. Thus all orbits of $G(f)$ on this set are regular, with point-stabilisers $G(f) \cap G_\alpha = I$ for $\alpha \in \Omega \setminus \Omega(f)$. □

Proof We can now prove Lemma 2. The maximal subgroups M of G are given by Proposition 7.

(1) This part is trivial, since F and M are the stabilisers in G of ∞ and ∞g, and G is doubly transitive on Ω.
(2) If $M = G(f)^g \in \mathscr{G}(f)$ then Lemma 3 shows that M acts doubly transitively on $\Omega(f)g$, and semiregularly on its complement. Thus $M \cap F = F(f)^g$ or I as $\infty \in \Omega(f)g$ or not, that is, as $g \in G(f)F$ or not.
(3) Each $M \in \mathscr{B}_0$ is the subgroup $G_{\{\alpha,\beta\}}$ of G preserving an unordered pair $\{\alpha, \beta\} \subset \Omega$. If $\infty \notin \{\alpha, \beta\}$ then since $G_{\alpha,\beta,\infty} = I$ we have $|M \cap F| \leq 2$, whereas if $\infty = \alpha$ or β then $M \cap F = G_{\alpha,\beta} \in \mathscr{A}_0$.
(4) If $M \in \mathscr{B}_i$ for $i = 1$ or 2 then $M = N_G(A) \cong A \rtimes C_4$ for some $A \in \mathscr{A}_i$; since A acts without fixed points, $M \cap F$ is isomorphic to a subgroup of C_4, so it is in $\mathscr{C}_m$ for $m = 1, 2$ or 4. □

5.4 Subgroups H of F

We first prove Theorem 6 for subgroups $H \in \mathscr{I}$ which are contained in groups in $\mathscr{F}$. Replacing H with a conjugate, we may assume that $H \leq F$.

5.4.1 Preliminaries

Here we record some observations and simplifications which will be used in the proof.

(a) Lemma 2 shows that each $M \in \mathscr{M}(H)$ satisfies $M \cap F \in \mathscr{X}_M$ where $\mathscr{X}_M = \mathscr{F}(f_M)$ for some f_M dividing e (depending on M), or $\mathscr{X}_M = \mathscr{A}_0$, or $\mathscr{X}_M = \mathscr{C}_m$ for some m dividing 4. If $\mathscr{X}_M = \mathscr{C}_m$ for some m, then since $H \leq M \cap F$ we have $H \in \mathscr{C}_1 \cup \mathscr{C}_2 \cup \mathscr{C}_4$, so H is as required, i.e. conjugate to an element of $\mathscr{S}$; we may therefore assume that for each $M \in \mathscr{M}(H) \setminus \{F\}$ we have $\mathscr{X}_M = \mathscr{F}(f_M)$ for some f_M or $\mathscr{X}_M = \mathscr{A}_0$, with $M \in \mathscr{G}(f_M)$ or $\mathscr{F}$ respectively.
(b) Since F is solvable, Hall's theorems [12] imply that a Hall $2'$-subgroup A of H is contained in one of the Hall $2'$-subgroups of F. These are the conjugates of

A_0, and Q permutes them regularly by conjugation, so by conjugating H with a suitable element of Q we may assume that $A \leq A_0$.

(c) If H has even order it contains an involution. The involutions in F (the non-identity elements of Z) are all conjugate under A_0, so in this case, by conjugating H with an element of A_0 we may also assume that H contains $z := (0, 1)$.

(d) If any $M \in \mathscr{M}(H) \setminus \{F\}$ satisfies $M = G(f_M)^g \in \mathscr{G}(f_M)$ for some $g \in G$, then since $M \cap F \in \mathscr{F}(f_M)$ we have $g \in G(f_M)F$ by Lemma 2. Without loss of generality we can therefore choose this conjugating element g to be in F. Then

$$F(f_M)^g = (G(f_M) \cap F)^g = G(f_M)^g \cap F = M \cap F.$$

Thus $z \in M \cap F$, so $z^{g^{-1}} \in F(f_M)$. Since $F = A_0 Q$ we can write $g = ab$ where $a \in A_0$ and $b \in Q$. Since z is in the centre Z of Q we have

$$z^{g^{-1}} = (z^{b^{-1}})^{a^{-1}} = z^{a^{-1}},$$

so $z^{a^{-1}} \in F(f_M)$. Since $z \in F(1)$ and A_0 acts regularly by conjugation on the involutions in Z, this implies that $a \in A_0(f_M)$. Thus $g = ab$ with $a \in F(f_M)$, so each $M \in \mathscr{M}(H) \setminus \{F\}$ satisfies

$$M \cap F = F(f_M)^g = F(f_M)^b$$

for some f_M dividing e, with $b \in Q$.

(e) We claim that if $f \mid h \mid e$ then the set

$$Q(f, h) := \{g \in Q \mid Q(f)^g \leq Q(h)\}$$

is the union of the cosets $(\alpha, 0)Z$ of Z in Q where $\alpha \in \mathbb{F}(h)$. Clearly this set consists of complete cosets of Z in Q. The elements $(\alpha, 0)$ where $\alpha \in \mathbb{F}$ are representatives of these cosets, since there is an epimorphism $(\alpha, \beta) \mapsto \alpha$ from Q to the additive group of $\mathbb{F}$, with kernel Z. Therefore it suffices to show that $g := (\alpha, 0) \in Q(f, h)$ if and only if $\alpha \in \mathbb{F}(h)$.
If $\alpha \in \mathbb{F}(h)$ then $g \in Q(h)$; since $Q(f) \leq Q(h)$ we have $Q(f)^g \leq Q(h)$ and hence $g \in Q(f, h)$. For the converse, note that $(1, 0) \in Q(f)$. A simple calculation shows that

$$(1, 0)^g = (1, \alpha + \alpha^\theta),$$

so if $g \in Q(f, h)$ then $\alpha + \alpha^\theta \in \mathbb{F}(h)$. The function $\phi : x \mapsto x + x^\theta$ maps each subfield $\mathbb{K}$ of $\mathbb{F}$ into itself. Composing ϕ with itself gives a quadratic polynomial

$$\phi^2 : x \mapsto (x + x^\theta) + (x + x^\theta)^\theta = x + x^{\theta^2} = x + x^2$$

defined over the prime field, so if $\beta \in \mathbb{K}$ then any element of $\phi^{-2}(\beta)$ has degree at most 2 over $\mathbb{K}$. Since e is odd, $\mathbb{F}$ contains no quadratic extensions, so $\phi^{-2}(\mathbb{K}) \subseteq \mathbb{K}$ and hence $\phi^{-1}(\mathbb{K}) \subseteq \mathbb{K}$. In particular, since $\phi(\alpha) \in \mathbb{F}(h)$ we have $\alpha \in \mathbb{F}(h)$.

We can now start the case-by-case analysis of maxint subgroups $H \leq F$.

5.4.2 Subgroups $H \leq F$ which are not 2-groups

First suppose that H is not a 2-group, or equivalently the Hall $2'$-subgroup $A \leq A_0$ of H is not the identity subgroup, so that $C_G(A) = A_0$. For each $M \in \mathscr{M}$, A is contained in a maximal cyclic subgroup A_M of M, which has order $2^{f_M} - 1$ if $M \in \mathscr{G}(f)$, and order $2^e - 1$ if $M \in \mathscr{F}$. This subgroup A_M centralises A, so it is contained in A_0; thus $A = A_0(f_M)$, where we take $f_M = e$ when $M \in \mathscr{F}$, since this is the unique subgroup of A_0 of order $2^{f_M} - 1$. Now H is the intersection of these subgroups M, so A is the intersection of the corresponding subgroups A_M; it therefore has the form $A = A_0(f)$ where f is the highest common factor of the divisors f_M of e.

If H has odd order then this gives $H = A \in \mathscr{A}_0(f)$, one of the types allowed for in Theorem 6, so we may assume that H has even order. As noted in Sect. 5.4.1(c), this allows us to assume that $z \in H$. This also implies that each $M \in \mathscr{M}(H) \setminus \{F\}$ is in $\mathscr{G}(f_M)$ for some f_M, for otherwise $M \in \mathscr{F}$ and so H is a subgroup of a two-point stabiliser $M \cap F$, which has odd order. As shown in Sect. 5.4.1(d), it follows that such subgroups M satisfy $M \cap F = F(f_M)^b$ for some $b \in Q$.

We have $A_0(f_M) = A \leq A_M \leq M \cap F = F(f_M)^b$, so $A_0(f_M)$ and $A_0(f_M)^{b^{-1}}$ are both point-stabilisers in $F(f_M)$; because $F(f_M)$ acts as a Frobenius group on $\Omega(f)$, its kernel $Q(f_M)$ permutes these point-stabilisers regularly by conjugation, so $A_0(f_M) = A_0(f_M)^{b^{-1}c}$ for some $c \in Q(f_M)$. Thus the element $b^{-1}c$ of Q normalises $A_0(f_M)$, so it also normalises $C_G(A_0(f_M)) = A_0$. However, Q permutes the conjugates of A_0 regularly by conjugation (since it is also a Frobenius group), so $b^{-1}c = 1$ and hence $b = c \in Q(f_M)$.

Thus $M \cap F = F(f_M)^b = F(f_M)$ for each $M \in \mathscr{M}(H) \setminus \{F\}$, so H, being the intersection of such subgroups $F(f_M)$, together with F, has the form $F(f) \cap F = F(f)$ for some divisor f of e, giving $H \in \mathscr{F}(f)$ as required.

5.4.3 Subgroups $H \leq F$ which are 2-groups

Now suppose that H is a 2-group, so $H \leq Q$. By Sect. 5.4.1(a), for each $M \in \mathscr{M}(H) \setminus \{F\}$ either $M \cap F \in \mathscr{F}(f_M)$ for some f_M dividing e, or $M \cap F \in \mathscr{A}_0$, with $M \in \mathscr{G}(f_M)$ or $\mathscr{F}$ respectively. We may assume that $H \neq I$, so H has even order and hence (as in Sect. 3.4.2) the second possibility cannot arise. Thus $M \cap F = F(f_M)^b$ for some $b \in Q$, as shown in Sect. 5.4.1(d). As Q is normal in F, and is a Sylow 2-subgroup of F, we have $M \cap Q = Q(f_M)^b$; comparing centres, we see that

$M \cap Z = Z(f_M)^b = Z(f_M)$ since $Z(f_M)$, being central in Q, is normalised by b. Thus if $H \le Z$ then

$$H = \bigcap_{M \in \mathcal{M}(H)} M = \bigcap_{M \in \mathcal{M}(H)} (M \cap Z) = \bigcap_{M \in \mathcal{M}(H)} Z(f_M) = Z(f),$$

where f is the highest common factor of the integers f_M, so $H \in \mathcal{Z}(f)$.

We may therefore assume that $H \not\le Z$, so H contains an element of order 4. Since F has a single conjugacy class of cyclic subgroups of order 4, we may assume that H contains the subgroup $Q(1) = \{1, z, y^{\pm 1}\}$ where $y := (1, 0)$. Then $Q(1) \le H \le M \cap Q = Q(f_M)^b$, so by Sect. 3.4.1(e) we have $b^{-1} \in (\alpha, 0)Z$ for some $\alpha \in \mathbb{F}(f_M)$. This shows that $M \cap Q = Q(f_M)^b = Q(f_M)$ for each $M \in \mathcal{M}(H) \setminus \{F\}$, so taking the intersection over all such M gives $H = Q(f) \in \mathcal{Q}(f)$ where f is the highest common factor of the integers f_M.

5.5 *Subgroups H of B_i*

Now suppose that H is a subgroup of a group in $\mathcal{B}_i$ for some $i = 0$, 1 or 2, and is maxint. Without loss of generality we may assume that $H \le B_i$.

5.5.1 Subgroups *H* of B_0

Suppose that $H \le B_0$. The subgroup $A := H \cap A_0 = H \cap F$ is maxint, since H is, it is contained in F, and it has odd order, so by an argument in Sect. 3.4.2 we see that $A = A_0(f)$ for some f dividing e. Now H contains A with index at most 2, so either $H = A_0(f)$, or H is a dihedral subgroup of B_0 conjugate (since $|A_0|$ is odd) to $B_0(f)$. Thus H is in $\mathcal{A}_0(f)$ or $\mathcal{B}_0(f)$.

5.5.2 Subgroups *H* of B_i for $i = 1$ or 2

Suppose that $H \le B_i$ where $i = 1$ or 2. Let $A := H \cap A_i$. If $|A| = 1$ then H is isomorphic to a subgroup of $B_i/A_i \cong C_4$, so H is in $\mathcal{C}_m$ for some $m = 1, 2$ or 4. We may therefore assume that $|A| > 1$. Any subgroup $M \in \mathcal{M}(H) \setminus \{B_i\}$ contains A, which has order dividing $q^2 + 1$, so it follows from the classification of the maximal subgroups in Proposition 7 that M must be in $\mathcal{B}_i$ or in $\mathcal{G}(f_M)$ for some f_M dividing e. The first possibility can be dismissed, since distinct subgroups in $\mathcal{B}_i$ have intersections of order dividing 4, so $M \in \mathcal{G}(f_M)$. We can now argue as in Sect. 3.4.2, by considering subgroups centralising A, to show that $A = A_i(f)$ for some f dividing e.

Now $|H : A|$ divides $|B_i : A_i| = 4$. If $|H : A| = 1$ then $H = A_i(f) \in \mathcal{A}_i(f)$, as required. If $|H : A| = 4$ then H is a subgroup of B_i of order $4a_i(f)$; all subgroups of this order are conjugate in B_i to $B_i(f)$, so $H \in \mathcal{B}_i(f)$. We will show that the remaining case $|H : A| = 2$, where $|H| = 2a_i(f)$, cannot arise.

In a Suzuki group $Sz(q)$, any subgroup K of order $2m$, where m divides q^2+1, is contained in a unique subgroup K^* of order $4m$. (This is because $A_i < KA_i < B_i$ for $i = 1$ or 2, up to conjugacy, and the complements for A_i in B_i have mutually trivial intersections.) Applying this to the subgroup $K = H$, firstly as a subgroup of G, and then as a subgroup of each of the Suzuki subgroups $M \cong G(f)$ in $\mathscr{M}(H)$, we see that there is a subgroup $H^* \leq B_i$, containing H with index 2, such that $H^* \leq M$ for all $M \in \mathscr{M}(H)$. Lemma 1 then shows that $H \notin \mathscr{I}$.

This completes the proof of Theorem 6. □

6 Size of Conjugacy Classes

An important step in proving the statement of the Möbius function of G in Theorem 5 is to determine the number of conjugates of each subgroup $H \in \mathscr{S}$, equal to the index in G of its normaliser $N_G(H)$. The orders of some of these normalisers are noted in Table 1.

Theorem 9 *Let f divide e. Then*

1. $N_G(G(f)) = G(f)$ *and* $|\mathscr{G}(f)| = |G|/|G(f)|$;
2. $N_G(F(f)) = F(f)$ *and* $|\mathscr{F}(f)| = |G|/|F(f)|$ *if* $f > 1$;
3. $|N_G(Q(f))| = 2^{e+f}(2^f-1)$ *and* $|\mathscr{Q}(f)| = |G|/2^{e+f}(2^f-1)$ *if* $f > 1$;
4. $N_G(Z(f)) = QA_0(f)$ *and* $|\mathscr{Z}(f)| = |G|/2^{2e}(2^f-1)$ *if* $f > 1$;
5. $N_G(B_i(f)) = B_i(f)$ *and* $|\mathscr{B}_i(f)| = |G|/|B_i(f)|$ *if* $i = 1$, *or if* $i = 0$ *or 2 and* $f > 1$;
6. $N_G(A_i(f)) = B_i$ *and* $|\mathscr{A}_i(f)| = |G|/|B_i|$ *if* $i = 1$, *or if* $i = 0$ *or 2 and* $f > 1$;
7. $|N_G(B_2(1))| = 2q$ *and* $|\mathscr{B}_2(1)| = q(q^2+1)(q-1)/2$;
8. $|N_G(B_0(1))| = q^2$ *and* $|\mathscr{B}_0(1)| = (q^2+1)(q-1)$.

Proof (1) Let $H = G(f)$ where f divides e. If $f > 1$, then since $N_G(G(f))$ contains $G(f)$ it cannot be solvable, so by Proposition 8 it must be conjugate to $G(h)$ for some multiple h of f. Since $G(h)$ is simple, we must have $h = f$ and $N_G(H) = H$, giving $|\mathscr{G}(f)| = |G|/|H|$. The case $f = 1$ is dealt with in (5), since $G(1)$ is conjugate to $B_1(1)$.

(4) It is convenient to prove (4) before (2) and (3). Let $f > 1$. Any element of G normalising $Z(f)$ must fix its unique fixed point ∞, so $N_G(Z(f)) \leq F$. By Sect. 4.2(4), $F = QA_0$. Now $Z(f)$ is centralised by Q since it lies in the centre Z of Q, and Sect. 4.2(5) implies that $N_G(Z(f)) \cap A_0 = A_0(f)$, so $N_G(Z(f)) = QA_0(f)$, of order $|Q|.|A_0(f)| = q^2(2^f-1) = 2^{2e}(2^f-1)$.

(3) Any element of G normalising $Q(f)$ must normalise its characteristic subgroup $Z(f)$, so $N_G(Q(f)) \leq N_G(Z(f)) = QA_0(f)$. Now $A_0(f) \leq F(f) \leq N_G(Q(f))$, and Sect. 3.4.1(e) shows that $N_Q(Q(f)) = \bigcup_\alpha (\alpha, 0)Z$ with the union over all $\alpha \in \mathbb{F}(f)$, so $N_G(Q(f))$ has order $|Z|.2^f.|A_0(f)| = 2^{e+f}(2^f-1)$.

(2) Clearly $N_G(F(f)) \leq N_G(Q(f)) \leq QA_0(f)$, and $A_0(f) \leq N_G(F(f))$. Since Z acts semi-regularly on $\Omega \setminus \{\infty\}$, it acts semi-regularly by conjugation on the

subgroups of F in the conjugacy class $\mathscr{A}_0$, so $N_G(F(f)) \cap Z = Z(f)$. Hence, using the proof of part (3), we see that $N_G(F(f)) \leq F \cap G(f) = F(f)$. Thus $F(f)$ is self-normalising.
(5, 6) See Sect. 4.3.2.
(7, 8) For $i = 0$ and 2 the subgroups in $\mathscr{B}_i(1)$ are cyclic groups of orders 2 and 4 respectively, so they are contained in Sylow 2-subgroups of G. There are $q^2 + 1$ Sylow 2-subgroups, each conjugate to Q and containing $q - 1$ subgroups of order 2, and containing $(q^2 - q)/2$ of order 4. Since distinct Sylow 2-subgroups have trivial intersection, there are $(q^2 + 1)(q - 1)$ and $(q^2 + 1)(q^2 - q)/2$ such subgroups in G. In each case such subgroups are all conjugate, so their normalisers have order q^2 and $2q$. □

7 Calculating Values of μ_G

We can now complete the proof of Theorem 5 by calculating $\mu_G(H)$ for each subgroup $H \in \mathscr{S}$. In order to use Eq. (13) for this (see Sect. 4), we first need to know, for each pair of subgroups $H, K \in \mathscr{S}$, the number $N(H; K)$ of conjugates in G of K containing H. If $M(H; K)$ denotes the number of conjugates in G of H contained in K, then a simple double counting argument gives

$$M(H; K)M(K; G) = M(H; G)N(H; K) \tag{15}$$

for all $H, K \in \mathscr{S}$. This allows $N(H; K)$ to be determined from the values of the function M. Now $M(H; G) = |\mathscr{H}|$ and $M(K; G) = |\mathscr{K}|$, where $\mathscr{H}$ and $\mathscr{K}$ are the conjugacy classes of subgroups of G containing H and K, so these values are given by Theorem 9. The values of $M(H; K)$ for $K \neq G$ can be found by using arguments similar to those used in proving Theorem 9, so details are omitted.

The non-zero values of $N(H; K)$ resulting from (15) are given in Table 2, where the rows and columns are indexed by the subgroups H and K respectively; the row corresponding to the identity subgroup $H = A_0(1) = A_2(1)$ is omitted since in this case $N(H; K) = |\mathscr{K}|$, given by Theorem 9 for all $K \in \mathscr{S}$. The table is split into two parts, the second part giving further entries for the last six rows of the first part. We assume that f divides h and that $f > 1$ unless otherwise stated. Thus $G(1)$ is represented by its conjugate $B_1(1)$, while $F(1)$ and $Q(1)$ are represented by $B_2(1)$, and $Z(1)$ by $B_0(1)$ (see Sect. 4.3.3 and the comments in Sect. 4.4).

Given Table 2, one can systematically use Eq. (13) to calculate $\mu_G(H)$ for each $H \in \mathscr{S}$, starting with $H = G(f)$ in the first row, and working downwards through the table. For instance, if $H = G(f)$ then the subgroups $K \in \mathscr{S}$ with $N(H; K) \neq 0$ are those of the form $K = G(h)$ where $f \mid h \mid e$; under inclusion, these form a lattice isomorphic to the lattice $\Lambda(e/f)$ of all divisors h/f of e/f, with $\mu_G(K) = 1$ when $h = e$, so we find that $\mu_G(H) = \mu(e/f)$, as in Table 1. Next, if $H = F(f)$ we consider the subgroups $K = G(h)$ and $F(h)$ where $f \mid h \mid e$; these form a lattice

Table 2 Values of $N(H;K)$ where $H, K \in \mathscr{S}$

	$G(h)$	$F(h)$	$Q(h)$	$Z(h)$	$B_0(h)$	$A_0(h)$
$G(f)$	1					
$F(f)$	1	1				
$Q(f)$	2^{e-h}	2^{e-h}	1			
$Z(f)$	$2^{2(e-h)}$	$2^{2(e-h)}$	2^{e-h}	1		
$B_0(f)$	1				1	
$A_0(f)$	$\frac{(2^e-1)}{(2^h-1)}$	$\frac{2(2^e-1)}{(2^h-1)}$			$\frac{(2^e-1)}{(2^h-1)}$	1
$B_1(f), f \geqslant 1$	1					
$B_2(f)$	1					
$A_1(f), f \geqslant 1$	$\frac{a_1(e)}{a_1(h)}$					
$A_2(f)$	$\frac{a_2(e)}{a_2(h)}$					
$B_2(1) \cong C_4$	2^{e-h}	2^{e-h}	1			
$B_0(1) \cong C_2$	$2^{2(e-h)}$	$2^{2(e-h)}$	2^{e-h}	1	2^{2e-1}	
	$B_1(h)$	$B_2(h)$	$A_1(h)$	$A_2(h)$	$B_2(1)$	$B_0(1)$
$B_1(f), f \geqslant 1$	1					
$B_2(f)$		1				
$A_1(f), f \geqslant 1$	$\frac{a_1(e)}{a_1(h)}$		1			
$A_2(f)$		$\frac{a_2(e)}{a_2(h)}$		1		
$B_2(1) \cong C_4$	2^{e-1}	2^{e-1}			1	
$B_0(1) \cong C_2$	$2^{2(e-1)}$	$2^{2(e-1)}$			2^{e-1}	1

isomorphic to $A(2e/f)$ since e is odd, giving $\mu_G(H) = \mu(2e/f) = -\mu(e/f)$. Similar arguments show that if $f > 1$ then $\mu_G(B_i(f)) = -\mu(e/f)$ for $i = 0, 1, 2$, and $\mu_G(Q(f)) = \mu_G(Z(f)) = \mu_G(A_i(f)) = 0$ for $i = 1, 2$. This process continues until $\mu_G(H)$ is evaluated for all $H \in \mathscr{S}$. The method is essentially the same as that described fully in [5, Sect. 4] for the groups $G = L_2(2^e)$, so the remaining details are omitted.

Now $\mu_G(H) = 0$ whenever $H = Q(f), Z(f), A_1(f)$ or $A_2(f)$ for any $f > 1$, so let $\mathscr{T}$ denote the remaining set of subgroups $H \in \mathscr{S}$, namely

$$G(f),\ F(f),\ B_i(f)\ (i = 0, 1, 2),\ A_0(f),\ B_2(1),\ B_0(1),\ A_0(1), \qquad (16)$$

where $1 < f \mid e$. This is a set of representatives for the conjugacy classes in Table 1, and every subgroup H of G with $\mu_G(H) \neq 0$ is in one of these classes. This, together with the values of $|H|$, $|N_G(H)|$ and $\mu_G(H)$ determined earlier, justifies the entries in Table 1 and in particular proves Theorem 5. □

Each conjugacy class in Table 1 contains $|G|/|N_G(H)|$ subgroups, and $|\mathrm{Aut}\, G| = e|G|$ by Sect. 4.2(2), so Eq. (1), counting the normal subgroups N of a finitely generated group Γ with $\Gamma/N \cong G$, can be reformulated as in Eq. (5), that is,

$$n_\Gamma(G) = \frac{1}{e} \sum_{H \in \mathscr{T}} \frac{\mu_G(H)|\mathrm{Hom}(\Gamma, H)|}{|N_G(H)|}.$$

Table 1 gives the values of $\mu_G(H)$ and $|N_G(H)|$, so in order to apply this equation to a particular group Γ one needs only to count the homomorphisms $\Gamma \to H$ for each $H \in \mathscr{T}$.

8 Enumerations

We can now use the values of the Möbius function μ_G given in Table 1 to enumerate regular objects with automorphism group $G = Sz(q)$ in various categories $\mathfrak{C}$ described in Sect. 2. Formulae (19) and (23) for maps have been found in equivalent form by Hubard and Leemans [15] and, in the context of polytopes, by Kiefer and Leemans [22]. All the other enumerations in this section, concerning regular and orientably regular hypermaps, k-valent maps, self-dual maps, and surface coverings, are new.

Before starting we record in Table 3 the number $|H|_k$ of elements of order $k = 2, 4$ or 5 in each subgroup H in Table 1, information needed later.

Table 3 Values of $|H|_k$ for $k = 2, 4$ and 5

Conjugacy class of H	$\lvert H\rvert_2$	$\lvert H\rvert_4$	$\lvert H\rvert_5$
$\mathscr{G}(f),\ 1 < f \mid e$	$(2^f-1)(2^{2f}+1)$	$2^f(2^{2f}+1)(2^f-1)$	$2^{2f}(2^f-1)a_2(f)$
$\mathscr{F}(f),\ 1 < f \mid e$	2^f-1	$2^f(2^f-1)$	0
$\mathscr{B}_0(f),\ 1 < f \mid e$	2^f-1	0	0
$\mathscr{A}_0(f),\ 1 < f \mid e$	0	0	0
$\mathscr{B}_1(f),\ 1 < f \mid e$	$a_1(f)$	$2a_1(f)$	4
$\mathscr{B}_2(f),\ 1 < f \mid e$	$a_2(f)$	$2a_2(f)$	0
$\mathscr{B}_2(1)$	1	2	0
$\mathscr{B}_0(1)$	1	0	0
$\mathscr{A}_0(1)$	0	0	0

8.1 Orientably Regular Hypermaps

If $\mathfrak{C}$ is the category $\mathfrak{H}^+$ of oriented hypermaps, we take Γ to be the free group F_2 of rank 2. Then $|\mathrm{Hom}(\Gamma, H)| = |H|^2$ for each subgroup $H \leq G$, so

$$r_{\mathfrak{H}^+}(G) = n_{F_2}(G) = \frac{1}{|\mathrm{Aut}\, G|} \sum_{H \leq G} \mu_G(H)|H|^2.$$

Now $|\mathrm{Aut}\, G| = e|G|$, so using the information in Table 1 about the subgroups H of G, their orders, numbers of conjugates, and values of $\mu_G(H)$, we obtain, after some routine algebra,

$$r_{\mathfrak{H}^+}(G) = n_{F_2}(G) = \frac{1}{e} \sum_{f|e} \mu\left(\frac{e}{f}\right) 2^f(2^{4f} - 2^{3f} - 9) \sim q^5/e. \tag{17}$$

(Here we have used the fact that $\Sigma_{f|e}\mu(e/f) = 0$ for $e > 1$ to eliminate a constant term in the summation.) Formula (17) gives the number of orientably regular hypermaps $\mathscr{O}$ with orientation-preserving automorphism group $\mathrm{Aut}_{\mathfrak{H}^+}\mathscr{O} \cong G = Sz(q)$, where $q = 2^e$ for some odd $e > 1$. It also gives the number of regular dessins d'enfants with automorphism group G, the number of normal subgroups of the free group F_2 with quotient group G, and the number of orbits of Aut G on ordered pairs of generators of G. The dominant term in the summation on the right-hand side is the leading term 2^{5e}, so $r_{\mathfrak{H}^+}(G) \sim q^5/e \sim |G|/e$ as $e \to \infty$. (More generally, results of Dixon [4], Kantor and Lubotzky [21], and Liebeck and Shalev [24] on probabilistic generation imply that for all non-abelian finite simple groups, $r_{\mathfrak{H}^+}(G) \sim |G|/|\mathrm{Out}\, G|$ as $|G| \to \infty$.)

8.2 Regular Hypermaps

If $\mathfrak{C}$ is the category $\mathfrak{H}$ of all hypermaps, then Γ is the free product $C_2 * C_2 * C_2$. Since G cannot be generated by fewer that three involutions, we can restrict attention to smooth homomorphisms and epimorphisms, those that map the three free factors of Γ faithfully into G. For each H the number of such homomorphisms $\Gamma \to H$ is $(|H|_2)^3$, where $|H|_2$ is the number of involutions in H. The values of $|H|_2$ for the nine families of conjugacy classes of subgroups H in Table 1 are given in Table 3, so after some algebra we obtain

$$r_{\mathfrak{H}}(G) = \frac{1}{e} \sum_{f|e} \mu\left(\frac{e}{f}\right) 2^f(2^{3f} - 2^{2f+1} + 2^{f+1} - 5) \sim q^4/e. \tag{18}$$

This is the number of regular hypermaps with automorphism group G, and also, by Proposition 1, the number of reflexible hypermaps in $\mathscr{R}_{\mathfrak{H}^+}(G)$. Subtracting the formula in Eq. (18) from that in (17) therefore gives the number of chiral hypermaps in $\mathscr{R}_{\mathfrak{H}^+}(G)$; note that these predominate.

8.3 Orientably Regular Maps

For the category $\mathfrak{M}^+$ of oriented maps we take $\Gamma = C_\infty * C_2$. As in the case of hypermaps we may restrict the summation to smooth homomorphisms. There are $|H||H|_2$ such homomorphisms $\Gamma \to H$, so we obtain

$$r_{\mathfrak{M}^+}(G) = \frac{1}{e}\sum_{f|e}\mu\left(\frac{e}{f}\right)2^f(2^{2f}-2^f-3) \sim q^3/e. \tag{19}$$

(This is equivalent to the formula obtained by Hubard and Leemans in [15, Theorem 15].) The k-valent maps in $\mathscr{R}_{\mathfrak{M}^+}(G)$ correspond to the torsion-free normal subgroups in $\mathscr{N}_\Gamma(G)$, where Γ is the Hecke group $C_k * C_2$ (see [14]). To count these we consider smooth homomorphisms $\Gamma = C_k * C_2 \to H$ for relevant subgroups H of G. There are $|H|_k|H|_2$ of these, so with $k = 4$ and $k = 5$ for example, Table 3 gives

$$r_{\mathfrak{M}_4^+}(G) = \frac{1}{e}\sum_{f|e}\mu\left(\frac{e}{f}\right)2^f(2^f-2) \sim q^2/e \tag{20}$$

and

$$r_{\mathfrak{M}_5^+}(G) = \frac{1}{e}\sum_{f|e}\mu\left(\frac{e}{f}\right)(2^f-1)a_2(f) \sim q^2/e, \tag{21}$$

where $a_2(f) = 2^f - \chi(f)2^{(f+1)/2} + 1$.

One can apply similar arguments for odd $k > 5$. For instance, G contains elements of order 7 if and only if e is divisible by 3, in which case they form three conjugacy classes, represented by elements of A_0. It follows that the number of normal subgroups of $\Gamma = C_7 * C_2$ with quotient group $Sz(q)$ is

$$\frac{3}{e}\sum_{3|f|e}\mu\left(\frac{e}{f}\right)(2^{2f}-2). \tag{22}$$

A map $\mathscr{M} \in \mathscr{R}_{\mathfrak{M}^+}(G)$ of type $\{n, n\}$, represented by a generating triple (x, y, z) for G of type $(n, 2, n)$, is self-dual if and only if G has an automorphism transposing x and z. Such an automorphism has order 2 and is therefore inner, induced by conjugation by an involution $i \in G$. This is equivalent to G having a generating triple (xi, i, x^{-1}) of type $(4, 2, n)$, corresponding to a map $\mathscr{M}^*$ of type $\{n, 4\}$ in $\mathscr{R}_{\mathfrak{M}^+}(G)$. If $\mathscr{M}$ corresponds to a subgroup $N \in \mathscr{N}_\Delta(G)$, where $\Delta := \Delta(n, 2, n)$, then its median map $\mathscr{M}^\dagger$ corresponds to N as an element of $\mathscr{N}_{\Delta^*}(G \times C_2)$, where $\Delta^* := \Delta(4, 2, n)$ contains $\Delta(n, 2, n)$ with index 2, and $\mathscr{M}^* = \mathscr{M}^\dagger/C_2$. (The median of any map $\mathscr{M}$ is a map on the same surface, with vertices at the edge-centres of $\mathcal{M}$, joined by an edge if they are on consecutive edges of a face of $\mathcal{M}$.) The correspondence $\mathcal{M} \mapsto \mathcal{M}^*$ is a bijection, so the number of self-dual maps $\mathcal{M} \in \mathcal{R}_{\mathfrak{M}^+}(G)$ is equal to the number $r_{\mathfrak{M}_4^+}(G)$ of 4-valent maps $\mathcal{M}^* \in \mathcal{R}_{\mathfrak{M}^+}(G)$, given in (20).

8.4 Regular Maps

For the category $\mathfrak{M}$ of all maps we take $\Gamma = V_4 * C_2$. In this case we may restrict attention to homomorphisms which embed the direct factors as subgroups V and C, such that the generator of C commutes with only the identity element of V. The only subgroups $H \leq G$ containing such subgroups V and C are those conjugate to some $G(f)$, with V and C central subgroups of distinct Sylow 2-subgroups of H. Since $G(f)$ has $2^{2f}+1$ Sylow 2-subgroups, and their centres are elementary abelian groups of order 2^f, one easily obtains

$$r_{\mathfrak{M}}(G) = \frac{1}{e}\sum_{f|e}\mu\left(\frac{e}{f}\right)(2^f-1)(2^f-2) = \frac{1}{e}\sum_{f|e}\mu\left(\frac{e}{f}\right)2^f(2^f-3) \sim q^2/e. \tag{23}$$

Note that for this result we use $\mu_G(H)$ only for subgroups $H \cong G(f)$, so it is feasible to obtain it without invoking the whole of the Möbius function. By an argument similar to that used in Sect. 8.2 for hypermaps, Proposition 1 implies that this is also the number of reflexible maps in $\mathcal{R}_{\mathfrak{M}^+}(G)$, all of them inner reflexible. Subtracting (23) from (19) gives the number of chiral maps (see also [15, Theorem 16]), and as before these predominate.

The formulae (23) are the same for the group $G = L_2(2^e)$ (see [6]). At first this may seem surprising, since $Sz(q)$ is much larger than $L_2(q)$. However, the distribution of involutions in these two groups is similar, and the above proof can be applied, with only minor changes, to $L_2(q)$. The fact that (19) also gives $r_{\mathfrak{H}}(L_2(2^e))$ and $r_{\mathfrak{H}^+}(L_2(2^e))$ seems to be more accidental.

This proof gives a natural interpretation for the first formula in (23). There is a unique inner automorphism of the subgroup $H = G(f)$ sending the image of the generator R_1 of Γ to $r_1 = \tau$, and sending R_0 and R_2 to a pair of distinct elements r_0, r_2 of the form $(0, \beta)$, in the notation of Sect. 5.1. Then $(2^f-1)(2^f-2)$ is the number of choices for such an ordered pair, the Möbius inversion picks out those triples (r_i) which generate G, and division by e is explained by the number of orbits of $\operatorname{Out} G$, acting on these triples, being the same as that for $\operatorname{Gal} \mathbb{F}_q \cong C_e$ acting on the coefficients β.

This parametrisation of maps also allows one to determine their types $\{m, n\}$, since m and n are the orders of r_ir_1 for $i = 0$ and 2. A matrix $(0, \beta)\tau$ has characteristic polynomial

$$p(\lambda) = \lambda^4 + \beta^\theta\lambda^3 + \beta^2\lambda^2 + \beta^\theta\lambda + 1, \tag{24}$$

so its order, as an element r_ir_1 of G, is the least common multiple of the multiplicative orders of the roots of p. Clearly $\lambda = 1$ is not a root, so m and n cannot be equal to 2 or 4, since elements of G of these orders are unipotent, with all eigenvalues equal to 1. Thus m and n are both odd. For example, if we take $\beta = 1$ then the roots of p are the primitive 5th roots of 1, so r_ir_1 has order 5. Specific examples are considered in Sect. 9.

None of these regular maps is self-dual. If one were, G would have an automorphism fixing r_1 and transposing r_0 and r_2. This would be induced by an element of G centralising the involutions r_0r_2 and r_1; however, these lie in distinct Sylow 2-subgroups of G, so their centralisers have trivial intersection. This explains why (23) gives twice the number of abstract regular polytopes with automorphism group G computed by Kiefer and Leemans in [22] (for instance 71576170 rather than 35788085 in their Example 2 (continued), with $e = 15$): by [23] such polytopes all have rank 3, so they are regular maps, and they were enumerated in [22] by counting *unordered* triples $\{r_0, r_1, r_2\}$ of involutions generating G and satisfying $(r_0r_2)^2 = 1$, corresponding to dual pairs of regular maps and polytopes.

8.5 Surface Coverings

In order to apply Theorem 3 to count regular surface coverings with covering group G, one needs to know the degrees of the irreducible complex characters of the subgroups H in Table 1. The irreducible characters of the Suzuki groups $G(f)$ are described in [30] and [16, Sect. XI.5], and the degrees for the other subgroups H are easily found; they are given in Table 4, where $s := 2^f$, $t := \sqrt{2s} = 2^{(f+1)/2}$, $k_i := (a_i(f) - 1)/4$ for $i = 1, 2$, and the notation $d^{\{k\}}$ denotes k characters of degree d.

With this information, Theorem 3 gives $|\mathrm{Hom}(\Gamma, H)|$ for each H in Table 1, where Γ is the fundamental group Π_g of an orientable surface $\mathscr{S}_g$ of genus g, and then

Table 4 Degrees of irreducible characters of subgroups $H \leq G$

Conjugacy class of H	Conditions on f	Degrees of irreducible characters of H
$\mathscr{G}(f)$	$1 < f \mid e$	$1, s^2, (s-1)t/2^{\{2\}}, (s^2+1)^{\{(s-2)/2\}}$, $(s-1)a_1(f)^{\{k_2\}}, (s-1)a_2(f)^{\{k_1\}}$
$\mathscr{F}(f)$	$1 < f \mid e$	$1^{\{s-1\}}, s-1, (s-1)t/2^{\{2\}}$
$\mathscr{B}_0(f)$	$1 < f \mid e$	$1^{\{2\}}, 2^{\{(s-2)/2\}}$
$\mathscr{A}_0(f)$	$1 < f \mid e$	$1^{\{s-1\}}$
$\mathscr{B}_1(f)$	$1 < f \mid e$	$1^{\{4\}}, 4^{\{k_1\}}$
$\mathscr{B}_2(f)$	$1 < f \mid e$	$1^{\{4\}}, 4^{\{k_2\}}$
$\mathscr{B}_2(1)$		$1^{\{4\}}$
$\mathscr{B}_0(1)$		$1^{\{2\}}$
$\mathscr{A}_0(1)$		1

$n_\Gamma(G)$ in Eq. (5) gives $r_g(G)$, the number of regular coverings of $\mathscr{S}_g$ with covering group G. The general formulae are very unwieldy, but in Sect. 9 we will give a simple example.

9 The Smallest Simple Suzuki Group

The smallest of the simple Suzuki groups is the group $G = G(3) = Sz(8)$ of order $29120 = 2^6.5.7.13$. Putting $e = 3$ in the enumerative formulae given above, we find that G is the automorphism group of 1054 regular hypermaps, of which 14 are maps; all of these are non-orientable. Similarly, it is the orientation-preserving automorphism group of 9534 orientably regular hypermaps, of which 142 are maps. By Proposition 1, 1054 of these orientably regular hypermaps, and 14 of these orientably regular maps, are reflexible; these are the orientable double covers of the regular hypermaps and maps associated with G, so they are all inner reflexible.

Theorem 2 and the character table of G in [2, 30] can be used to find how many of these orientably regular maps and hypermaps there are for each type. For instance, they show that G contains $2^6.3.7.13.331$ triples (x, y, z) of type $(5, 5, 5)$ satisfying $xyz = 1$; of these, $2^6.3.7.13$ generate the $2^4.7.13$ Sylow 5-subgroups, while the remaining $2^6.3.7.13.330 = 66|\mathrm{Aut}\, G|$ generate G, so there are 66 hypermaps of type $(5, 5, 5)$ in $\mathscr{R}_{\mathfrak{H}^+}(G)$.

Each element of G has order 1, 2, 4, 5, 7 or 13, so the entries in each type must come from this list. Moreover, since G is non-solvable, any map arising must have type $\{m, n\}$ with $m, n \geq 4$, other than $\{4, 4\}$. As shown in [19], it follows from Theorem 2 that there are four maps of type $\{4, 5\}$ in $\mathscr{R}_{\mathfrak{M}^+}(G)$, forming two chiral pairs. In fact, repeated use of Theorem 2 shows that the distribution of the maps in $\mathscr{R}_{\mathfrak{M}^+}(G)$ into types is as in Table 5, which is symmetric in m and n by the duality of maps. The number of self-dual maps of type $\{n, n\}$ is equal to the number of maps of type $\{n, 4\}$ in this table.

One can use the argument at the end of Sect. 8.4 to determine the types $\{m, n\}$ of the 14 regular maps in $\mathscr{R}_{\mathfrak{M}}(G)$. Taking $\beta = 1$ gives an element $r_i r_1$ of order 5. Of the six remaining elements $\beta \in \mathbb{F}_8^*$, three have minimal polynomial $t^3 + t + 1$ over $\mathbb{F}_2$, and three have $t^3 + t^2 + 1$. In the first case p splits into four linear factors, with roots $\beta + 1$, β^2, $\beta^2 + \beta$ and $\beta^2 + \beta + 1$ all of order 7, so that $r_i r_1$ has order 7. In

Table 5 Number of orientably regular maps of type $\{m, n\}$ in $\mathscr{R}_{\mathfrak{M}^+}(Sz(8))$

$m \setminus n$	4	5	7	13	Total
4	0	4	8	4	16
5	4	4	13	9	30
7	8	13	26	15	62
13	4	9	15	6	34
Total	16	30	62	34	142

the second case, p is irreducible over $\mathbb{F}_8$, and its roots in its splitting field $\mathbb{F}_{2^{12}}$ have order 13, so $r_i r_1$ has order 13. By considering the action of $\mathrm{Gal}\,\mathbb{F}_8 \cong C_3$ on distinct ordered pairs of elements $\beta \in \mathbb{F}_8^*$, we find that the number of regular maps of each type $\{m, n\}$ is as in Table 6; there are none with $m = 4$ or $n = 4$ since elements of order 4 are not conjugate to their inverses. This table also gives the types of the 14 reflexible maps in $\mathcal{R}_{\mathfrak{M}^+}(G)$, so subtracting its entries from the corresponding entries in Table 5 gives the number of chiral maps of each type in $\mathcal{R}_{\mathfrak{M}^+}(G)$.

One can also enumerate regular surface coverings with covering group G. If we put $f = e = 3$ in Table 4, so that $s = 8$ and $t = 4$, we find from Eq. (5) and Theorem 3 that the number $r_g(G)$ of regular coverings of an orientable surface of genus g with covering group G is

$$\frac{1}{3}\Big\{29120^n(1 + 2.14^{-n} + 3.35^{-n} + 64^{-n} + 3.65^{-n} + 91^{-n})$$
$$-448^n(7 + 7^{-n} + 2.14^{-n}) - 14^n(2 + 3.2^{-n}) + 7^{n+1}$$
$$-52^n(4 + 3.4^{-n}) - 20^n(4 + 4^{-n}) + 8.4^n + 2.2^n - 1\Big\}$$

where $n = 2g - 2$ is the negative of the Euler characteristic of the surface. When $g = 1$ there are no coverings, as one should expect since the fundamental group Π_1 is abelian, and when $g = 2$ there are 286063776. As $g \to \infty$ we have $r_g(G) \sim |G|^n/3 = 847974400^{g-1}/3$.

In the non-orientable case, in addition to the degrees of the irreducible characters χ of the subgroups H in Table 1, we also need to know their Frobenius-Schur indicators c_χ. For $H = G$ these are given in [2]: we have $c_\chi = 1$ for all χ except the two characters of degree 14, which satisfy $c_\chi = 0$ since they take values $\pm 2i$ on elements of order 4. The indicators for proper subgroups $H < G$ are easily found, either directly or from the character table for G in [2], [16, Sect. XI.5] or [30]. It follows from Eq. (5) and Theorem 4 that the number $r_g^-(G)$ of regular coverings of a non-orientable surface of genus $g \geq 1$ with covering group G is

$$\frac{1}{3}\Big\{29120^n(1 + 3.35^{-n} + 64^{-n} + 3.65^{-n} + 91^{-n})$$
$$-448^n(1 + 7^{-n}) - 14^n(2 + 3.2^{-n}) + 7^n$$
$$-52^n(2 + 3.4^{-n}) - 20^n(2 + 4^{-n}) + 4.4^n + 2.2^n - 1\Big\}$$

where $n = g - 2$, the negative of the Euler characteristic. We have $r_g^-(G) = 0$ when $g \leq 2$, while $r_3^-(G) = 11004$. As $g \to \infty$, $r_g^-(G) \sim 29120^{g-2}/3$.

Table 6 Number of regular maps of type $\{m, n\}$ in $\mathscr{R}_{\mathfrak{M}}(Sz(8))$

$m \setminus n$	4	5	7	13	Total
4	0	0	0	0	0
5	0	0	1	1	2
7	0	1	2	3	6
13	0	1	3	2	6
Total	0	2	6	6	14

10 Postscript

The calculations described in Sects. 5–7 were carried out by the first author, Martin Downs, in the early 1990s, but were never published. Recent interest in combinatorial and geometric actions of the Suzuki groups motivated the authors to revisit these calculations, and to provide various applications of them. Despite suffering from a long-term illness, Martin was deeply involved in the preparation of this paper, right up to its submission. Unfortunately, he died before it could be published. As a good friend and a valued colleague, he will be greatly missed.

Acknowledgments The authors are grateful to Dimitri Leemans for some very helpful comments on enumeration with Suzuki groups, and to Nikos Kanakis for help in preparing the TeX file.

References

1. E. Breuillard, B. Green and T. Tao, Suzuki groups as expanders, *Groups Geom. Dyn.* 5 (2011), 281–299.
2. J. H. Conway, R. T. Curtis, S. P. Norton, R. A. Parker and R. A. Wilson, *ATLAS of Finite Groups*, Clarendon Press, Oxford, 1985.
3. H. S. M. Coxeter and W. O. J. Moser, *Generators and Relations for Discrete Groups*, 4th ed., Springer-Verlag, Berlin–Heidelberg–New York, 1980.
4. J. D. Dixon, The probability of generating the symmetric group, *Math. Z.* 110 (1969), 199–205.
5. M. L. N. Downs, The Möbius function of $PSL_2(q)$, with application to the maximal normal subgroups of the modular group, *J. London Math. Soc.* 43 (1991), 61–75.
6. M. L. N. Downs and G. A. Jones, Enumerating regular objects with a given automorphism group, *Discrete Math.* 64 (1987), 299–302.
7. M. L. N. Downs and G. A. Jones, The Möbius function of the Suzuki groups, with applications to enumeration, arXiv.math:1404.5470 [GR].
8. F. G. Frobenius, Über Gruppencharaktere, *Sitzber. Königlich Preuss. Akad. Wiss. Berlin*, (1896), 985–1021.
9. F. G. Frobenius and I. Schur, Über die reellen Darstellungen der endlichen Gruppen, *Sitzber. Königlich Preuss. Akad. Wiss. Berlin* (1906), 186–208.
10. E. Girondo and G. González-Diez, *Introduction to Compact Riemann Surfaces and Dessins d'Enfants*, London Math. Soc. Student Texts 79, Cambridge University Press, Cambridge, 2011.

11. A. Grothendieck, Esquisse d'un Programme, in: P. Lochak and L. Schneps (Eds.), *Geometric Galois Actions I, Around Grothendieck's Esquisse d'un Programme*, London Math. Soc. Lecture Note Ser. 242, Cambridge University Press, Cambridge, 1997, pp. 5–48.
12. P. Hall, A note on soluble groups, *J. London Math. Soc.* 3 (1928), 98–105.
13. P. Hall, The Eulerian functions of a group, *Q. J. Math.* 7 (1936), 134–151.
14. E. Hecke, Über die Bestimmung Dirichletscher Reihen durch ihre Funktionalgleichung, *Math. Ann.* 112 (1936), 664–699.
15. I. Hubard and D. Leemans, Chiral polytopes and Suzuki simple groups, in eds. R. Connelly, A. I. Weiss and W. Whiteley, *Rigidity and Symmetry*, Fields Inst. Commun. 70 (2014), 155–175.
16. B. Huppert and N. Blackburn, *Finite Groups III*, Springer-Verlag, Berlin–Heidelberg–New York, 1982.
17. G. A. Jones, Enumeration of homomorphisms and surface-coverings, *Q. J. Math.* 46 (1995), 485–507.
18. G. A. Jones, Combinatorial categories and permutation representations, *Ars Math. Contemp.* 10 (2016), 237–254.
19. G. A. Jones and S. A. Silver, Suzuki groups and surfaces, *J. London Math. Soc.* (2) 48 (1993), 117–125.
20. G. A. Jones and D. Singerman, Maps, hypermaps and triangle groups., in: L. Schneps (Ed.), *The Grothendieck Theory of Dessins d'Enfants (Luminy, 1993)*, London Math. Soc. Lecture Note Ser. 200, Cambridge Univ. Press, Cambridge, 1994, pp. 115–145.
21. W. M. Kantor and A. Lubotzky, The probability of generating a finite classical group, *Geom. Dedicata* 36 (1990), 67–87.
22. A. Kiefer and D. Leemans, On the number of abstract regular polytopes whose automorphism group is a Suzuki simple group $Sz(q)$, *J. Combin. Theory Ser. A* 117 (2010), 1248–1257.
23. D. Leemans, Almost simple groups of Suzuki type acting on polytopes, *Proc. Amer. Math. Soc.* 134 (2006), 3649–3651 (electronic).
24. M. W. Liebeck and A. Shalev, The probability of generating a finite simple group, *Geom. Dedicata* 56 (1995), 103–113.
25. H. Lüneburg, *Die Suzukigruppen und ihre Geometrien*, Springer-Verlag, Berlin–New York, 1965.
26. A. D. Mednykh, Determination of the number of nonequivalent coverings over a compact Riemann surface, *Dokl. Akad. Nauk SSSR* 239 (1978), 269–271 (Russian); *Soviet Math. Dokl.* 19 (1978), 318–320 (English).
27. J. R. Munkres, *Topology* (2nd ed.), Prentice Hall, Upper Saddle River NJ, 2000.
28. J-P. Serre, *Topics in Galois Theory* (2nd ed.), A. K. Peters, Wellesley MA, 2008.
29. M. Suzuki, A new type of simple groups of finite order, *Proc. Nat. Acad. Sci. U.S.A.* 46 (1960), 868–870.
30. M. Suzuki, On a class of doubly transitive groups, *Ann. of Math.* (2) 75 (1962), 105–145.
31. R. A. Wilson, *The Finite Simple Groups*, Graduate Texts in Math. 251, Springer, London, 2009.

More on Strongly Real Beauville Groups

Ben Fairbairn

Abstract Beauville surfaces are a class of complex surfaces defined by letting a finite group G act on a product of Riemann surfaces. These surfaces possess many attractive geometric properties several of which are dictated by properties of the group G. A particularly interesting subclass are the 'strongly real' Beauville surfaces that have an analogue of complex conjugation defined on them. In this survey we discuss these objects and in particular the groups that may be used to define them. *En route* we discuss several open problems, questions and conjectures and in places make some progress made on addressing these.

1 Introduction

Roughly speaking (precise definitions will be given in the next section), a Beauville surface is a complex surface $\mathcal{S}$ defined by taking a pair of complex curves, i.e. Riemann surfaces, $\mathcal{C}_1$ and $\mathcal{C}_2$ and letting a finite group G act freely on their product to define $\mathcal{S}$ as a quotient $(\mathcal{C}_1 \times \mathcal{C}_2)/G$. These surfaces have a wide variety of attractive geometric properties: they are surfaces of general type; their automorphism groups [45] and fundamental groups [17] are relatively easy to compute (being closely related to G); they are rigid surfaces in the sense of admitting no nontrivial deformations [8] and thus correspond to isolated points in the moduli space of surfaces of general type [34].

Much of this good behaviour stems from the fact that the surface $(\mathcal{C}_1 \times \mathcal{C}_2)/G$ is uniquely determined by a particular pair of generating sets of G known as a 'Beauville structure'. This converts the study of Beauville surfaces to the study of groups with Beauville structures, i.e. Beauville groups.

Beauville surfaces were first defined by Catanese in [17] as a generalisation of an earlier example of Beauville [12, Exercise X.13(4)] (native English speakers

B. Fairbairn (✉)
Department of Economics, Mathematics and Statistics, Birkbeck, University of London, Malet Street, London WC1E 7HX, UK
e-mail: b.fairbairn@bbk.ac.uk

J. Širáň and R. Jajcay (eds.), *Symmetries in Graphs, Maps, and Polytopes*, Springer Proceedings in Mathematics & Statistics 159,
DOI 10.1007/978-3-319-30451-9_6

may find the English translation [13] somewhat easier to read and get hold of) in which $\mathscr{C}_1 = \mathscr{C}_2$ and the curves are both the Fermat curve defined by the equation $X^5 + Y^5 + Z^5 = 0$ being acted on by the group $(\mathbb{Z}/5\mathbb{Z}) \times (\mathbb{Z}/5\mathbb{Z})$ (this choice of group may seem somewhat odd at first, but the reason will become clear later). Bauer, Catanese and Grunewald went on to use these surfaces to construct examples of smooth regular surfaces with vanishing geometric genus [9]. Early motivation came from the consideration of the 'Friedman-Morgan speculation'—a technical conjecture concerning when two algebraic surfaces are diffeomorphic which Beauville surfaces provide counterexamples to. More recently, they have been used to construct interesting orbits of the absolute Galois group Gal $(\overline{\mathbb{Q}}/\mathbb{Q})$ (connections with Gothendeick's theory of *dessins d'enfant* make it possible for this group to act on the set of all Beauville surfaces). Indeed one of the more impressive applications of these surfaces is the recent proof by González-Diez and Jaikin-Zapirain in [36] that Gal$(\overline{\mathbb{Q}}/\mathbb{Q})$ acts faithfully on the set of regular dessins by showing that it acts regularly on the set of Beauville surfaces.

Furthermore, Beauville's original example has also recently been used by Galkin and Shinder in [32] to construct examples of exceptional collections of line bundles.

Like any survey article, the topics discussed here reflect the research interests of the author. Slightly older surveys discussing related geometric and topological matters are given by Bauer et al. in [10, 11]. Other notable works in the area include [6, 23, 46, 53, 58]. Whilst this article is largely expository in nature we also report incremental progress on various different problems that will appear here. Indeed, this work can be naturally viewed as a sequel to the author's earlier article [24], though the reader will lose little if they have neither read nor have a copy of [24] to hand.

We remark that throughout we shall use the standard 'Atlas' notation for finite groups and related concepts as described in [20], excepting that we will occasionally deviate to minimise confusion with similar notation for geometric concepts. In particular, given two groups A and B we use the following notation.

- We write $A \times B$ for the direct product of A and B, that is, the group whose members are ordered pairs (a, b) with $a \in A$ and $b \in B$ such that for two pairs $(a, b), (a', b') \in A \times B$ we have the multiplication $(a, b)(a', b') = (aa', bb')$. Given a positive integer k we write A^k for the direct product of k copies of A.
- We write $A \cdot B$ for the extension of A by B, that is, a group with a normal subgroup isomorphic to A whose quotient is B (such groups are not necessarily direct products—for instance $\mathrm{SL}_2(5) = 2 \cdot \mathrm{PSL}_2(5)$).
- We write $A : B$ for a semi-direct product of A and B, also known as a split extension A and B, that is, there is a homomorphism $\phi\colon B \to Aut(A)$ with elements of this group being ordered pairs (b, a) with $a \in A$ and $b \in B$ such that for $(b, a), (b', a') \in A : B$ we have the multiplication $(b, a)(b', a') = (bb', a^{\phi(b')}a')$.
- We write $A \wr B$ for the wreath product of A and B, that is, if B is a permutation group on n points then we have the split extension $A^n : B$ with B acting in a way that permutes the n copies of A.

In Sect. 2 we provide preliminary information and in particular give specific definitions for the concepts we have only talked about very vaguely until now. In Sect. 3 we will discuss the case of the finite simple groups. In Sects. 4, 5 and 6 we will discuss families of groups closely related to these such as characteristically simple groups and almost simple groups. Finally, in Sect. 7 we will conclude with a brief discussion of the question of which of the abelian and nilpotent groups are strongly real Beauville groups.

2 Preliminaries

We give the main definition.

Definition 1 A surface $\mathcal{S}$ is a **Beauville surface of unmixed type** if

- the surface $\mathcal{S}$ is isogenous to a higher product, that is, $\mathcal{S} \cong (\mathcal{C}_1 \times \mathcal{C}_2)/G$ where $\mathcal{C}_1$ and $\mathcal{C}_2$ are complex algebraic curves of genus at least 2 and G is a finite group acting faithfully on $\mathcal{C}_1$ and $\mathcal{C}_2$ by holomorphic transformations in such a way that it acts freely on the product $\mathcal{C}_1 \times \mathcal{C}_2$, and
- each $\mathcal{C}_i/G$ is isomorphic to the projective line $\mathbb{P}_1(\mathbb{C})$ and the corresponding covering map $\mathcal{C}_i \to \mathcal{C}_i/G$ is ramified over three points.

There also exists a concept of Beauville surfaces of mixed type in which the action of G interchanges the two curves $\mathcal{C}_1$ and $\mathcal{C}_2$ but these are much harder to construct and we shall not discuss these here. (For further details of the mixed case, the most up-to-date information at the time of writing may be found in the work of the author and Pierro in [27].)

In the first of the above conditions the genus of the curves in question needs to be at least 2. It was later proved by Fuertes, González-Diez and Jaikin-Zapirain in [30] that in fact we can take the genus as being at least 6. The second of the above conditions implies that each $\mathcal{C}_i$ carries a regular dessin in the sense of Grothendieck's theory of *dessins d'enfants* (children's drawings) [39]. Furthermore, by Belyĭ's Theorem [14] this ensures that $\mathcal{S}$ is defined over an algebraic number field in the sense that when we view each Riemann surface as being the zeros of some polynomial we find that the coefficients of that polynomial belong to some number field. Equivalently they admit an orientably regular hypermap [47], with G acting as the orientation-preserving automorphism group. A modern account of *dessins d'enfants* and proofs of Belyĭ's theorem may be found in the recent book of Girondo and González-Diez [35].

These constructions can also be described instead in terms of uniformisation and using the language of Fuchsian groups [38, 56].

What makes this class of surfaces so good to work with is the fact that all of the above definition can be 'internalised' into the group. It turns out that a group G can be used to define a Beauville surface if and only if it has a certain pair of generating sets known as a Beauville structure.

Definition 2 Let G be a finite group. For $x, y \in G$ let

$$\Sigma(x, y) := \bigcup_{i=1}^{|G|} \bigcup_{g \in G} \{(x^i)^g, (y^i)^g, ((xy)^i)^g\}.$$

An **unmixed Beauville structure** for the group G is a set of pairs of elements $\{\{x_1, y_1\}, \{x_2, y_2\}\} \subset G \times G$ with the property that $\langle x_1, y_1 \rangle = \langle x_2, y_2 \rangle = G$ such that

$$\Sigma(x_1, y_1) \cap \Sigma(x_2, y_2) = \{e\}.$$

If G has a Beauville structure we say that G is a **Beauville group**. Furthermore we say that the structure has **type**

$$((o(x_1), o(y_1), o(x_1 y_1)), (o(x_2), o(y_2), o(x_2 y_2))).$$

In some parts of the literature authors have defined the above structure in terms of so-called 'spherical systems of generators of length 3', meaning $\{x, y, z\} \subset G$ with $xyz = e$, but we omit $z = (xy)^{-1}$ from our notation in this survey. (The reader is warned that this terminology is a little misleading since the underlying geometry of Beauville surfaces is hyperbolic thanks to the below constraint on the orders of the elements.) Furthermore, many earlier papers on Beauville structures add the condition that for $i = 1, 2$ we have that

$$\frac{1}{o(x_i)} + \frac{1}{o(y_i)} + \frac{1}{o(x_i y_i)} < 1,$$

but this condition was subsequently found to be unnecessary following Bauer et al. investigation of the wall-paper groups in [7]. A triple of elements and their orders satisfying this condition are said to be hyperbolic. Geometrically, the type gives us considerable amounts of geometric information about the surface: the Riemann-Hurwitz formula

$$g(\mathscr{C}_i) = 1 + \frac{|G|}{2}\left(1 - \frac{1}{o(x_i)} - \frac{1}{o(y_i)} - \frac{1}{o(x_i y_i)}\right)$$

tells us the genus of each of the curves used to define the surface $\mathscr{S}$ and by a theorem of Zeuthen-Segre this in turn gives us the Euler number of the surface $\mathscr{S}$ since

$$e(\mathscr{S}) = 4\frac{(g(\mathscr{C}_1) - 1)(g(\mathscr{C}_2) - 1)}{|G|}$$

which in turn gives us the holomorphic Euler-Poincaré characteristic of $\mathscr{S}$ since $4\chi(\mathscr{S}) = e(\mathscr{S})$ (see [17, Theorem 3.4]). On a more practical and group theoretic note, the type is often useful for verifying that the critical condition that

$\Sigma(x_1, y_1) \cap \Sigma(x_2, y_2) = \{e\}$ is satisfied since this will clearly hold whenever the number $o(x_1)o(y_1)o(x_1y_1)$ is coprime to the number $o(x_2)o(y_2)o(x_2y_2)$.

Furthermore, if a group can be generated by a pair of elements of orders a and b whose product has order c then G is a homomorphic image of the triangle group

$$\Delta(a, b, c) = \langle x, y, z \mid x^a = y^b = z^c = xyz = e \rangle.$$

Homomorphic images of the triangle group $\Delta(2, 3, 7)$ are known as Hurwitz groups. In several places in the literature, knowing that a particular group is a Hurwitz group has proved useful for deciding its status as a Beauville group. For a discussion of known results on Hurwitz groups see the excellent surveys of Conder [18, 19] and for a more historically oriented discussion see the brief account given by Murray Macbeath in [50].

The abelian Beauville groups were essentially classified by Catanese in [17, p. 24.] and the full argument is given explicitly in [7, Theorem 3.4] where the following is proved.

Theorem 1 *Let G be an abelian group. Then G is a Beauville group if, and only if, $G = (\mathbb{Z}/n\mathbb{Z}) \times (\mathbb{Z}/n\mathbb{Z})$ where $n > 1$ is coprime to* 6.

This explains why Beauville's original example used the group $(\mathbb{Z}/5\mathbb{Z}) \times (\mathbb{Z}/5\mathbb{Z})$—it is the smallest abelian Beauville group.

Given any complex surface $\mathcal{S}$ it is natural to consider the complex conjugate surface $\overline{\mathcal{S}}$. In particular, it is natural to ask whether or not these two surfaces are biholomorphic.

Definition 3 Let $\mathcal{S}$ be a complex surface. We say that $\mathcal{S}$ is **real** if there exists a biholomorphism $\sigma : \mathcal{S} \to \overline{\mathcal{S}}$ such that σ^2 is the identity map.

(We remark that strictly speaking the above definition is not quite right, it being impossible to compose σ with itself. It is more accurate to talk of the composition $\sigma \circ \overline{\sigma}$ where $\overline{\sigma} : \overline{\mathcal{S}} \to \mathcal{S}$.)

As is often the case with Beauville surfaces, the above geometric condition can be translated into purely group theoretic terms.

Definition 4 Let G be a Beauville group and let $X = \{\{x_1, y_1\}, \{x_2, y_2\}\}$ be a Beauville structure for G. We say that G and X are **strongly real** if there exists an automorphism $\phi \in \mathrm{Aut}(G)$ and elements $g_i \in G$ for $i = 1, 2$ such that

$$g_i\phi(x_i)g_i^{-1} = x_i^{-1} \text{ and } g_i\phi(y_i)g_i^{-1} = y_i^{-1}$$

for $i = 1, 2$.

In practice we can always replace one generating pair by some generating pair that is conjugate to it and so we can take $g_1 = g_2 = e$ and this is often what is done in practice. (We take this opportunity to point out that the definition of strongly real

Beauville structure as given by the author in [23, Definition 5.2] is slightly incorrect since the indices of the g_is appearing there are mixed up.)

In [7] Bauer et al. show that a Beauville surface is real if, and only if, the corresponding Beauville group and structure are strongly real. This all comes from the study of the following related concept in the theory of Riemann surfaces. In Singerman's nomenclature of [52], a Riemann surface with a function behaving like the function σ in Definition 3 are said to be symmetric. The relationship with automorphisms of the corresponding group critically depends on the main result of [52]. The reader is warned, however, that some notable errors in [52] were subsequently found and are corrected by Jones et al. in [48]. More specifically, the condition that an automorphism like the above exists is sufficient but it is not necessary. This is corrected by Jones et al. by giving a complete list of conditions that are both necessary and sufficient in [48, Theorem 1.1].

Question 1. Are there interesting strongly real Beauville surfaces arising from the conditions given in [48, Theorem 1.1] but not [52, Theorem 2]?

We remark that symmetric Riemann surfaces are also connected to the theory of Klein surfaces. Real algebraic curves and compact Klein surfaces are equivalent in the same way that the categories of complex algebraic curves and compact Riemann surfaces are equivalent. Indeed, just as a compact, connected, orientable surface admits the structure of a complex analytic manifold of dimension 1 (this is, a Riemann surface structure) then a compact connected surface that is not necessarily orientable admits the structure of a complex *dianalytic* manifold of dimension 1, that is, a Klein surface structure. See [51] for an introductory discussion and [16] for a recent survey of these surfaces.

By way of immediate easy examples, note that the function $x \mapsto -x$ is an automorphism of any abelian group and so every Beauville group given by Theorem 1 is an example of a strongly real Beauville group. More generally the following question is the main subject of this article.

Question 2. Which groups are strongly real Beauville groups?

3 The Finite Simple Groups

Naturally, a necessary condition for being a strongly real Beauville group is being a Beauville group. Furthermore, a necessary condition for being a Beauville group is being 2-generated: we say that a group G is 2-generated if there exist two elements $x, y \in G$ such that $\langle x, y \rangle = G$. It is an easy exercise for the reader to show that the alternating groups A_n for $n \geq 3$ are 2-generated (see the work of Miller in [49]). In [54] Steinberg proved that all of the simple groups of Lie type are 2-generated and in [1] Aschbacher and Guralnick used cohomological methods to show that the larger of the sporadic simple groups are 2-generated, the smaller ones having dealt with by numerous previous authors. We thus have that all of the non-abelian finite simple groups are 2-generated making them natural candidates for Beauville groups. This lead Bauer, Catanese and Grunewald to conjecture that aside from A_5, which is easily seen to not be a Beauville group, every non-abelian finite simple group is a Beauville group—see [7, Conjecture 1] and [8, Conjecture 7.17]. This suspicion was later proved correct [25, 26, 33, 40], indeed the full theorem proved by the author, Magaard and Parker in [26] is actually a more general statement about quasisimple groups (recall that a group G is quasisimple if it is generated by its commutators and the quotient by its center $G/Z(G)$ is a simple group.). A sketch of the proof of this Theorem is given by the author in [23, Sect. 3].

Having found that almost all of the non-abelian finite simple groups are Beauville groups, it is natural to ask which of the finite simple groups are strongly real Beauville groups. In [7, Sect. 5.4] Bauer et al. wrote

> There are 18 finite simple nonabelian groups of order $\leq$15,000. By computer calculations we have found strongly [real] Beauville structures on all of them with the exceptions of A_5, $PSL_2(7)$, A_6, A_7, $PSL_3(3)$, $U_3(3)$ and the Mathieu group M_{11}.

On the basis of these computations they conjectured that all but finitely many of the non-abelian finite simple groups are strongly real Beauville groups. Several authors have worked on this conjecture and consequently many special cases are now known to be true.

- In [29] Fuertes and González-Diez showed that the alternating groups A_n $(n \geq 7)$ and the symmetric groups S_n $(n \geq 5$, cf Sect. 5) are strongly real Beauville groups by explicitly writing down permutations for their generators and the automorphisms and applying some of the classical theory of permutation groups to show that their elements had the properties they claimed. Subsequently the alternating group A_6 was also shown to be a strongly real Beauville group.
- In [31] Fuertes and Jones prove that the simple groups $PSL_2(q)$ for prime powers $q > 5$ and the quasisimple groups $SL_2(q)$ for prime powers $q > 5$ are strongly real Beauville groups. As with the alternating and symmetric groups, these results are proved by writing down explicit generators, this time combined with a celebrated theorem usually (but historically inaccurately) attributed to Dickson for the maximal subgroups of $PSL_2(q)$. General lemmas for lifting Beauville structures from a group to its covering groups are also used.
- Settling the case of the sporadic simple groups makes no impact on the Bauer, Catanese and Grunewald's original conjecture, there being only 26 of them. Nonetheless, for reasons we shall return to below, in [22] the author determined which of the sporadic simple groups are strongly real Beauville groups, including the '27th sporadic simple group', the Tits group ${}^2F_4(2)'$. Of all the sporadic simple groups only the Mathieu groups M_{11} and M_{23} are not strongly real. For all of the other sporadic groups smaller than the Baby Monster group $\mathbb{B}$ explicit words in the 'standard generators' [57] for a strongly real Beauville structure are given. For the Baby Monster group $\mathbb{B}$ and Monster group $\mathbb{M}$ character theoretic methods are used.
- In [24, Theorem 2] the author verifies the conjecture for the Suzuki groups ${}^2B_2(2^{2n+1})$, again making use of knowledge of the subgroup structure of these

groups and writing down explicit matrices in the natural 4 dimensional representations of these groups.
- In unpublished calculations, the author has pushed the original computations of Bauer, Catanese and Grunewald to every non-abelian finite simple group of order at most 100,000,000.

Many of the smaller groups seem to require the use of outer automorphisms to make their Beauville structures strongly real, which explains much of the above difficulty in finding strongly real Beauville structures in certain groups. Slightly larger groups have enough conjugacy classes for inner automorphisms to be used instead. Consequently, it seems that 'small' non-abelian finite simple groups fail to be strongly real if they have too few conjugacy classes (as is the case with A_5 and as we would intuitively expect) or if they have no outer automorphisms—a phenomenon that is extremely rare but is true of both of the groups M_{11} and M_{23}. We are thus lead to assert the following somewhat bolder strengthening of the above, which was previously asserted by the author in [23, Conjecture 5.5] and [24, Conjecture 1].

Conjecture 1 *All non-abelian finite simple groups apart from A_5, M_{11} and M_{23} are strongly real Beauville groups.*

4 Characteristically Simple Groups

Another class of finite groups that has recently been studied from the viewpoint of Beauville constructions, and appears to be fertile ground for providing further examples of strongly real Beauville groups, are the characteristically simple groups that we define as follows (the definition commonly given is somewhat different from the below but in the case of finite groups it can easily be shown that below is equivalent to it).

Definition 5 A finite group G is said to be **characteristically simple** if G is isomorphic to some direct product H^k where H is a finite simple group.

For example, as we saw in Theorem 1, if $p > 3$ is prime then the abelian Beauville groups isomorphic to $\mathbb{Z}_p \times \mathbb{Z}_p$ are characteristically simple.

Characteristically simple Beauville groups have recently been investigated by Jones in [24, 43, 44] where the following conjecture is discussed.

Conjecture 2 *Let G be a finite non-abelian characteristically simple group. Then G is a Beauville group if and only if it is a 2-generator group not isomorphic to A_5.*

In particular, the main results of [43, 44] verify this conjecture in the cases where H is any of the alternating groups; the linear groups $PSL_2(q)$ and $PSL_3(q)$; the unitary groups $PSU_3(q)$; the Suzuki groups ${}^2B_2(2^{2n+1})$; the small Ree groups ${}^2G_2(3^{2n+1})$ and the sporadic simple groups. Furthermore, as discussed in the previous section this conjecture is true for all simple groups in the case $k = 1$.

For large values of k, the group H^k will not be 2-generated despite the fact that H will be as discussed in Sect. 3. The values of k for which H^k is 2-generated were first investigate by Hall in [41] where various techniques for calculating, or at least bounding, these were investigated. The values of k for which H^k is 2-generated can be surprisingly large. For example, a special case of the results alluded to here is the somewhat amusing fact that

$$A_5 \times A_5 \times A_5 \times A_5 \times A_5 \times A_5 \times A_5 \times A_5 \times A_5 \times A_5 \\ \times A_5 \times A_5 \times A_5 \times A_5 \times A_5 \times A_5 \times A_5 \times A_5 \times A_5$$

is a Beauville group, despite the fact that A_5 itself is not a Beauville group.

In general, the full automorphism group of H^k will be the wreath product $\mathrm{Aut}(H) \wr S_k$ where S_k is the kth symmetric group acting on the product by permuting the groups H. This bounteous supply of automorphisms makes it likely that characteristically simple Beaville groups are in general strongly real.

The question of which characteristically simple Beauville groups are strongly real was first investigated by the author in [24, Sect. 3]. More specifically the following conjecture was investigated.

Conjecture 3 *If H is a finite simple group of order greater than 3, then the group $H \times H$ is a strongly real Beauville group.*

In many cases previously known strongly real Beauville structures of simple groups H can be used to provide strongly real Beauville structures for the groups $H \times H$. In particular we have the following.

Theorem 2 *If H is any of the following groups, then $H \times H$ is a strongly real Beauville group.*

(a) The cyclic groups of prime order greater than 3;
(b) The alternating groups A_n for $n \geq 5$;
(c) The linear groups $PSL_2(q)$ for prime powers $q > 5$;
(d) The Suzuki groups ${}^2B_2(2^{2n+1})$;
(e) All simple groups of order at most 100,000,000;
(f) The sporadic simple groups.

The proofs of several of these cases relied on the following general construction.

Theorem 3 *Let G be a strongly real Beauville group with strongly real Beauville structure $\{\{x_1, y_1\}, \{x_2, y_2\}\}$ such that the numbers $o(x_1)o(y_1)o(x_1y_1)$ is coprime to $o(x_2)o(y_2)o(x_2y_2)$. Furthermore, suppose that there is an automorphism $\phi \in Aut(G)$ such that*

$$\phi(x_1) = x_1^{-1}, \phi(y_1) = y_1^{-1}, \phi(x_2) = x_2^{-1} \textit{ and } \phi(y_2) = y_2^{-1}.$$

Then the group $G \times G$ is a strongly real Beauville group.

Proof See [24, Theorem 3]. □

The reader may be somewhat suspicious of the cases of the alternating group A_5 as well as the Mathieu groups M_{11} and M_{23} given these groups non-status as strongly real Beauville groups, in addition to asking why we are not being ambitious in considering larger direct products. Each of the groups $A_5 \times A_5$, $M_{11} \times M_{11}$ and $M_{23} \times M_{23}$ are indeed strongly real Beauville groups, the automorphisms used to invert the elements of the structures being ones that interchanges the two factors of the direct products. The fact that none of the corresponding simple groups in these cases are strongly real means that this automorphism cannot be extended to higher products so in particular none of the groups $A_5 \times A_5 \times A_5$, $M_{11} \times M_{11} \times M_{11}$ and $M_{23} \times M_{23} \times M_{23}$ strongly real despite the fact that the automorphisms making the product of two copies of the simple groups strongly real can be adapted to make each of the groups $A_5 \times A_5 \times A_5 \times A_5$, $M_{11} \times M_{11} \times M_{11} \times M_{11}$ and $M_{23} \times M_{23} \times M_{23} \times M_{23}$ are strongly real. In short, any precise statement concerning which higher products of simple groups are strongly real must be much more complicated.

Despite the remarks made in the previous paragraph we have the following.

Lemma 1

(a) Let $n \geq 11$ be odd and let $k \leq (n-6)/2$ be positive integers. Then A_n^k is a strongly real Beauville group.
(b) Let $n \geq 12$ be an even integer and let $k \leq (n-8)/4$. Then A_n^k is a strongly real Beauville group.

Proof See [24, Lemmas 5 and 6]. □

More generally the following question seems natural.

Question 3. Given a finite simple group H for which values of k is the characteristically simple group H^k a strongly real Beauville group?

By way of a partial answer to this question the author has computed values of k such that H^r is a strongly real Beauville group for every $r \leq k$ for every simple group of order at most 100,000 (with the exception of the alternating group A_5 and the Mathieu group M_{11} for which we have already shown that $k = 0$ is the largest

Table 1 Values of k such that every a simple group H with $|H| < 100{,}000$ the group H^r is a strongly real Beauville group for every $r \leq k$

H	k	H	k	H	k	H	k	H	k	H	k
A_5	0	$L_2(7)$	2	A_6	2	$L_2(8)$	4	$L_2(11)$	4	$L_2(13)$	4
$L_2(17)$	6	A_7	14	$L_2(19)$	6	$L_2(16)$	2	$L_3(3)$	14	$U_3(3)$	6
$L_2(23)$	2	$L_2(25)$	10	M_{11}	0	$L_2(27)$	12	$L_2(29)$	12	$L_2(31)$	14
A_8	18	$L_3(4)$	4	$U_4(2)$	6	$^2B_2(8)$	52	$L_2(32)$	2	$L_2(41)$	18
$L_2(43)$	18	$L_2(47)$	22	$L_2(49)$	22	$U_3(4)$	28	$L_2(53)$	48	M_{12}	16

value). The best known values of k are listed in Table 1. We do not claim that these values are best possible, merely lower bounds on the correct value, and it is likely that these may be improved upon. The author hopes to push these computations further in the future. Furthermore, the author is happy to provide details of the computations done on request.

5 Almost Simple Groups

We first recall the definition of almost simple groups.

Definition 6 Let G be a group. Recall that we say G is **almost simple** if there exists a simple group S such that $S \leq G \leq \mathrm{Aut}(S)$.

For example, any simple group is almost simple, as are the symmetric groups.

As briefly mentioned earlier in [29] Fuertes and González-Diez considered which of the symmetric groups are strongly real Beauville groups. We wish to take this opportunity to correct a minor error in [29] that as far as the author is aware has not previously been corrected in the literature. In [29, Proposition 9] permutations providing strongly real Beauville structures for the smallest symmetric groups are given. In particular, in the case $n = 8$ the following permutations are given to provide the x_1 and y_1 elements of a Beauville structure for S_8:

$$x_1 := (2,7,3)(5,8,6),\ y_1 := (1,2,3,4,5,6).$$

Despite the claim that these generate the whole group, it is easy to see that these permutations preserve the partition $\{\{1,4\},\{2,5\},\{3,6\},\{7,8\}\}$ and therefore they do not even generate a primitive group, let alone the whole of the symmetric group! Since the other half of the Beauville structure given in [29, Proposition 9] is

$$x_2 := (7,8),\ y_2 := (1,2,3,4,5,6,7)$$

(these are easily seen to be inverted by $(1,6)(2,5)(3,4)$) we complete this Beauville structure by setting

$$x_1 := (1,2,3,4,5,6),\ y_1 := (8,7,6,5,4).$$

It is easy to verify that these permutations generate the whole of S_8 and are both inverted by conjugation by $(1,3)(4,6)(7,8)$.

We briefly digress by noting that in [21] the author has recently generalised Fuertes and González-Diez's results to reflection groups more generally. (Recall that we may view the symmetric group S_{n+1} as the Coxeter group of type A_n—see [42].) For the irreducible Coxeter groups we have the following.

Theorem 4 *Every finite irreducible Coxeter group is a strongly real Beauville group aside from the groups of type:*

(a) A_n *for* $n \leq 3$*;*
(b) B_n *for* $n \leq 4$*;*
(c) D_n *for* $n \leq 4$*;*
(d) F_4, H_3 *and*
(e) $I_2(k)$ *for any* k.

From this we can deduce the complete classification of strongly real Beauville Coxeter groups.

Corollary 1 *No product of three or more irreducible Coxeter groups is a Beauville group. Furthermore,* $K_1 \times K_2$ *is a strongly real Beauville group if* K_1 *and* K_2 *are strongly real irreducible Coxeter Beauville groups not of type* B_n.

As a consequence of the proof of this theorem we have the following.

Corollary 2 *An irreducible Coxeter group is a Beauville group if and only if it is a strongly real Beauville group.*

The first place the more general question of which almost simple groups are (strongly real) Beauville groups was the author's discussion given in [24, Sect. 5] where the following conjecture is asserted.

Conjecture 4 *A split extension of a simple group is a Beauville group if, and only if, it is a strongly real Beauville group.*

There are multiple 'warning shots' to be fired here—there are infinitely many almost simple groups that are not even 2-generated, let alone Beauville groups, the smallest example being $\mathrm{PSL}_4(9)$ whose outer automorphism group is $2 \times \mathrm{D}_8$ (and more generally, if p is an odd prime and r is an even positive integer then $Aut(PSL_4(p^r))$ is not 2-generated). Even among the almost simple groups that are 2-generated, many are not Beauville groups—for example the almost simple groups ${}^2\mathrm{B}_2(2^{2n+1}) : 3$ where $n \equiv 1 \pmod 3$ are never Beauville groups since for any generating pair $x, y \in {}^2\mathrm{B}_2(2^{2n+1}) : 3$ we have that $\Sigma(x, y)$ contains elements from the only class of elements of order 3.

6 The Groups $G : \langle g \rangle \times G : \langle g \rangle$

After characteristically simple groups, the next most natural direct products to try and deal with are products of almost simple groups. Alas, any group of the form $G : H$ where H is not cyclic will have the property that $G : H \times G : H$ will not be 2-generated, so the best we can hope for is that the groups $G : \langle g \rangle \times G : \langle g \rangle$ are Beauville. (This observation was also extremely useful when proving the above Corollary 1.)

In [24, Lemma 7] the author proves that for $n \geq 5$ the groups $S_n \times S_n$ are strongly real. Besides symmetric groups there are infinitely many simple groups with non-trivial outer automorphisms, indeed it is unusual for a simple group to have no outer automorphisms. It is natural to consider groups in which G is not necessarily the alternating group.

Question 4. For which simple groups G is the group $G : \langle g \rangle \times G : \langle g \rangle$ a strongly real Beauville group?

It is easy to see that if G is a sporadic simple group with a non-trivial outer automorphism (namely one of the groups M_{12}, M_{22}, J_2, HS, J_3, McL, He, Suz, O'N, Fi_{22}, HN and Fi_{24}) then the strongly real Beauville structures obtained for the groups $G : 2$ in [24, Sect. 5] provide further examples of groups of this kind. As far as the author is aware this class of groups has not been investigated more generally elsewhere in the literature.

7 Abelian and Nilpotent Groups

Recall that the abelian Beauville groups were classified in Theorem 1 and that an immediate corollary of this result is that every abelain Beauville group is strongly real.

Theorem 1 has been put to great use by González-Diez et al. in [37] where several structural results concerning the surfaces defined by abelian Beauville groups are proved. For each abelian Beauville group they describe all the surfaces arising from that group, enumerate them up to isomorphism and impose constraints on their automorphism groups. As a consequence they show that all such surfaces are defined over $\mathbb{Q}$.

After the abelian groups, the next most natural class of finite groups to consider are the nilpotent groups. In [2, Lemma 1.3] Barker, Boston and the author note the following easy Lemma.

Lemma 2 *Let G and G' be Beauville groups and let $\{\{x_1, y_1, \}, \{x_2, y_2\}\}$ and $\{\{x_1', y_1', \}, \{x_2', y_2'\}\}$ be their respective Beauville structures. Suppose that*

$$gcd(o(x_i), o(x_i')) = gcd(o(y_i), o(y_i')) = 1$$

for $i = 1, 2$. Then $\{\{(x_1, x_1'), (y_1, y_1')\}, \{(x_2, x_2'), (y_2, y_2')\}\}$ is a Beauville structure for the group $G \times G'$.

Recall that a finite group is nilpotent if, and only if, it is isomorphic to the direct product of its Sylow subgroups. It thus follows that Lemma 2, and its easy to prove converse, reduces the study of nilpotent Beauville groups to that of Beauville p-groups. Note that Theorem 1 gives us infinitely many examples of Beauville p-groups for every prime $p > 3$: simply let n be any power of p. Early examples of Beauville 2-groups and 3-groups were constructed by Fuertes et al. in [30] where a Beauville group of order 2^{12} and another of order 3^{12} were constructed. Even earlier than this, two (mixed) Beauville 2-groups of order 2^8 arose as part of a classification due to Bauer et al. in [9] of certain classes of surfaces of general type, which give rise to examples of (unmixed) Beauville 2-groups of order 2^7.

More recently, in [2] Barker, Boston and the author classified the Beauville p-groups of order at most p^4 and made substantial progress on the cases of groups of order p^5 and p^6. More recently still in [55] Stix and Vdovina have constructed infinite series of Beauville p-groups. In particular this gives the first examples of non-abelian Beauville p-groups of arbitrarily large order and any prime $p \geq 5$. To do this they make use of the theory of pro-p groups and in doing so provide generalisations of examples from [2]. The first explicit construction of an infinite family of Beauville 3-groups was recently given by Fernández-Alcober and Gül in [28] where they consider homomorphic images of the famous Nottingham group as well as providing other general constructions for Beauville p-groups. In doing so they settled several conjectures made in [2]. The earliest explicit infinite family of Beauville 2-groups have been constructed by Barker et al. in [3–5] where, again, more general constructions are also considered. The most comprehensive survey on Beauville p-groups is given by Boston in [15].

Up until now, however, the only example of Beauville p-groups that have been explicitly shown to be strongly real have been the abelian ones. Below we give what is, as far as the author is aware, the very first example of non-abelian strongly real Beauville p-group.

Lemma 3 *There exist strongly real Beauville 2-groups.*

Proof Consider the group

$$G = \langle u, v \mid (u^i v^j)^4,\ i, j = 0, 1, 2, 3 \rangle.$$

Straightforward computations verify that $|G| = 2^{14}$ and that $\{\{u, v\}, \{uvu, vuv\}\}$ is a Beauville structure. Moreover, the function mapping $u \leftrightarrow u^{-1}$ and $v \leftrightarrow v^{-1}$ is an automorphism since it simply permutes the relations appearing the above presentation. This automorphism of this group clearly inverts all of the elements in the above Beauville structure so we have a strongly real Beauville structure and thus a strongly real Beauville group. □

As far as the author is aware, the above is an isolated example—replacing 4 with a higher power of 2 or replacing 2 with a larger prime does not appear to produce a finite group. The utility of the above group stems from the fact that it admits an unusually easy to write down presentation and in particular a presentation that makes it unusually easy to explicitly write down an automorphism of the group. The closest to a further example the author has been able to find is the group

$$G = \langle u, v \mid u^8, v^8, [u^2, v^2], (u^i v^j)^4,\ i, j = 1, 2, 3 \rangle$$

which can easily be shown to be a strongly real Beauville group of order 2^{13} by an argument similar to the above.

Whilst constructing infinitely many examples of strongly real Beauville p-groups is difficult, constructing infinitely many strongly real nilpotent Beauville groups from the above is easy: as noted earlier any nilpotent group is isomorphic to the direct product of its Sylow subgroups. This fact combined with the strongly real abelian Beauville p-groups given to us by Theorem 1 provides an infinite supply of examples.

Clearly much work on the following question remains to be done.

Question 5.

(a) Are there infinitely many strongly real Beauville p-groups?
(b) What proportion of the 2-generated p-groups that are Beauville groups are strongly real?

Since p-groups in general tend to have large automorphism groups it seems likely in the opinion of the author that there are infinite families of strongly real Beauville p-groups.

Acknowledgments The author wishes to express his deepest gratitude to the organisers of the 2014 installment of the conferences on Symmetries in Graphs, Maps, and Polytopes hosted by The Open University and in particular to Professor Jozef Širáň for making this publication possible. The author wishes to thank the anonymous referees for their lengthy and in-depth commentary they provided on this submission.

References

1. M. Aschbacher and R. Guralnick "Some applications of the first cohomology group" J. Algebra 90 (1984), no. 2, 446–460.
2. N. W. Barker, N. Bosten and B. T. Fairbairn "A note on Beauville p-groups" Exp. Math., 21(3): 298–306 (2012) doi:10.1080/10586458.2012.669267.
3. N. W. Barker, N. Boston, N. Peyerimhoff and A. Vdovina "An infinite family of 2-groups with mixed Beauville structures" Int. Math. Res. Notices., 2014 doi:10.1093/imrn/rnu045. arXiv:1304.4480.
4. N. W. Barker, N. Boston, N. Peyerimhoff and A. Vdovina "Regular algebraic surfaces isogenous to a higher product constructed from group representations using projective planes" preprint 2011 arXiv:1109.6053.
5. N. W. Barker, N. Boston, N. Peyerimhoff and A. Vdovina "New examples of Beauville surfaces" Monatsh. Math. 166 (2012), no. 3-4, pp. 319–327 doi:10.1007/s00605-011-0284-6.
6. I. Bauer "Product-Quotient Surfaces: Results and Problems" preprint 2012 arXiv:1204.3409.
7. I. Bauer, F. Catanese and F. Grunewald "Beauville surfaces without real structures" in Geometric methods in algebra and number theory pp. 1–42, Progr. Math., 235, Birkhuser Boston, Boston, MA, 2005.
8. I. Bauer, F. Catanese and F. Grunewald "Chebycheff and Belyi Polynomials, Dessins d'Enfants, Beauville Surfaces and Group Theory" Mediterr. J. math. 3 (2006), 121–146.
9. I. Bauer, F. Catanese and F. Grunewald "The classification of surfaces with $p_g = q = 0$ isogenous to a product of curves" Pre Apple. Math. Q. 4 (2008), no. 2, Special Issue: In Honor of Fedor Bogomolov. Part 1, 547–586.
10. I. Bauer, F. Catanese and R. Pignatelli "Surfaces of General Type with Geometric Genus Zero: A Survey" in Complex and differential geometry 1–48, Springer Proc. Math., 8, Springer-Verlag, Heidelberg, 2011.

11. I. C. Bauer, F. Catanese and R. Pignatelli "Complex surfaces of general type: some recent progress" in Global Aspects of Complex Geometry, 1–58, Springer, Berlin, 2006.
12. A. Beauville "Surfaces algébriques complexes" (Astérisque 54 1978).
13. A. Beauville "Complex Algebraic Surfaces" (London Mathematical Society Student Texts 34, Cambridge University Press, Cambridge, 1996).
14. G. V. Belyĭ "On Galois extensions of a maximal cyclotomic field" Math. USSR Izvestija 14 (1980), 247–256.
15. N. Boston "A Survey of Beauville p-Groups" in *Beauville Surfaces and Groups, Springer Proceedings in Mathematics & Statistics, Vol. 123* (eds I. Bauer, S. Garion and A. Vdovina), Springer-Verlag (2015) pp. 35–40.
16. E. Bujalance, F. J. Cirre, J. J. Etayo, G. Gromadzki and E. Martínez "A Survey on the Minimum Genus and Maximum Order Problems for Bordered Klein Surfaces" in Proceedings of Groups St Andrews 2009 London Mathematical Society Lecture Note Series 387' (eds. C. M. Campbell, M. R. Quick, E. F. Robertson, C. M. Roney-Dougal, G. C. Smith and G. Traustason) Cambridge University Press, Cambridge, (2011) pp. 161–182.
17. F. Catanese "Fibered surfaces, varieties isogenous to a product and related moduli spaces" Amer. J. Math. 122 (2000), no. 1, 1–44.
18. M. D. E. Conder "Hurwitz groups: a brief survey" Bull. Amer. Math. Soc. 23 (1990), 359–370.
19. M. D. E. Conder "An update on Hurwitz groups" Groups Complexity Cryptology, Volume 2, Issue 1 (2010) 35–49.
20. J. H. Conway, R. T. Curtis, S. P. Norton, R. A. Parker and R. A. Wilson, "Atlas of Finite Groups" (Clarendon Press, Oxford) 1985.
21. B. T. Fairbairn "Coxeter groups as Beauville groups" Monatshefte für Mathematik 178:4 pp. 1–17 (2015) 10.1007/s00605-015-0848-y.
22. B. T. Fairbairn "Some Exceptional Beauville Structures" J. Group Theory, 15(5), pp. 631–639 (2012) arXiv:1007.5050 doi:10.1515/jgt-2012-0018.
23. B. T. Fairbairn, "Recent work on Beauville surfaces, structures and groups" in 'Groups St Andrews 2013 London Mathematical Society Lecture Note Series 422' (eds. C. M. Campbell, M. R. Quick, E. F. Robertson and C. M. Roney-Dougald) Cambridge University Press, Cambridge (2015).
24. B. T. Fairbairn, "Strongly Real Beauville Groups" in *Beauville Surfaces and Groups, Springer Proceedings in Mathematics & Statistics, Vol. 123* (eds I. Bauer, S. Garion and A. Vdovina), Springer-Verlag (2015) pp. 41–61.
25. B. T. Fairbairn, K. Magaard and C. W. Parker "Generation of finite simple groups with an application to groups acting on Beauville surfaces" Proc. London Math. Soc. (2013) 107 (4): 744-798. doi:10.1112/plms/pds097.
26. B. T. Fairbairn, K. Magaard and C. W. Parker "Corrigendum to Generation of finite simple groups with an application to groups acting on Beauville surfaces" Proc. London Math. Soc. (2013) 107 (5): 1220 doi:10.1112/plms/pdt037.
27. B. T. Fairbairn and E. Pierro "New Examples of Mixed Beauville Groups" J. Group Theory 18(5), pp. 761–795 (2015).
28. Gustavo A. Fernández-Alcober and Şükran Gül "Beauville structures in finite p-groups" preprint 2015, arXiv:1507.02942.
29. Y. Fuertes and G. González-Diez "On Beauville structures on the groups S_n and A_n" Math. Z. 264 (2010), no. 4, 959–968.
30. Y. Fuertes, G. González-Diez and A. Jaikin-Zapirain "On Beauville surfaces" Groups Geom. Dyn. 5 (2011), no. 1, 107–119.
31. Y. Fuertes and G. Jones "Beauville surfaces and finite groups" J. Algebra 340 (2011) 13–27.
32. S. Galkin and E. Schinder "Exceptional collections of line bundles on the Beauville surface" Advances in Mathematics (2013) Vol. 224. No. 10 1033–1050 arXiv:1210.3339.
33. S. Garion, M. Larsen and A. Lubotzky "Beauville surfaces and finite simple groups" J. Reine Angew. Math. 666 (2012), 225–243.
34. S. Garion and M. Penegini "New Beauville surfaces, moduli spaces and finite groups" Comm. Algebra. 42, Issue 5 (2014), 2126–2155 arXiv:0910.5402.

35. E. Girondo and G. González-Diez "Introduction to Compact Riemann Surfaces and Dessins d'Enfants" (London Mathematical Society Student texts 79) Cambridge University Press, Cambridge 2011.
36. G. González-Diez and A. Jaikin-Zapirain "The absolute Galois group acts faithfully on regular dessins and on Beauville surfaces". Proc. London Math. Soc. (2015) 111(4): 775–796. doi:10.1112/plms/pdv041.
37. G. González-Diez, G. A. Jones and D. Torres-Teigell "Beauville surfaces with abelian Beauville group" Math. Scand. 114 (2014), no. 2, 191–204 arXiv:1102.4552.
38. G. González-Diez and D. Torres-Teigell "An introduction to Beauville surfaces via uniformization, in Quasiconformal mappings, Riemann surfaces, and Teichmüller spaces" 123–151, Contemp. Math., 575, Amer. Math. Soc., Providence, RI, 2012.
39. A. Grothendieck "Esquisse d'un Programme" in Geometric Galois Actions 1. Around Grothendieck's Esquisse d'un Programme, eds P. Lochak and L. Schneps, London Math. Soc. Lecture Note Ser. 242, Cambridge University Press, 1997, pp. 5–84.
40. R. Guralnick and G. Malle "Simple groups admit Beauville structures" J. Lond. Math. Soc. (2) 85 (2012), no. 3, 694–721.
41. P. Hall "The Eulerian functions of a group" Quarterly Journal of Mathematics 7 (1936), 134–151.
42. J. E. Humphreys "Reflection groups and Coxeter groups Cambridge Studies in Advanced Mathematics 29" Cambridge University Press, Cambridge, 1997.
43. G. A. Jones "Characteristically simple Beauville groups, I: cartesian powers of alternating groups" in *Geometry, Groups and Dynamics* (eds C. S. Aravinda, W. M. Goldman et. al.) Contemp. Math. 639, pp. 289–306 (2015) arXiv:1304.5444v1.
44. G. A. Jones "Characteristically simple Beauville groups, II: low rank and sporadic groups" in *Beauville Surfaces and Groups, Springer Proceedings in Mathematics & Statistics, Vol. 123* (eds I. Bauer, S. Garion and A. Vdovina), Springer-Verlag (2015) pp. 97–120 arXiv:1304.5450v1.
45. G. A. Jones "Automorphism groups of Beauville surfaces" J. Group Theory. Volume 16, Issue 3, Pages 353–381, doi:10.1515/jgt-2012-0049, 2013 arXiv:1102.3055.
46. G. A. Jones "Beauville surfaces and groups: a survey" in 'Rigidity and Symmetry, Fields Institute Communications vol. 70' (eds. R. Connelly, A. I. Weiss and W. Whiteley) pp. 205–226, Springer 2014.
47. G. A. Jones and D. Singerman "Belyi functions, hypermaps and Galois groups" Bull. Lond. Math. Soc. 28 (1996) 561–590.
48. G. A. Jones, D. Singerman and P. D. Watson "Symmetries of quasiplatonic Riemann surfaces" Revista Mathemática Iberoamericana Volume 31, Issue 4, 2015, pp. 1403–1414 doi:10.4171/RMI/873. arXiv:1401.2575.
49. G. A. Miller "On the groups generated by two operators" Bull. Amer. Math. Soc. Volume 7, Number 10 (1901) 424–426.
50. A. Murray Macbeath "Hurwitz Groups and Surfaces" in 'The Eightfold Way: The Beauty of Klein's Quartic Curve' (ed. S. Levy) MSRI Publications, 35, Cambridge University Press, Cambridge (1998) pp.103–114.
51. F. Schaffhauser "Lectures on Klein surfaces and their fundamental groups" Advanced Courses in Mathematics — CRM Barcelona, to appear http://matematicas.uniandes.edu.co/~florent/resources/papers/Barcelona.pdf.
52. D. Singerman "Symmetries of Riemann surfaces with large automorphism group" Math. Ann. 210 (1974) 17–32.
53. J. Širáň "How symmetric can maps on surfaces be?" in 'Surveys in Combinatorics 2013' (Simon R. Blackburn, Stefanie Gerke and Mark Wildon eds.), London Mathematical Society Lecture Note Series 409 (Cambridge University Press, Cambridge, 2013), 161–238.
54. R. Steinberg "Generators for simple groups" Canad. J. Math., 14 (1962), pp. 277–283.
55. J. Stix and A. Vdovina "Series of p-groups with Beauville structure" Monatshefte der Mathematik, 2015, doi:10.1007/s00605-015-0805-9. arXiv:1405.3872.

56. D. Torres-Teigell "Triangle groups, dessins d'enfants and Beauville surfaces" PhD thesis, Universidad Autonoma de Madrid, 2012.
57. R. A. Wilson, "Standard generators for sporadic simple groups" J. Algebra 184 (1996), no. 2, 505–515.
58. J. Wolfart "ABC for polynomials, dessins d'enfants and uniformization — a survey" Elementare und analytische Zahlentheorie, Schr. Wiss. Ges. Johann Wolfgang Goethe Univ. Frankfurt am Main, 20, Franz Steiner Verlag Stuttgart, Stuttgart, 313–345 (2006) http://www.math.uni-rankfurt.de/~wolfart/.

On Pentagonal Geometries with Block Size 3, 4 or 5

Terry S. Griggs and Klara Stokes

Abstract Let PLS(k, r) be a partial linear space which is both uniform, i.e. every line has the same cardinality $k \geq 2$, and regular, i.e. every point is incident with the same number $r \geq 1$ of lines. In a recent paper (J. Combin. Des. 21 (2013), 163–179), Ball, Bamberg, Devillers & Stokes introduced the concept of a pentagonal geometry PENT(k, r) as a PLS(k, r) in which all the points not collinear with any given point are themselves collinear. They also determined the existence spectrum for $k = 1$ or 2 and $r = k$ or $k+1$. In this paper we prove that the existence spectrum for PENT(3, r) is $r \equiv 0$ or 1 (mod 3) except $r = 4$ or 6. We also prove that there exists a PENT(4, r) for $r \equiv 1 \pmod 8$ and a PENT(5, r) for $r \equiv 1 \pmod 5$, $r \neq 6$, apart from nine possible exceptions. Further we construct an infinite class of pentagonal geometries PENT$(2^m, 2^{m+1} + 1)$, $m \geq 1$, and a PENT(6, 13).

1 Introduction

In [2], the authors introduced the concept of a pentagonal geometry and developed the theory of this structure. The framework within which this is done is that of a *partial linear space*. This is an ordered pair $(V, \mathscr{L})$ where V is a set of elements, usually called *points*, of cardinality v and $\mathscr{L}$ is a family of subsets of V, usually called *lines* or *blocks*, such that every pair of distinct points is contained in at most one line. We will only be concerned with partial linear spaces which are both *uniform*, i.e. every line has the same cardinality $k \geq 2$, and *regular*, i.e. every point is incident with the same number $r \geq 1$ of lines. Denote such a partial linear space by PLS(k, r). A *pentagonal geometry*, PENT(k, r), is a partial linear space in which for all points x,

T.S. Griggs (✉)
Department of Mathematics and Statistics, The Open University, Walton Hall, Milton Keynes MK7 6AA, UK
e-mail: terry.griggs@open.ac.uk

K. Stokes
School of Engineering Science, University of Skövde, Box 408, 54128 Skövde, Sweden
e-mail: klara.stokes@his.se

J. Širáň and R. Jajcay (eds.), *Symmetries in Graphs, Maps, and Polytopes*, Springer Proceedings in Mathematics & Statistics 159,
DOI 10.1007/978-3-319-30451-9_7

the points not collinear with x are themselves collinear. We call this line the *opposite line* to x and denote it by x^{opp}. If two points x and y have the same opposite line $x^{opp} = y^{opp} = l$, then $z^{opp} = m$ for all points $z \in l$ where m is the line joining x and y. Similarly $w^{opp} = l$ for all $w \in m$. We will call such a pair of lines (l, m) an *opposite line pair*. The pentagon is the geometry PENT(2, 2) and the Desargues configuration is PENT(3, 3). When $r = 1$, PENT(k, 1) consists of two disjoint lines, each of cardinality k. We will say that this is a *degenerate* pentagonal geometry.

A number of basic lemmas about pentagonal geometries are proved in [2], of which the following will be important for this paper.

Lemma 1 *A pentagonal geometry PENT(k, r) has* $rk - r + k + 1$ *points and* $(rk - r + k + 1)r/k$ *lines. Thus a necessary condition for existence is that k divides* $r(r-1)$.

Lemma 2 *If there exists a pentagonal geometry PENT(k, r) with* $r > 1$, *then* $r \geq k$.

Lemma 3 *A pentagonal geometry PENT(k, r) with* $1 < r < 3k$ *has either*

(i) no opposite line pair, or
(ii) $r = 2k + 1$ *and the points are partitioned into opposite line pairs.*

An important concept in the theory of partial linear spaces is that of the *leave* or *deficiency graph*. This is the graph G whose vertex set is V with two points x and y being adjacent if and only if they are not collinear. The following result is also proved in [2].

Lemma 4 *The deficiency graph G of a pentagonal geometry PENT(k, r) is the disjoint union of complete bipartite graphs* $K_{k,k}$ *(one for each opposite line pair) and* G' *where* G' *is a k-regular graph of girth at least 5, not necessarily connected.*

With the aid of this latter result, the authors of [2] were able to relate the existence of a pentagonal geometry PENT(k, k) or PENT(k, $k+1$) to that of a *Moore graph* of girth 5, i.e. k-regular graph with $k^2 + 1$ vertices. Such graphs exist only for $k = 2$ (pentagon), 3 (Petersen graph), 7 (Hoffman-Singleton graph) and possibly 57. Thus they were able to prove the two following theorems.

Theorem 1 *A pentagonal geometry PENT(k, k) exists only for k = 2, 3, 7 and possibly 57.*

Theorem 2 *A pentagonal geometry PENT(k, k+1) exists only for k = 2, 6 and possibly 56.*

The existence spectrum of pentagonal geometries with block size 2 was also determined. From Lemma 1, the number of points in a PENT(2, r) is $r+3$. So, from Lemma 4 we have the following theorem which is taken from [2].

Theorem 3 *A pentagonal geometry PENT(2, r) is a complete graph on r + 3 vertices from which a union of disjoint cycles, none of size 3, spanning the vertex set has been deleted.*

Thus the number of non-isomorphic pentagonal geometries PENT(2, r) is equal to the number of decompositions, say $q(r)$, of $r+3$ into integers greater than or equal to 4. This is sequence A008484 in the On-line Encyclopedia of Integer Sequences (OEIS) and, as observed there, can easily be expressed in terms of the partition function, (sequence A000041 in OEIS). Recall that for all $n \geq 1$, the *partition function* $p(n)$ is the number of ways in which n can be expressed as a sum of positive integers, called *parts*, where the order of the parts is immaterial. By convention $p(0) = 1$ and $p(n) = 0$ for $n < 0$.

We express this result formally as a theorem.

Theorem 4 *The number $q(r)$ of non-isomorphic pentagonal geometries PENT(2, r) is $p(r+3) - p(r+2) - p(r+1) + p(r-1) + p(r-2) - p(r-3)$.*

Values of $q(r)$ for $2 \leq r \leq 21$ are given in the table below.

r	2	3	4	5	6	7	8	9	10	11
$q(r)$	1	1	1	2	2	3	3	5	5	7
r	12	13	14	15	16	17	18	19	20	21
$q(r)$	8	11	12	16	18	24	27	34	39	50

The aim of this paper is to extend the results of [2] by considering pentagonal geometries with block size 3, 4 or 5. The main result is the proof that the existence spectrum for PENT(3, r) is $r \equiv 0$ or 1 (mod 3) except $r = 4$ or 6. The case $r \equiv 2$ (mod 3) is prohibited by Lemma 1. The necessary condition for the existence of PENT(4, r) as given by Lemma 1 is $r \equiv 0$ or 1 (mod 4). We also prove that there exists a PENT(4, r) for $r \equiv 1$ (mod 8). Further progress on the existence spectrum is inhibited by the lack of any examples in the residue classes 0, 4 or 5 (mod 8). However we indicate how the construction of certain small examples would lead, via our methods, to further infinite linear classes. The necessary condition for the existence of PENT(5, r) as given by Lemma 1 is $r \equiv 0$ or 1 (mod 5). We prove the existence of pentagonal geometries in the latter residue class apart from $r = 6$ and possibly nine further values. We also establish some results on pentagonal geometries PENT(k, $2k+1$).

2 Constructions

A partial linear space PLS(k, r) is said to have *deficiency one* if every point is collinear with every other point except one, called its *antipodal point*. In [2], the authors construct pentagonal geometries by taking the union of k copies of a partial linear space of deficiency one with k points on each line and replacing the disconnected configuration consisting of the k copies of the same line together with the points on these lines by an affine plane of order k. This is a product construction and all opposite lines of the resulting pentagonal geometry occur in pairs. This gives the following theorem.

Theorem 5 *Let PLS(k, r) be a partial linear space of deficiency one where k is a power of a prime. Then there exists a pentagonal geometry PENT($k, kr + 1$).*

In the next two sections we use this theorem to prove the existence of infinite linear classes of pentagonal geometries with block size 3 or 4.

The authors of [2] also give a second construction.

Theorem 6 *Let PENT(k, r) be a, (possibly degenerate), pentagonal geometry. If there exists a set of $k - 2$ mutually orthogonal Latin squares (MOLS) of order $(k - 1)r + k + 1$, then there exists a pentagonal geometry PENT($k, rk + k + 1$).*

We prove a generalisation of this theorem by replacing the set of mutually orthogonal Latin squares by an appropriate group divisible design.

A *k-group divisible design*, k-GDD, is an ordered triple $(V, \mathcal{G}, \mathcal{B})$ where V is a set of *points* of cardinality v, $\mathcal{G}$ is a partition of V into *groups* and $\mathcal{B}$ is a family of subsets of V, called *lines* or *blocks*, each of cardinality k, such that every pair of distinct points is contained in either precisely one group or one block, but not both. If $v = a_1g_1 + a_2g_2 + \ldots + a_sg_s$ and if there are a_i groups of cardinality g_i, $i = 1, 2, \ldots, s$, then the k-GDD is said to be of *type* $g_1^{a_1} g_2^{a_2} \ldots g_s^{a_s}$. Thus the existence of $k - 2$ MOLS of order n is equivalent to the existence of a group divisible design of type n^k.

For group divisible designs with uniform group size we have the following result.

Theorem 7 *Let PENT(k, r) be a (possibly degenerate) pentagonal geometry. If there exists a k-GDD of type $((k-1)r+(k+1))^u$, then there exists a pentagonal geometry PENT($k, ru + (k + 1)(u - 1)/(k - 1)$).*

Proof Let the point set of the pentagonal geometry PENT(k, r) be $V = \{1, 2, \ldots, (k - 1)r + (k + 1)\}$. Define disjoint sets $V^{(i)} = \{x^{(i)} : x \in V\}$, $i = 1, 2, \ldots, u$, and construct pentagonal geometries PENT(k, r) on each of these sets. Adjoin the blocks of a k-GDD of type $((k - 1)r + (k + 1))^u$ where the sets $V^{(i)}$ form the groups of the design. We obtain a pentagonal geometry with block size k and in which every point occurs in $r + ((k - 1)r + (k + 1))(u - 1)/(k - 1) = ru + (k + 1)(u - 1)/(k - 1)$ blocks. □

We can also prove an extension of Theorem 7 in which all group sizes except one have the same cardinality.

Theorem 8 *Let PENT(k, r) and PENT(k, s) be (possibly degenerate) pentagonal geometries. If there exists a k-GDD of type $((k-1)r+(k+1))^u((k-1)s+(k+1))^1$, then there exists a pentagonal geometry PENT($k, (r + (k + 1)/(k - 1))u + s$).*

Proof Let the point set of the pentagonal geometry PENT(k, r) be $V = \{1, 2, \ldots, (k - 1)r + (k + 1)\}$. Define disjoint sets $V^{(i)} = \{x^{(i)} : x \in V\}$, $i = 1, 2, \ldots, u$, and construct pentagonal geometries PENT(k, r) on each of these sets. In addition construct a pentagonal geometry PENT(k, s) on the set $W = \{1^{(w)}, 2^{(w)}, \ldots, ((k - 1)s + (k + 1))^{(w)}\}$. Adjoin the blocks of a k-GDD of type $((k - 1)r + (k +$

$1))^u((k-1)s+(k+1))^1$ where the sets $V^{(i)}$ and W form the groups of the design. We obtain a pentagonal geometry with block size k. A point $x \in V^{(i)}$ occurs in $r+(((k-1)r+(k+1))(u-1)+(k-1)s+(k+1))/(k-1) = (r+(k+1)/(k-1))u+s$ blocks and a point $y \in W$ also occurs in $s+((k-1)r+(k+1))u/(k-1) = (r+(k+1)/(k-1))u+s$ blocks. □

3 Block Size 3

From Lemma 1, the number of points in a PENT$(3, r)$ is $2r+4$ and the necessary condition for existence is $r \equiv 0$ or 1 (mod 3). We prove that this condition is also sufficient with the exception of the two values $r = 4$ and $r = 6$, where the pentagonal geometries do not exist.

In arithmetic set density terms, this means that 1/3 of the possible spectrum can be constructed using Theorem 5. PLS$(3, r)$ of deficiency one are readily obtained from Steiner triple systems. Recall that a *Steiner triple system* of order v, STS(v), is an ordered pair $(V, \mathscr{B})$ where V is a set of *points* of cardinality v and $\mathscr{B}$ is a family of *lines* or *blocks*, each of cardinality 3, such that every pair of distinct points is contained in precisely one block. It is very well known that STS(v) exist if and only if $v \equiv 1$ or 3 (mod 6), [12], and there is an extensive theory and literature on the systems, see for example [6]. Given an STS(v), then by selecting any point and deleting all blocks through that point, a PLS$(3, r)$ of deficiency one is obtained for all $r \equiv 0$ or 2 (mod 3). It then follows immediately that there exist pentagonal geometries PENT$(3, 9t+1)$ and PENT$(3, 9t+7)$, $t \geq 0$.

However using Theorems 7 and 8, we can obtain the entire existence spectrum. First we need the following two results on the existence of 3-GDDs, the first due to Hanani, [10], and the second to Colbourn, Hoffman & Rees, [5].

Proposition 1 **(Hanani)** *There exists a 3-GDD of type g^u, $u \geq 3$, if and only if*

1. *$g \equiv 1$ or 5 (mod 6) and $u \equiv 1$ or 3 (mod 6), or*
2. *$g \equiv 2$ or 4 (mod 6) and $u \equiv 0$ or 1 (mod 3), or*
3. *$g \equiv 3$ (mod 6) and $u \equiv 1$ (mod 2), or*
4. *$g \equiv 0$ (mod 6) with no constraint on u.*

Proposition 2 **(Colbourn, Hoffman & Rees)** *There exists a 3-GDD of type $g^u m^1$ if and only if the following conditions are all satisfied.*

1. *if $g > 0$, then $u \geq 3$, or $u = 2$ and $m = g$, or $u = 1$ and $m = 0$, or $u = 0$;*
2. *$m \leq g(u-1)$ or $gu = 0$;*
3. *$g(u-1)+m \equiv 0$ (mod 2) or $gu = 0$;*
4. *$gu = 0$ (mod 2) or $m = 0$;*
5. *$\frac{1}{2}g^2u(u-1)+gum \equiv 0$ (mod 3).*

We are now in a position to prove the main theorem.

Theorem 9 ***(Existence theorem for block size 3)*** *The existence spectrum of pentagonal geometries PENT(3, r) is $r \equiv 0$ or $1 \pmod 3$, except $r = 4$ or 6.*

Proof In Theorem 7, let $k = 3$ and $r = 1$. There exists a pentagonal geometry PENT(3, 1) and a 3-GDD of type 6^t, $t \geq 3$. Hence there exists a pentagonal geometry PENT(3, $3t - 2$), $t \geq 3$. We have already observed that there exists a PENT(3, 1) and from Theorem 2 there is no PENT(3, 4).
In Theorem 8, let $k = 3$, $r = 1$ and $s = 3$. There exist pentagonal geometries PENT(3, 1) and PENT(3, 3) and a 3-GDD of type $6^t 10^1$, $t \geq 3$. Hence there exists a pentagonal geometry PENT(3, $3t + 3$), $t \geq 3$. Again we have already observed that a PENT(3, 3) exists and it was shown in [2] that there is no PENT(3, 6). A PENT(3, 9) with one opposite line pair is also given in [2]. □

The systems constructed in the above theorem contain the maximum number of opposite line pairs among all pentagonal geometries with the same parameters. From Lemma 4 the deficiency graph of a pentagonal geometry with $k = 3$ is a union of (i) complete bipartite graphs $K_{3,3}$ (corresponding to opposite line pairs) and (ii) cubic graphs of girth at least 5, of which the smallest is the Petersen graph with 10 vertices. In the pentagonal geometries PENT(3, $3t - 2$), $t = 1$ or $t \geq 3$, the $6t$ points are partitioned into t opposite line pairs. In the pentagonal geometries PENT(3, $3t + 3$), $t = 0$ or $t \geq 3$, the $6t + 10$ points are partitioned into t opposite line pairs and 10 points which form a PENT(3, 3). This just leaves PENT(3, 9) to be considered. By the same argument as above, there are at most two opposite line pairs. But this is not possible as the following lemma shows.

Lemma 5 *There is no pentagonal geometry PENT(3, 9) with two opposite line pairs.*

Proof A pentagonal geometry PENT(3, 9) has 22 points and 66 lines. Suppose that there are two opposite line pairs. Call the points of one of the opposite line pairs type A and the points of the other opposite line pair type B. The remaining 10 points are type C. There are two lines of type AAA and two lines of type BBB (the opposite line pairs). The remaining 62 lines are of type ABC, ACC, BCC or CCC. Of the remaining pairs to be covered, 36 are of type AB, 60 are of type AC and 60 are of type BC. This leaves $(62 \times 3) - (36 + 60 + 60) = 30$ further pairs which must be of type CC. So there are 36 lines of type ABC, 12 lines of type ACC, 12 lines of type BCC and 2 lines of type CCC. Now consider a point of type C. Its opposite line is of type CCC and since there are no opposite line pairs other than those of types AAA or BBB there are at least as many lines of type CCC as points of type C giving a contradiction. □

So for all admissible r we have an example of a pentagonal geometry PENT(3, r) which is extremal in the sense that it contains the maximum number of opposite line pairs; although zero in the case of PENT(3, 3). It would therefore be of interest to have examples at the other extreme, i.e. with no opposite line pairs. We can use the PENT(3, 3) and our constructional methods using 3-GDDs to obtain infinite linear classes.

Theorem 10 *There exists a pentagonal geometry PENT(3, r) with no opposite line pair for all $r \equiv 3$ or 13 (mod 15).*

Proof In Theorem 7, let $k = 3$ and $r = 3$. There exists a 3-GDD of type 10^u, $u = 3t$ or $3t + 1$, $t \geq 1$. Hence there exists a pentagonal geometry PENT(3, $5u - 2$) for the same values of u. □

We conclude this section with an enumeration result. Trivially PENT(3, 1) is unique and it is easily seen that the Desargues configuration is the unique PENT(3, 3). There is no PENT(3, r) for $r = 4$ or 6. The next value to consider is $r = 7$. From Lemma 3, a pentagonal geometry PENT(3, 7) either has no opposite line pair or the points are partitioned into opposite line pairs. The former possibility was eliminated in [2] by computer search. For the latter possibility the 18 points are partitioned into three opposite line pairs (l_0, m_0), (l_1, m_1), (l_2, m_2). The remaining 36 lines come from a group divisible design with block size 3 and of type 6^3, (i.e. a Latin square of side 6), whose groups are the points of $l_i \cup m_i$, $i = 0, 1, 2$. Thus the number of non-isomorphic PENT(3, 7)s is the number of paratopy or main classes of Latin squares of side 6. There are precisely 12 of these and they are listed in [4]. The deficiency graph of every pentagonal geometry PENT(3, 7) is disconnected, being three copies of the bipartite graph $K_{3,3}$.

4 Block Sizes 4 and 5

Necessary and sufficient conditions for the existence of 4-GDDs with uniform block size was determined in [3].

Proposition 3 (Brouwer, Schrijver & Hanani) *There exists a 4-GDD of type g^u, $u \geq 4$, if and only if*

1. *$g \equiv 1$ or 5 (mod 6) and $u \equiv 1$ or 4 (mod 12), or*
2. *$g \equiv 2$ or 4 (mod 6) and $u \equiv 1$ (mod 3), $(g, u) \neq (2, 4)$, or*
3. *$g \equiv 3$ (mod 6) and $u \equiv 0$ or 1 (mod 4), or*
4. *$g \equiv 0$ (mod 6) with no constraint on u, $(g, u) \neq (6, 4)$.*

From Lemma 1, the number of points in a PENT(4, r) is $3r + 5$ and the necessary condition for existence is $r \equiv 0$ or 1 (mod 4). In arithmetic set density terms we can determine 1/4 of the possible spectrum again by using Theorem 5. From the above there exists a 4-GDD of type 2^{3t+1}, $t \geq 2$. This is a PLS(4, $2t$) of deficiency one and so it follows that there exist pentagonal geometries PENT(4, $8t + 1$), $t \geq 2$. This leaves the existence of a PENT(4, 9) still in doubt but using Theorem 7 the existence of this geometry can also be shown.

Theorem 11 *There exists a pentagonal geometry PENT(4, r) for all $r \equiv 1$ (mod 8).*

Proof In Theorem 7 let $k = 4$ and $r = 1$. There exists a pentagonal geometry PENT(4, 1) and a 4-GDD of type 8^{3t+1}, $t \geq 1$. Hence there exists a PENT(4, $8t + 1$), $t \geq 1$. □

Necessary and sufficient conditions for the existence of 4-GDDs in which all group sizes except one have the same cardinality are not known. However there do exist very many 4-GDDs of this type, see [7]. But it is not the lack of suitable group divisible designs which prevents further progress on the existence spectrum of PENT(4, r)s. As the reader can easily check, putting $k = 4$, $u \equiv 0 \pmod 3$ and $r, s \equiv 1 \pmod 8$ in Theorem 8, and supposing that the relevant GDDs exist, yields further geometries in the residue class 1 (mod 8). In order to obtain pentagonal geometries in the residue classes 0, 4 or 5 (mod 8) by this method we need at least one geometry in one of these classes. To illustrate this point we present two putative constructions.

Construction 1 Suppose that there exists a PENT(4, $4s$) for some $s \geq 2$. Now let $k = 4$ and $r = 4s$ in Theorem 7. There exists a 4-GDD of type $(12s + 5)^u$ where $u = 12t + 1, t \geq 1$ or $u = 12t + 4, t \geq 0$. In the former case we obtain pentagonal geometries PENT(4, $48st + 4s + 20t$), i.e. further infinite linear subclasses of systems in the residue class 0 (mod 4) and in the latter case PENT(4, $48st + 16s + 20t + 5$), systems in the residue class 1 (mod 4).

Construction 2 First we prove the existence of a relevant 4-GDD. From [8], see also [7], there exists a 4-GDD of type $2^{3w}11^1$ for $w \geq 4$. Now in this group divisible design inflate each point by a factor 4, i.e. replace each point by four points and in addition replace each block by a 4-GDD of type 4^4. This is a standard procedure known as Wilson's fundamental construction. We obtain a 4-GDD of type $8^{3w}44^1$ for $w \geq 4$. Now suppose that there exists a PENT(4, 13). Let $r = 1$ and $s = 13$ in Theorem 8. Then there exists PENT(4, $8w + 13$) for $w \geq 4$, which deals with the entire residue class 5 (mod 8) apart from three possible exceptions.

Turning now to block size 5, the number of points in a PENT(5, r) is $4r + 6$ and the necessary condition for existence is $r \equiv 0$ or $1 \pmod 5$. Here we are able to determine 1/2 of the possible existence spectrum, specifically we can deal with the residue class 1 (mod 5) apart from nine possible exceptions. First we need the following result from [9].

Proposition 4 (Ge & Ling) *There exists a 5-GDD of type* 10^{2t+1}, $t \geq 2$, *possibly apart from when* $t \in \{2, 3, 7, 11, 13, 16, 17, 19, 23\}$.

We now have the following theorem.

Theorem 12 *There exists a pentagonal geometry PENT(5, r) for all* $r \equiv 1 \pmod 5$, $r \neq 6$, *possibly apart from* $r \in \{11, 16, 36, 56, 66, 81, 86, 96, 116\}$.

Proof In Theorem 7, let $k = 5$ and $r = 1$. There exists a pentagonal geometry PENT(5, 1) and a 5-GDD of type 10^{2t+1}, $t \geq 2$, possibly apart from the values of t given in Proposition 4. Hence there exists a pentagonal geometry PENT(5, $5t + 1$), $t \geq 2$, again possibly apart from these values. There is no PENT(5, 6) by Theorem 2. □

5 The Case $r = 2k + 1$

This section is concerned with pentagonal geometries PENT(k, r) in which $r = 2k + 1$. From Lemma 3, such a pentagonal geometry either has no opposite line pair or the points are partitioned into opposite line pairs. Consider the latter option. We prove the following characterization theorem.

Theorem 13 *There exists a pentagonal geometry PENT*$(k, 2k+1)$ *whose points are partitioned into opposite line pairs if and only if there exists a set of* $k-2$ *mutually orthogonal Latin squares (MOLS) of side* $2k$.

Proof The number of points in a PENT$(k, 2k + 1)$ is $2k^2$ and the number of lines is $4k^2 + 2k$. Put $r = 1$ in Theorem 6. There exists a pentagonal geometry PENT$(k, 1)$. Hence if there exists a set of $k - 2$ MOLS of order $2k$, then there exists a pentagonal geometry PENT$(k, 2k + 1)$. Further, the k copies of PENT$(k, 1)$ contained in the PENT$(k, 2k + 1)$ will form a partition of the $2k^2$ points into k opposite line pairs. Conversely, if there exists a PENT$(k, 2k+1)$ whose points are partitioned into opposite line pairs, then these line pairs will form k copies of the degenerate pentagonal geometry PENT$(k, 1)$ accounting for $2k$ lines in total. The remaining $4k^2$ lines must then form a set of $k - 2$ MOLS of side $2k$. □

The theoretical maximum number of MOLS of side n is $n - 1$, called a *complete set*. These are known to exist when n is a power of a prime and are equivalent to the existence of a projective plane of order n. We are interested in the case where n is even. Thus when $k = 2^m$, $m \geq 1$, there exist $2^{m+1} - 1 > 2^m - 2$ MOLS of side 2^{m+1}. We therefore have the following infinite class of pentagonal geometries as stated in the theorem below.

Theorem 14 *There exists a pentagonal geometry PENT*$(2^m, 2^{m+1}+1)$ *for all* $m \geq 1$ *whose points are partitioned into opposite line pairs.*

From the table given in [1], the only other value of k for which it is known that the number of MOLS of side $2k$ is at least $k - 2$ is $k = 6$. There exists a set of 5 MOLS of side 12, [11]. So we have the following result.

Theorem 15 *There exists a pentagonal geometry PENT*$(6, 13)$ *whose points are partitioned into opposite line pairs.*

We conclude this section with the remark that the existence of a pentagonal geometry PENT(5, 11) with points partitioned into opposite line pairs, which is one of the "missing" values in Theorem 12, is equivalent to the existence of three MOLS of side 10, one of the most intriguing unanswered questions about Latin squares.

6 Concluding Remarks

As we stated in the introduction, pentagonal geometries with block size 2 were characterised in [2]. The main result of the present paper has been to completely determine the existence spectrum for block size 3. In doing this, we have not been unduly concerned with the structure of the geometries. Nevertheless as we observed, the pentagonal geometries given in Theorem 9 contain the maximum number of opposite line pairs. Then in Theorem 10 we constructed infinite linear classes of PENT$(3, r)$ with no opposite line pair. It would be good to complete the existence spectrum of such pentagonal geometries. In order to do this using the group divisible design constructions it is necessary to find examples other than those in the residue classes 3 or 13 (mod 15) already dealt with.

Also the group divisible design constructions of Theorems 7 and 8 yield pentagonal geometries whose deficiency graph is not connected. The only known pentagonal geometries PENT$(3, r)$ with connected deficiency graph are those for $r = 1$ ($K_{3,3}$), $r = 3$ (Petersen graph) and an example for $r = 13$ given in [2]. It would be of interest to have more examples, indeed infinite families. This would appear to be a priority for future research but new ideas for the constructional methods would seem to be needed.

We have also proved the existence of pentagonal geometries PENT$(4, r)$ for all $r \equiv 1 \pmod 8$. But the situation for $r \equiv 0, 4$ or $5 \pmod 8$ remains completely open. As we indicated, the obstacle is that currently not a single PENT$(4, r)$ is known to exist in these residue classes. Our constructions in Sect. 4 show that the existence of for example a PENT(4, 8), PENT(4, 12) and/or PENT(4, 13) would lead to further infinite classes. The construction of these geometries or proof of their non-existence is also of prime importance.

For $k \geq 4$, although we have been able to construct pentagonal geometries PENT(k, r) for $k = 5$ and $r \equiv 1 \pmod 5$ apart from nine possible exceptions, further progress seems limited because of the lack of knowledge of appropriate group divisible designs. Nevertheless it would be good to deal with these nine "missing" values but that is more a problem of constructing the relevant group divisible designs than the pentagonal geometries.

References

1. R. J. R. Abel, C. J. Colbourn & J. H. Dinitz, Mutually Orthogonal Latin Squares (MOLS), Handbook of Combinatorial Designs, second edition (ed. C. J. Colbourn & J. H. Dinitz), Chapman and Hall/CRC Press, 160–193, 2007.
2. S. Ball, J. Bamberg, A. Devillers & K. Stokes, An alternative way to generalise the pentagon, *J. Combin. Des.* **21** (2013), 163–179.
3. A. E. Brouwer, A. Schrijver & H. Hanani, Group divisible designs with block size 4, *Discrete Math.* **20** (1977), 1–10.

4. C. J. Colbourn, J. H. Dinitz & I. M. Wanless, Latin Squares, Handbook of Combinatorial Designs, second edition (ed. C. J. Colbourn & J. H. Dinitz), Chapman and Hall/CRC Press, 135–152, 2007.
5. C. J. Colbourn, D. G. Hoffman & R. Rees, A new class of group divisible designs with block size three, *J. Combin. Theory Ser. A* **59** (1992), 73–89.
6. C. J. Colbourn & A. Rosa, *Triple Systems*, Clarendon Press, Oxford, 1999.
7. G. Ge, Group divisible designs, Handbook of Combinatorial Designs, second edition (ed. C. J. Colbourn & J. H. Dinitz), Chapman and Hall/CRC Press, 255–260, 2007.
8. G. Ge & A. C. H. Ling, Group divisible designs with block size four and group type $g^u m^1$ for small g, *Discrete Math.* **285** (2004), 97–120.
9. G. Ge & A. C. H. Ling, Asymptotic results on the existence of 4-RGDDs and uniform 5-GDDs, *J. Combin. Des.* **13** (2005), 222–237.
10. H. Hanani, Balanced incomplete block designs and related designs, *Discrete Math.* **11** (1975), 255–369.
11. D. M. Johnson, A. L. Dumage & N. S. Mendelsohn, Orthomorphisms of groups of orthogonal latin squares, *Canad. J. Math.* **13** (1961), 356–372.
12. T. P. Kirkman, On a problem in combinations, *Cambridge and Dublin Math. J.* **2** (1847), 191–204.

The Grothendieck-Teichmüller Group of a Finite Group and G-Dessins d'enfants

Pierre Guillot

Abstract For each finite group G, we define the *Grothendieck-Teichmüller group* of G, denoted $\mathrm{GT}(G)$, and explore its properties. The theory of dessins d'enfants shows that the inverse limit of $\mathrm{GT}(G)$ as G varies can be identified with a group defined by Drinfeld and containing $\mathrm{Gal}(\overline{\mathbb{Q}}/\mathbb{Q})$. We give, in particular, an identification of $\mathrm{GT}(G)$, in the case when G is simple and non-abelian, with a certain very explicit group of permutations that can be analyzed easily. With the help of a computer, we obtain precise information for $G = PSL_2(\mathbb{F}_q)$ when $q \in \{4, 7, 8, 9, 11, 13, 16, 17, 19\}$, and we treat A_7, $PSL_3(\mathbb{F}_3)$ and M_{11}. In the rest of the paper we give a conceptual explanation for the technique which we use in our calculations. It turns out that the classical action of the Grothendieck-Teichmüller group on dessins d'enfants can be refined to an action on "G-dessins", which we define, and this elucidates much of the first part.

1 Introduction

Suppose that Γ is a finite group, generated by two distinguished elements x and y, and such that

(i) Γ has an automorphism θ such that $\theta(x) = y$ and $\theta(y) = x$,
(ii) Γ has an automorphism δ such that $\delta(x) = z$ and $\delta(y) = y$, where z is the element such that $xyz = 1$.

In this situation we define a subgroup $A(\Gamma) \subset Aut(\Gamma)$ as follows: an element $\varphi \in Aut(\Gamma)$ belongs to $A(\Gamma)$, by definition, when

1. $\varphi(x)$ is a conjugate of x^k for some integer k,
2. φ commutes with θ and δ in $Out(\Gamma)$.

(It follows that $\varphi(y)$ is a conjugate of y^k, and likewise for z). The image of $A(\Gamma)$ in $Out(\Gamma)$ will be denoted by $\mathscr{A}(\Gamma)$.

P. Guillot (✉)
IRMA, 7 rue Descartes, 67084 Strasbourg, France
e-mail: guillot@math.unistra.fr

J. Širáň and R. Jajcay (eds.), *Symmetries in Graphs, Maps, and Polytopes*, Springer Proceedings in Mathematics & Statistics 159,
DOI 10.1007/978-3-319-30451-9_8

For any finite group G at all, we shall see that there is a way to construct a group $\overline{G}$ satisfying (i) and (ii), so that it can play the role of Γ (and moreover $\overline{\overline{G}} = \overline{G}$). Thus it makes sense to define $\mathrm{GT}(G) := \mathscr{A}(\overline{G})$. We call it the *Grothendieck-Teichmüller group* of G, and the present paper is dedicated to the study of its properties. We start with a few words of motivation and background.

How $\mathrm{GT}(G)$ varies with G is a discussion which we postpone; for the time being, we take it for granted that it is possible to form the inverse limit

$$\mathrm{GT} := \lim_{G} \mathrm{GT}(G).$$

In [2] we proved the central (for us) result that there is a monomorphism

$$\mathrm{Gal}(\overline{\mathbb{Q}}/\mathbb{Q}) \longrightarrow \mathrm{GT}.$$

Thus GT, with its very brief definition, gives a group-theoretic angle to the study of the absolute Galois group $\mathrm{Gal}(\overline{\mathbb{Q}}/\mathbb{Q})$ of the field $\mathbb{Q}$. A very first step towards understanding GT is to provide information on $\mathrm{GT}(G)$ for some individual choices of G, and this is what we propose to do here.

As an aside, the reader will probably find it useful to know that

$$\lim_{G} Out(\overline{G}) \cong Out(\hat{F}_2),$$

where $\hat{F}_2$ is the profinite completion of the free group F_2 on two generators. Thus GT can be seen as a certain subgroup of $Out(\hat{F}_2)$, and one can show that it can be lifted to a subgroup of $Aut(\hat{F}_2)$. Also, let us indicate that GT coincides with the group denoted $\widehat{\mathrm{GT}}_0$ by Drinfeld in [1] (we shall have nothing to say about the subgroup $\widehat{\mathrm{GT}} \subset \widehat{\mathrm{GT}}_0$, also considered by Drinfeld). All this, and more, is proved in [2].

A good deal of the present paper will in fact pertain to $\mathrm{GT}_1(G)$, which is the subgroup of $\mathrm{GT}(G)$ obtained by restricting condition (1) above to $k = 1$ only. One can show that there is a monomorphism

$$\mathrm{Gal}(\overline{\mathbb{Q}}/\mathbb{Q})' \longrightarrow \mathrm{GT}_1 := \lim_{G} \mathrm{GT}_1(G),$$

where $\mathrm{Gal}(\overline{\mathbb{Q}}/\mathbb{Q})'$ is the derived subgroup of $\mathrm{Gal}(\overline{\mathbb{Q}}/\mathbb{Q})$. So GT_1 can potentially give us information on $\mathrm{Gal}(\overline{\mathbb{Q}}/\mathbb{Q})'$ just like GT can give us information on $\mathrm{Gal}(\overline{\mathbb{Q}}/\mathbb{Q})$, and of course the abelianization $\mathrm{Gal}(\overline{\mathbb{Q}}/\mathbb{Q})/\mathrm{Gal}(\overline{\mathbb{Q}}/\mathbb{Q})' \cong \hat{\mathbb{Z}}^\times$ is well-understood. Here $\hat{\mathbb{Z}}^\times$ is the group of units in the profinite completion of $\mathbb{Z}$.

The following simple example should illuminate the situation. If $G = C_n$, the cyclic group of order n, we have $\overline{C_n} \cong C_n \times C_n$ with its canonical pair of generators. Then $\mathrm{GT}(C_n) \cong (\mathbb{Z}/n)^\times$, directly from the definition, while $\mathrm{GT}_1(C_n)$ is trivial. Letting n vary, we can take the inverse limit and obtain

$$\mathrm{Gal}(\overline{\mathbb{Q}}/\mathbb{Q}) \longrightarrow \lim_{n} \mathrm{GT}(C_n) \cong \hat{\mathbb{Z}}^\times.$$

In turn, this homomorphism can be identified with the celebrated *cyclotomic character*, whose kernel is $\mathrm{Gal}(\overline{\mathbb{Q}}/\mathbb{Q})'$. In a sense, consideration of cyclic groups accounts for what is abelian in $\mathrm{Gal}(\overline{\mathbb{Q}}/\mathbb{Q})$, and we must turn to non-abelian groups and their GT to proceed further.

⋆⋆⋆

The Grothendieck-Teichmüller group is strongly related to the theory of *dessins d'enfants*, which are the object of many papers in these Proceedings (some information on dessins is given below in this Introduction, and more is said in Sect. 6). On the one hand one uses dessins in order to construct the homomorphism from $\mathrm{Gal}(\overline{\mathbb{Q}}/\mathbb{Q})$ into GT and show that it is injective. On the other hand, the group GT can be used to shed light on the action of $\mathrm{Gal}(\overline{\mathbb{Q}}/\mathbb{Q})$ on (isomorphism classes) of dessins. Indeed, when trying to predict whether two dessins belong to the same Galois orbit, one starts by checking a few combinatorial properties which they must have in common: the same number of black vertices, the same number of white vertices, the same number of faces, and the same "monodromy group", for example.

All these are subsumed by the following statement: the action of $\mathrm{Gal}(\overline{\mathbb{Q}}/\mathbb{Q})$ on those dessins with monodromy group G factors via

$$\mathrm{Gal}(\overline{\mathbb{Q}}/\mathbb{Q}) \longrightarrow \mathrm{GT} \longrightarrow \mathrm{GT}(G).$$

For simplicity, say that one is interested in *regular* dessins, those with "maximal symmetry". Then the regular dessins with monodromy group (or automorphism group) G are in bijection with those normal subgroups N of $\overline{G}$ such that $\overline{G}/N \cong G$, and the action of $\mathrm{Gal}(\overline{\mathbb{Q}}/\mathbb{Q})$ factors through the natural action of $\mathrm{GT}(G)$ on these. The combinatorial features above can be recovered from this, and more. This motivates the computation of $\mathrm{GT}(G)$ for a single group G individually.

An example of a finer statement which one can make about the Galois action is the following: if $\mathrm{GT}_1(G) = 1$, then $\mathrm{Gal}(\overline{\mathbb{Q}}/\mathbb{Q})'$ acts trivially on the set of dessins with monodromy G. In different terms, the *moduli field* of such a dessin, that is, the number field F whose fixed subgroup in $\mathrm{Gal}(\overline{\mathbb{Q}}/\mathbb{Q})$ is the stabilizer of the dessin, is an *abelian* extension of $\mathbb{Q}$. (The field F is strongly related to, though sometimes smaller than, the number fields over which the dessin can be defined.)

During the SIGMAP conference, Gareth Jones asked for examples of regular dessins with non-abelian moduli field. A hint for those trying to answer the question is thus that the monodromy group G must satisfy $\mathrm{GT}_1(G) \neq 1$. In the course of this paper we shall see that this rules out $G = A_5$ and $G = D_n$ when n is divisible by 4, among others.

⋆⋆⋆

Let us now describe the contents of the paper. It is in Sect. 2 where, after expanding on the definitions above, we prove that properties of G are reflected in properties of GT_1 (but not GT). For example we establish:

Theorem 1 *If G is a p-group for some prime p, then so is* $\mathrm{GT}_1(G)$*; if G is nilpotent, then so is* $\mathrm{GT}_1(G)$.

The group $\mathrm{GT}(G)/\mathrm{GT}_1(G)$ *is abelian, with exponent dividing that of* $(\mathbb{Z}/N)^\times$*, where N is the order of x or y in $\overline{G}$ (in particular, this exponent may not be a power of p when G is a p-group).*

In Sect. 4 we define a new group $\mathscr{S}(G)$. We hasten to add that when G is non-abelian and simple we shall prove that there is an isomorphism $\mathrm{GT}_1(G) \cong \mathscr{S}(G)$, so the material in that section can be seen at least as a study of the "simple case". However $\mathscr{S}(G)$ is defined for all G, and it is a much easier group to deal with than $\mathrm{GT}_1(G)$. It is described as the intersection, in a permutation group, of a Young subgroup and the centralizer of a few explicit permutations. The first virtue of $\mathscr{S}(G)$ is that it is easy to reason with, leading for example to the next result:

Theorem 2 *Let G be a finite, simple, non-abelian group, and let m be the size of the largest conjugacy class in G. A simple factor occurring in* $\mathrm{GT}_1(G)$ *must be isomorphic to either:*

- C_2,
- C_3,
- *a subquotient of* $Out(G)$,
- *an alternating group A_s where* $s \leq \frac{m^2}{|G|}$.

(We stress that the theorem mentions $Out(G)$, not $Out(\overline{G})$ which is much bigger and would make for a tautological statement.)

It is also easy to compute explicitly with $\mathscr{S}(G)$. The reader should keep in mind that a computer, unleashed after $\mathrm{GT}_1(G)$ by a direct, brute force approach, will not be able to finish its task within a day or without exceeding the memory on a group G whose order is much bigger than 32. Relying only on naive calculations, the author has yet to see a completed example for which the order of $\mathrm{GT}_1(G)$ is anything but $1, 2, 3, 4, 5, 6, 7$. By contrast, the machinery of $\mathscr{S}(G)$ has allowed us to treat, for example, the case of the Mathieu group M_{11} of order 7920, yielding:

Theorem 3 *The direct product of the simple factors of* $\mathrm{GT}_1(M_{11})$ *is*

$$C_2^{465} \times C_3^{46} \times A_5^{10} \times A_6^{9} \times A_7^{10} \times A_8^{4} \times A_9^{4} \times A_{10}^{5} \times A_{11}^{5} \times A_{12} \times A_{14}^{2} \times A_{15}^{4} \times A_{16} \times A_{17}^{3} \times A_{18}^{12} \times A_{19} \times A_{20}^{2} \times A_{23} \times A_{28} \times A_{31} \times A_{33}^{2}.$$

Accordingly, the order of $\mathrm{GT}_1(M_{11})$ *is* $2^{1141} \cdot 3^{407} \cdot 5^{165} \cdot 7^{98} \cdot 11^{43} \cdot 13^{34} \cdot 17^{23} \cdot 19^{8} \cdot 23^{5} \cdot 29^{3} \cdot 31^{3}$.

We also give a complete description of $\mathrm{GT}_1(PSL_2(\mathbb{F}_q))$ for $q \in \{4, 7, 8, 9, 11, 13, 16, 17, 19\}$, and we treat A_7 and $PSL_3(\mathbb{F}_3)$. In Sect. 5 we explain some of the practicalities of the implementation with the open-source computer algebra system GAP.

To move on with our outline, let $\mathscr{P}$ be the set of pairs $(g, h) \in G$ such that $\langle g, h\rangle = G$. We will see that there is a very natural action of $\mathrm{GT}(G)$ on the set $\mathscr{P}/Aut(G)$.

However, the development of the isomorphism between $\mathrm{GT}_1(G)$ and $\mathscr{S}(G)$ relies on the existence of an action of GT(G) on $\mathscr{P}_c$, the set of orbits in $\mathscr{P}$ under the action of the inner automorphisms only (the letter c is for "conjugation"). At first sight this appears rather mysterious, and the arguments are *ad hoc*. In Sect. 6 we give a conceptual explanation.

The key is to bring dessins d'enfants into the picture. Here we must recall that a dessin is essentially a bipartite graph drawn on a compact, oriented surface in such a way that the complement of the graph is a union of topological discs. The (isomorphism classes of) dessins d'enfants are in bijection with many other sets of (isomorphism classes of) objects, notably algebraic curves over $\overline{\mathbb{Q}}$ with a certain ramification property, or étale algebras over $\overline{\mathbb{Q}}(x)$, again with a ramification property. The group $\mathrm{Gal}(\overline{\mathbb{Q}}/\mathbb{Q})$ acts naturally on étale algebras, and this is turned into an action on dessins *via* the said bijection.

In [2] (which is our reference for dessins), we prove that dessins form a category $\mathfrak{Dessins}$, and that the aforementioned bijections can be refined into equivalences of categories. Such a refinement may not seem to bring much new information at first sight, but it is not so. Indeed, with this formalism it is completely straightforward to define the category $G\mathfrak{Dessins}$ of *G-dessins*, that is, dessins equipped with an action of a fixed group G; and we prove the following:

Theorem 4 *The group* $\mathrm{Gal}(\overline{\mathbb{Q}}/\mathbb{Q})$ *acts on the set of isomorphism classes of objects in* $G\mathfrak{Dessins}$*, for any group G.*

Moreover, suppose we consider those regular G-dessins X in $G\mathfrak{Dessins}$ *such that the action gives an isomorphism* $G \longrightarrow Aut(X)$*. Then the set of isomorphism classes of such objects is naturally in bijection with* $\mathscr{P}_c$*, and the latter is endowed with an action of* $\mathrm{Gal}(\overline{\mathbb{Q}}/\mathbb{Q})$.

(The word *regular* will be explained in the text.) It is now much more believable that GT should act on $\mathscr{P}_c$; given that the action of GT on dessins, when restricted to those dessins X such that $Aut(X) \cong G$, factors *via* GT(G), we should not be overly surprised by the discovery made in Sect. 4 that GT(G) does act on $\mathscr{P}_c$.

2 Generalities

We start by expanding on the definitions given in the Introduction. We define $\overline{G}$, the group GT(G) as a subgroup of $Out(\overline{G})$, explain the relationship with $\mathrm{Gal}(\overline{\mathbb{Q}}/\mathbb{Q})$, and prove the most basic properties.

2.1 The Group $\overline{G}$

Let G be a finite group. Whenever N is a subgroup of a group Γ, it will be convenient to say that N *has index G in* Γ when (i) N is normal in Γ and (ii) there is an isomorphism $\Gamma/N \cong G$.

Writing $F_2 = \langle x, y \rangle$ for the free group on two generators x and y, we call N_G the intersection of all the subgroups of F_2 having index G. There are finitely many of these, so the group $\overline{G} := F_2/N_G$ is finite. We usually write x and y for the images of the generators of F_2 in $\overline{G}$, since no confusion should arise.

The following lemma is almost trivial.

Lemma 1 *$\overline{G}$ has the following properties:*

1. *The intersection of all the subgroups of $\overline{G}$ having index G is trivial.*
2. *If Γ is any group such that the intersection of all its subgroups of index G is trivial, and if x' and y' are generators of Γ, then there is a homomorphism $\overline{G} \to \Gamma$ mapping x to x' and y to y'.*
3. *If x' and y' are generators for $\overline{G}$, then there is an automorphism of $\overline{G}$ mapping x to x' and y to y'.*

We turn to the description of a concrete "model" for $\overline{G}$. The key observation is that subgroups of F_2 of index G are in bijection with the orbits of $Aut(G)$ on the set $\mathscr{P}$ of pairs of generators for G; the bijection sends a pair (x', y') to the kernel of the map $F_2 \to G$ sending x to x' and y to y'.

Based on this, we select pairs $(x_1, y_1), \ldots, (x_r, y_r)$ forming a system of representatives for the orbits of $Aut(G)$, that is, with just one pair out of each orbit. (The number $r = r(G)$ was much studied in [4].) Consider then the subgroup $\tilde{G}$ of G^r generated by $x = (x_1, x_2, \ldots, x_r)$ and $y = (y_1, y_2, \ldots, y_r)$. Then it is straightforward to show that $\tilde{G}$ satisfies (2) of Lemma 1 (since the group Γ mentioned there embeds into G^r). This property clearly characterizes $\overline{G}$ as a group with distinguished generators, so there must be an isomorphism $\overline{G} \cong \tilde{G}$ identifying the two elements which we have both called x, and likewise for y. For most of this paper we will consider $\overline{G}$ to be the subgroup of G^r just defined.

Let p_i be the projection onto the ith factor of G^r, restricted to $\overline{G}$. It sends x to x_i and y to y_i, so it is surjective and its kernel K_i has index G. The various K_i's are distinct (by choice of the pairs (x_i, y_i)), so they must constitute the r different subgroups of index G in $\overline{G}$. In particular, they form a characteristic family of subgroups, that is, for any $\varphi \in Aut(\overline{G})$ we must have $\varphi(K_i) = K_{\sigma(i)}$ for some permutation $\sigma \in S_r$.

Finally, we note that $\overline{\overline{G}} = \overline{G}$. Indeed, if we try to construct the model for $\overline{\overline{G}}$ as we have just done with $\overline{G}$, then property (3) of Lemma 1 leaves us only one pair to consider; in other words, $r(\overline{G}) = 1$.

2.2 *The Group* GT(G)

By (3) of Lemma 1, the group $\overline{G}$ has an automorphism θ with $\theta(x) = y$ and $\theta(y) = x$; likewise, $\overline{G}$ possesses an automorphism δ with $\delta(x) = y^{-1}x^{-1}$ and $\delta(y) = y$.

Consider now the elements $\varphi \in Aut(\overline{G})$ satisfying

1. $\varphi(x)$ is a conjugate of x^k for some k prime to the order of $\overline{G}$,
2. φ commutes with θ and δ in $Out(\overline{G})$.

(It follows that $\varphi(y)$ is a conjugate of y^k, and likewise xy is a conjugate of $(xy)^k$.) These form a subgroup of $Aut(\overline{G})$, and its image in $Out(\overline{G})$ will be called $\mathrm{GT}(G)$.

Likewise, we can consider those automorphisms satisfying (1) for $k = 1$ only, as well as (2); they induce a normal subgroup $\mathrm{GT}_1(G)$ of $\mathrm{GT}(G)$.

A complication to keep in mind is that there is no well-defined map on $\mathrm{GT}(G)$ that would associate to φ the number k as above: the latter is not unique, and not even unique modulo the order of x, for some powers of x may well be conjugated to one another. In other words an element of $\mathrm{GT}_1(G)$ may have the property that $\varphi(x)$ is a conjugate of x^k for many values of $k \neq 1$.

It is however true that when $\varphi(x) \sim x^k$ and $\psi(x) \sim x^\ell$, then $\psi \circ \varphi(x) \sim x^{k\ell}$, where we write $a \sim b$ when a and b are conjugate. In particular, since φ^{-1} is a power of φ, we note that $\varphi^{-1}(x) \sim x^{k'}$ where $x^{kk'} \sim x$. If $\psi^{-1}(x) \sim x^{\ell'}$ with $x^{\ell\ell'} \sim x$, then the commutator $[\varphi, \psi]$ takes x to a conjugate of

$$x^{kk'\ell\ell'} = (x^{kk'})^{\ell\ell'} \sim x^{\ell\ell'} \sim x.$$

We have proved that all commutators in $\mathrm{GT}(G)$ must belong to $\mathrm{GT}_1(G)$. Thus we may state:

Lemma 2 *The group* $\mathrm{GT}(G)/\mathrm{GT}_1(G)$ *is an abelian group, of exponent dividing that of* $(\mathbb{Z}/N\mathbb{Z})^\times$*, where* N *is the order of* x *(or* y*) in* $\overline{G}$*.*

The statement about the exponent follows from the fact that $\varphi(x) \sim x^k$ for some $k \in (\mathbb{Z}/N\mathbb{Z})^\times$, whenever $\varphi \in \mathrm{GT}(G)$. So $\varphi^n(x) \sim x^{k^n} = x$ when $k^n = 1$ mod N, and then $\varphi^n \in \mathrm{GT}_1(G)$.

2.3 Inverse Limits

If N is a normal subgroup of F_2 of finite index, we can always find a G such that $N_G \subset N$: indeed it suffices to take $G = F_2/N$. From this one can show that

$$\lim F_2/N_G \cong \hat{F}_2,$$

where $\hat{F}_2$ is the profinite completion of F_2. Here the inverse limit is over the directed set of all the subgroups of the form N_G (with their inclusions). Details for this, and everything else in the next few paragraphs, are provided in [2].

When $N_G \subset N_H$, we have a map $\overline{G} \to \overline{H}$, whose kernel is the intersection of all the subgroups of $\overline{G}$ having index H. In particular, this kernel is a characteristic subgroup, and as a result we have an induced map

$$\mathrm{GT}(G) \longrightarrow \mathrm{GT}(H).$$

Thus it makes sense to talk about the inverse limit $\lim \mathrm{GT}(G)$. Again the indexing set for the limit is the set of the various subgroups N_G, but we prefer to write more suggestively

$$\lim_{G} \mathrm{GT}(G)$$

which we call GT. We also put

$$\mathrm{GT}_1 := \lim_{G} \mathrm{GT}_1(G).$$

2.4 *The Galois Group of* $\mathbb{Q}$

In [2] we prove the existence of a monomorphism

$$\Phi\colon \mathrm{Gal}(\overline{\mathbb{Q}}/\mathbb{Q}) \longrightarrow \mathrm{GT}$$

which is the motivation for the study of GT. Moreover, if $\lambda \in \mathrm{Gal}(\overline{\mathbb{Q}}/\mathbb{Q})$ and if $\varphi = \Phi(\lambda)$, then we can compute for any G an integer k such that $\varphi(x)$ and x^k are conjugate in $\overline{G}$: namely, let N be the order of x, let $\zeta = e^{\frac{2i\pi}{N}}$, and pick k such that $\lambda(\zeta) = \zeta^k$.

We write $\mathrm{Gal}(\overline{\mathbb{Q}}/\mathbb{Q})'$ for the derived subgroup of $\mathrm{Gal}(\overline{\mathbb{Q}}/\mathbb{Q})$ (the closed subgroup generated by the commutators). A celebrated result in number theory asserts that $\mathrm{Gal}(\overline{\mathbb{Q}}/\mathbb{Q})'$ is precisely the subgroup of elements acting trivially on all the roots of unity (this is essentially the Kronecker-Weber theorem, see [6], Chap. 5, Theorem 1.10). As a result, or simply as an application of Lemma 2, there is also a monomorphism

$$\mathrm{Gal}(\overline{\mathbb{Q}}/\mathbb{Q})' \longrightarrow \mathrm{GT}_1.$$

It is surprising that the lemma below seems hard to prove without appealing to $\mathrm{Gal}(\overline{\mathbb{Q}}/\mathbb{Q})$. It is never used on the sequel.

Lemma 3 *Let N be the order of x (or y) in $\overline{G}$. Then for any integer k prime to N, there is $\varphi \in \mathrm{GT}(G)$ such that $\varphi(x)$ is a conjugate of x^k.*

Proof Simply take $\varphi = \Phi(\lambda)$ where $\lambda \in \mathrm{Gal}(\overline{\mathbb{Q}}/\mathbb{Q})$ has the appropriate effect on roots of unity.

2.5 p-Groups and Nilpotent Groups

Proposition 1 *If G is a p-group, then so is* $\mathrm{GT}_1(G)$.

Proof First note that $\overline{G}$ is itself a p-group, being a subgroup of G^r. Let $A'(\overline{G})$ denote the preimage of $\mathrm{GT}_1(\overline{G})$ in $Aut(\overline{G})$. If $A'(\overline{G})$ is not a p-group, then it contains an element φ whose order is a prime $\ell \neq p$.

Consider the elementary abelian p-group $E = \overline{G}/\Phi(\overline{G})$, where $\Phi(\overline{G})$ is the Frattini subgroup of $\overline{G}$, generated by the images $\overline{x}$ and $\overline{y}$ of x and y. The induced action of φ on E is then trivial. It follows from [3], Corollary 3.29 (a result sometimes referred to as the Burnside basis theorem), that the action of φ on $\overline{G}$ is trivial, violating the assumption that the order of φ is ℓ. This contradiction shows that $A'(\overline{G})$ is a p-group, and so also is $\mathrm{GT}_1(G)$.

Proposition 2 *If G and H have coprime orders, we have $\overline{G \times H} \cong \overline{G} \times \overline{H}$ and* $\mathrm{GT}(G\times H) \cong \mathrm{GT}(G) \times \mathrm{GT}(H)$, *as well as* $\mathrm{GT}_1(G\times H) \cong \mathrm{GT}_1(G) \times \mathrm{GT}_1(H)$.

Proof We start with a remark. Whenever a group N has index $G \times H$ in a group Γ, then we can write $N = N' \cap N''$ where N' has index G and N'' has index H, clearly. Now suppose the orders of G and H are coprime, and let us prove the converse. If $N = N' \cap N''$ for such N' and N'', then Γ/N injects in $\Gamma/N' \times \Gamma/N'' \cong G \times H$, and its image surjects onto both G and H. Thus the order of Γ/N is divisible by both $|G|$ and $|H|$ and so by their product, so that $\Gamma/N \cong G \times H$, as we wished to show. Applying this remark to the subgroups of the free group F_2, we deduce that

$$N_{G\times H} = N_G \cap N_H. \tag{*}$$

(Recall that N_G is the intersection of the subgroups of index G, and likewise for N_H and $N_{G\times H}$.)

What is more, $\overline{G}$ and $\overline{H}$ also have coprime orders since they are subgroups of G^r and H^s respectively. Thus we may apply the remark again, and deduce from (*) that $N_{\overline{G\times H}}$ has index $\overline{G} \times \overline{H}$ (being the intersection of a group of index $\overline{G}$ and a group of index $\overline{H}$). This shows that there is an isomorphism $\overline{G \times H} \to \overline{G} \times \overline{H}$.

Next we note that an automorphism of $\overline{G} \times \overline{H}$ must be of the form $\alpha \times \beta$ where $\alpha \in Aut(\overline{G})$ and $\beta \in Aut(\overline{H})$. It follows easily that $\mathrm{GT}(G \times H) \cong \mathrm{GT}(G) \times \mathrm{GT}(H)$.

Corollary 1 *If G is nilpotent, then so is* $\mathrm{GT}_1(G)$.

Proof A finite group is nilpotent precisely when it is a direct product of p-groups.

3 An Elementary Example: Dihedral Groups

In this section we present a computation of $\mathrm{GT}_1(D_n)$ where D_n is the dihedral group of order $2n$ (with details only when n is odd). It is simple enough to be carried out "by hand" to the end, while by contrast the methods developped in the sequel ultimately rely on computers when put to practice. We believe that many features of $\mathrm{GT}_1(G)$ are already visible here.

So let s and t be involutions generating $G = D_n$, and let $R = st$, so that the $2n$ elements of G are the "rotations" R^m and the involutions sR^m, for $0 \leq m < n$. It is easily seen that a pair of generators (x_1, x_2) for G can be taken by an automorphism to one of (s, t), (R, t) or (s, R). As a result, $\overline{G}$ is the subgroup of G^3 generated by $x = (s, R, s)$ and $y = (t, t, R)$.

From now on, we assume that n is odd, and we proceed to prove that $\mathrm{GT}_1(G)$ has order 2. We shall state the corresponding results for other values of n below.

Observations

First we describe the group $\overline{G}$ a to some extent. To do so, we observe that the abelianization of G is C_2, so G^3 has projects onto $C_2^3 = \{(\pm 1, \pm 1, \pm 1)\}$, and looking at the images of x and y we see that $\overline{G}$ maps onto the subgroup of elements (a, b, c) with $abc = 1$. Thus the index of $\overline{G}$ in G^3 is at least 2. However, since $x^2 = (1, R^2, 1)$ and the order of R is odd, we see that $(1, R, 1) \in \overline{G}$; likewise, starting with y and xy, we see that $(1, 1, R)$ and $(R, 1, 1)$ are in $\overline{G}$. There is thus a subgroup $A \cong C_n^3 \subset \overline{G}$, and the order of $\overline{G}$ is a multiple of n^3. Finally, note that A is normal in G^3 and hence, also in $\overline{G}$, and the quotient $\overline{G}/A$ is easily seen to be $C_2 \times C_2$, so the order of $\overline{G}$ is $4n^3$ and its index in G^3 is just 2. In passing we have established a recipe for checking whether an element $(\alpha, \beta, \gamma) \in G^3$ belongs to $\overline{G}$: namely, this is the case if and only if there are an even number of involutions among α, β, γ.

It will be useful to know the centralizer $C_{\overline{G}}(y)$ of y in $\overline{G}$. First off, the centralizer in G^3 is $C_{G^3}(y) = C_2 \times C_2 \times C_n$ generated by $(t, 1, 1)$, $(1, t, 1)$ and $(1, 1, R)$, so it has order $4n$. Since the order of y is $2n$ (using that n is odd), and since there are elements in $C_{G^3}(y)$ which are not in $\overline{G}$, such as $(t, 1, 1)$, we conclude that $C_{\overline{G}}(y) = \langle y \rangle$.

Choices for φ

Now let $\varphi \in Aut(\overline{G})$ represent an element of $\mathrm{GT}_1(G)$. Composing with an inner automorphism if necessary, we may assume that $\varphi(y) = y$, and we know that $\varphi(x) = x'$ can be conjugated to x within $\overline{G}$, and so also within G^3. Put $x' = (s', R', s'')$, where s' and s'' are involutions and $R' = R^{\pm 1}$ is a rotation.

Now suppose ψ is another such automorphism of $\overline{G}$, with $\psi(y) = y$ and $\psi(x) = x''$, a conjugate of x. Then φ and ψ differ by an inner automorphism, or equivalently represent the same element in $\mathrm{GT}_1(G)$, if and only if x' can be conjugated to x'' by an element of $C_{\overline{G}}(y) = \langle y \rangle$.

Here we point out that all the involutions in G are conjugate, and indeed can be conjugated to one another using a power of R: using the notation a^b for $b^{-1}ab$, this follows from $(sR^i)^R = sR^{i+2}$ and the fact that the order of R is odd. Given that $y = (t, t, R)$, we can clearly conjugate x' by a power of y to obtain an element whose third coordinate is any involution we want, say s. In other words, we may assume that $s'' = s$ without loss of generality. Conjugating further by $y^n = (t, t, 1)$ if necessary, we may assume that $R' = R$, that is $x' = (s', R, s)$. Different choices for s' can only lead to different elements of $\mathrm{GT}_1(G)$.

We must have $s' = sR^m$ for an integer m (taken mod n). The next step is to show that there are only two possibilities for m.

The Condition Involving δ

This will be imposed by the condition stating that φ and δ must commute in $Out(\overline{G})$, by definition of $\mathrm{GT}_1(G)$. Recall that $\delta(y) = y = (t, t, R)$ and $\delta(x) = y^{-1}x^{-1} = (ts, tR^{-1}, R^{-1}s) = (R^{-1}, s, t)$. Pick a power of R, say R^p, such that $s^{R^p} = t$. As the element $(t, 1, R^p)$ commutes with y, and $(R, s, s)^{(t,1,R^p)} = (R^{-1}, s, t)$, we conclude that

$$\delta(a, b, c) = (b, a, c)^{(t,1,R^p)},$$

for any $(a, b, c) \in \overline{G}$. Indeed, both sides of this equation define homomorphisms $\overline{G} \to G^3$, and they agree on x and y. The attentive reader will notice that finding a simple expression for δ, replacing the definition in terms of the generators x and y, is a silent but major theme in all the rest of the paper, and the same applies to θ.

We are now able to compute $\delta(\varphi(x)) = \delta(x') = (R^{-1}, -, -)$ (what happens with the second and third coordinates turns out to be irrelevant for the sequel, and would be distracting to look at). On the other hand $\varphi(\delta(x)) = y^{-1}(x')^{-1} = (ts', -, -) = (R^{m-1}, -, -)$. And of course $\delta(\varphi(y)) = \varphi(\delta(y)) = y$.

The condition on φ thus states the existence of $c \in \overline{G}$ such that (i) $y^c = y$, that is c centralizes y, and (ii) $(R^{m-1}, -, -)^c = (R^{-1}, -, -)$. By the observation above, (i) implies $c \in \langle y \rangle$. The element c, in particular, is of the form $(1, -, -)$ or $(t, -, -)$.

Each possibility implies a value for m. Indeed if $c = (1, -, -)$, condition (ii) gives $R^{-1} = R^{m-1}$ so that $m = 0$. The case $c = (t, -, -)$ yields $R = R^{m-1}$ so that $m = 2$.

Existence

We know now that there can be at most two elements in $\mathrm{GT}_1(G)$: the identity and the class of a potential automorphism φ such that $\varphi(y) = y$ and $\varphi(x) = x' = (sR^2, R, s)$. To show that such an automorphism actually exists, we may simply consider conjugation by the element $(t, 1, 1) \in G^3$, which does not belong to $\overline{G}$.

We are left with the task of checking that φ really defines an element of $\mathrm{GT}_1(G)$, that is, it must be verified that φ and θ commute up to an inner automorphism. Recall that $\theta(x) = y$ and $\theta(y) = x$. Using that $x' = x(xy)^2$, a straightforward computation shows that we must find an element which simultaneously conjugates $y(yx)^2 = (tR^{-2}, t, R)$ to $y = (t, t, R)$ and $x = (s, R, s)$ to $x' = (sR^2, R, s)$. For this one may take $(R, 1, 1)$.

We have proved the first part of the following proposition:

Proposition 3 *If n is odd, then the group $\mathrm{GT}_1(D_n)$ has order 2.*

If $n = 2k$ and k is odd, then the group $\mathrm{GT}_1(D_n)$ also has order 2. If k is even, then the group $\mathrm{GT}_1(D_n)$ is trivial.

The rest of the proposition is left as a lengthy exercise. Note that when $n = 2k$, the group $\overline{G}$ has order $4k^3$ and so has index 16 in G^3.

Let us say a word about the image of GT_1 in $\mathrm{GT}_1(D_n)$. Let us use the notation s_n, t_n and R_n for the elements in D_n written s, t, R up to now. There is a homomorphism $D_{nm} \to D_n$ sending s_{nm} to s_n, t_{nm} to t_n, and R_{nm} to R_n. Clearly the induced homomorphism $D_{nm}^3 \to D_n^3$ maps $\overline{D}_{nm}$ onto $\overline{D}_n$. It is a general fact, already mentioned in Sect. 2.3, that in this situation there is a map $\mathrm{GT}_1(D_{nm}) \to \mathrm{GT}_1(D_n)$.

The projection map $\mathrm{GT}_1 \to \mathrm{GT}_1(D_n)$ thus factors through $\mathrm{GT}_1(D_{nm})$ for any m, in particular, through $\mathrm{GT}_1(D_{4n}) = 1$. As a result, the image of GT_1 in $\mathrm{GT}_1(D_n)$ is trivial, for all n. What amounts essentially to the same thing, the inverse limit $\lim_n \mathrm{GT}_1(D_n)$ makes sense here, but sadly, it is trivial.

4 The Case of Simple Groups

For any finite group G, we define a permutation group $\mathscr{S}(G)$. When G is simple and non-abelian, we proceed to show that there is an isomorphism $\mathrm{GT}_1(G) \cong \mathscr{S}(G)$. This is used to analyse the possible simple factors in $\mathrm{GT}_1(G)$ in this case.

4.1 Notation

Let G be a finite group (shortly to be assumed simple and non-abelian, but not at the moment). The following notation will be used throughout this section. Let us emphasize that we make some arbitrary *choices* at the same time.

Let $\mathscr{P}$ denote the set of pairs of elements (g, h) generating G. The group $Aut(G)$ acts on $\mathscr{P}$, and the set $\mathscr{P}/Aut(G)$ of orbits has cardinality r. It will be useful to also work with $\mathscr{P}_c$, the set of orbits under the sole action of the inner automorphisms. We see that $Out(G)$ acts freely on $\mathscr{P}_c$, and $\mathscr{P}_c/Out(G) = \mathscr{P}/Aut(G)$. Thus the set $\mathscr{P}_c$ has cardinality $r|Out(G)|$. (Please note that the actions considered here are on the left. In this section the composition on $Aut(G)$ is $\alpha\beta = \alpha \circ \beta$.)

For each $1 \leq i \leq r$ we choose a representative $(x_i, y_i) \in \mathscr{P}$ for the ith orbit in $\mathscr{P}/Aut(G)$, in some ordering.

The $Aut(G)$-orbit of $(g, h) \in \mathscr{P}$ will be denoted $[g, h]$, while its orbit under $Inn(G)$ will be written $[g, h]_c$ (the brackets will never denote commutators in this section). In this notation the action of $\alpha \in Out(G)$ on $[g, h]_c$ is $\alpha \cdot [g, h]_c = [\alpha(g), \alpha(h)]_c$. The elements of $\mathscr{P}_c$ are precisely enumerated as $[\alpha(x_i), \alpha(y_i)]_c$ for $\alpha \in Out(G)$ and $1 \leq i \leq r$. The following is immediate.

Lemma 4 *There is a bijection of sets*

$$\mathscr{P}_c \longrightarrow Out(G) \times \mathscr{P}/Aut(G),$$

sending $[\alpha(x_i), \alpha(y_i)]_c$ to the pair $(\alpha, [x_i, y_i])$. It is equivariant with respect to the $Out(G)$ actions, where on the right hand side the group $Out(G)$ acts trivially on $\mathscr{P}/Aut(G)$ and by left multiplication on itself.

Finally, each pair (x_i, y_i) determines a unique homomorphism $p_i \colon \overline{G} \to G$ sending x and y to x_i and y_i respectively (recall that x and y are the canonical generators of $\overline{G}$). The kernel of p_i will be written K_i.

Remark 1 In the literature on dessins d'enfants or related group-theoretical topics, one often works with triples (x, y, z) of elements generating a finite group G and satisfying $xyz = 1$. Our $\mathscr{P}$ can be identified with the set of such triples, clearly, and $\mathscr{P}_c$ can be thought of as the set of triples up to simultaneous (triple) conjugation. Likewise the rest of this section could be developed with this (hardly different) point of view.

4.2 An Action of $Out(\overline{G})$ on $\mathscr{P}_c$

First recall (from the discussion in Sect. 2.1) that the K_i's form a characteristic family of subgroups in $\overline{G}$; in other words, for any $\varphi \in Aut(\overline{G})$ and any i there is a $\sigma(i)$ such that $\varphi(K_i) = K_{\sigma(i)}$. The permutation $\sigma \in S_r$ thus obtained from φ may occasionally be denoted $\sigma(\varphi)$.

Next, the composition $\overline{G} \xrightarrow{\varphi} \overline{G} \xrightarrow{p_{\sigma(i)}} G$ factors through p_i, thus resulting in an automorphism $G \to G$ which we denote φ_i. There is the composition formula

$$(\psi \circ \varphi)_i = \psi_{\sigma(i)} \circ \varphi_i \quad \text{where} \quad \sigma = \sigma(\varphi).$$

Also observe that when φ is inner, the permutation $\sigma(\varphi)$ is the identity, and each φ_i is inner.

We can now define an action of $Aut(\overline{G})$ on $Out(G) \times \mathscr{P}/Aut(G)$ by setting

$$\varphi \cdot (\alpha, [x_i, y_i]) = (\alpha\varphi_i^{-1}, [x_{\sigma(i)}, y_{\sigma(i)}]).$$

(On the second factor this is the natural action on $\mathscr{P}/Aut(G)$, which can be identified with the set of the K_i's). In this expression we have written φ_i^{-1} for the class of this automorphism in $Out(G)$. It is clear that this is indeed an action and that it factors through $Out(\overline{G})$.

Crucially, we notice that the action just defined commutes with that of $Out(G)$ (by left multiplication on itself and trivially on $\mathscr{P}/Aut(G)$.)

By Lemma 4, we also have an action of $Out(\overline{G})$ on $\mathscr{P}_c$, which commutes with the natural action of $Out(G)$. We have in particular

$$\varphi \cdot [x_i, y_i]_c = [\varphi_i^{-1}(x_{\sigma(i)}), \varphi_i^{-1}(y_{\sigma(i)})]_c,$$

which we will use more often than the general expression

$$\varphi \cdot ([\alpha(x_i), \alpha(y_i)]_c) = [\alpha\varphi_i^{-1}(x_{\sigma(i)}), \alpha\varphi_i^{-1}(y_{\sigma(i)})]_c.$$

(Commutation with $Out(G)$ means that the first formula implies the second anyway.)
This discussion is summarized in the next proposition.

Proposition 4 *There is a homomorphism*

$$Out(\overline{G}) \longrightarrow C_S(Out(G)),$$

where $S = S(\mathscr{P}_c)$ is the symmetric group of the set $\mathscr{P}_c$, and $C_S(Out(G))$ is the centralizer of $Out(G)$ for the natural action. The corresponding action of $Out(\overline{G})$ on $\mathscr{P}_c$ satisfies in particular

$$\varphi \cdot [x_i, y_i]_c = [\varphi_i^{-1}(x_{\sigma(i)}), \varphi_i^{-1}(y_{\sigma(i)})]_c,$$

(Note that when $\varphi \in Out(\overline{G})$, or $\varphi \in Aut(\overline{G})$, we simply write $\varphi \cdot [g, h]_c$ for the action.)

Remark 2 The specific choices we have made for the elements x_i and y_i actually matter here. The curious reader may prove the following. Using the material below on simple groups, one can at least establish that when G is simple, the permutation $\sigma(\varphi)$ and the automorphisms φ_i are uniquely defined (once one has a numbering of the elements of $\mathscr{P}/Aut(G)$), and so the action on $Out(G) \times \mathscr{P}_c$ can be defined without making choices. However, even in this case, the bijection of Lemma 4 depends on choices.

We need to identify the permutations of $\mathscr{P}_c$ induced by certain specific elements of $Aut(\overline{G})$. We start with the automorphism θ of $\overline{G}$ which exchanges x and y. The next lemma is perhaps not surprising, but its proof requires some care.

Lemma 5 *For all $g, h \in G$, we have*

$$\theta \cdot [g, h]_c = [h, g]_c\,.$$

Proof Consider the following commutative diagram:

$$\begin{array}{ccc} \overline{G} & \xrightarrow{\theta} & \overline{G} \\ \downarrow & & \downarrow p_{\sigma(i)} \\ G & \xrightarrow{\theta_i} & G \end{array}$$

Recall that $\theta(x) = y$, $\theta(y) = x$, $p_i(x) = x_i$, $p_i(y) = y_i$, and likewise for $p_{\sigma(i)}$. Thus we see that $\theta_i(x_i) = y_{\sigma(i)}$ and $\theta_i(y_i) = x_{\sigma(i)}$, which we may profitably rewrite as $\theta_i^{-1}(y_{\sigma(i)}) = x_i$ and $\theta_i^{-1}(x_{\sigma(i)}) = y_i$.

Following the definitions, we see that

$$\theta \cdot [x_i, y_i]_c = [y_i, x_i]_c.$$

Thus the proposed formula is true at least when $[g, h]_c = [x_i, y_i]_c$ for some i. However, the map $[g, h]_c \mapsto [h, g]_c$ commutes with the action of $Out(G)$, as does $[g, h]_c \mapsto \theta \cdot [g, h]_c$, so these two maps have to agree.

In the exact same vein, we have

Lemma 6 *For all $g, h \in G$, we have*

$$\delta \cdot [g, h]_c = [h^{-1}g^{-1}, h]_c.$$

We leave the proof to the reader (recall that $\delta(x) = y^{-1}x^{-1}$ and $\delta(y) = y$.)

The next (and last) lemma involves the action of $G \times G$ on $\mathscr{P}_c$ by conjugation.

Lemma 7 *Let $\varphi \in Aut(\overline{G})$ be such that $\varphi(x)$ is conjugate to x, and $\varphi(y)$ is conjugate to y. Then the action of φ on $\mathscr{P}_c$ preserves the orbits of $G \times G$.*

Remark 3 (on our cavalier use of the word "orbit" here). While $G \times G$ acts on itself by conjugation, the action does not restrict to the subset $\mathscr{P}$. As a result it does not make sense to speak of the orbits of $G \times G$ on $\mathscr{P}$, let alone $\mathscr{P}_c$. However, it does make sense to ask whether two elements of $\mathscr{P}$ lie in the same $G \times G$-orbit (that is, orbit on $G \times G$); it also makes sense to ask whether two elements of $\mathscr{P}_c$ are the images of two elements of $\mathscr{P}$ in the same $G \times G$-orbit: very explicitly $[a_1, b_1]_c$ and $[a_2, b_2]_c$ are thus related if a_1 and a_2 are conjugate and b_1 and b_2 are conjugate, a relation which is well-defined.

This is how the notion of an "orbit" should be interpreted in the lemma and in related statements that follow.

In Sect. 4.5 we give an example where the elements of $\mathscr{P}_c$ lying in the same $G \times G$-"orbit" are grouped together into blocks; one of these blocks is of size 10 while G has order 168, showing that the blocks are not actual orbits.

Proof (Proof of Lemma 7) The action of $Out(G)$ on $\mathscr{P}_c$ preserves the $G \times G$-orbits, clearly, so it suffices to show that $\varphi \cdot [x_i, y_i]_c$ is in the same $G \times G$-orbit as $[x_i, y_i]_c$ for each index i.

By assumption $\varphi(x) = x^g$ so $\varphi_i(x_i) = p_{\sigma(i)}(\varphi(x)) = x_{\sigma(i)}^{g'}$ where $g' = p_{\sigma(i)}(g)$. It follows that $\varphi_i^{-1}(x_{\sigma(i)})$ is conjugate to x_i. Likewise, $\varphi_i(y_{\sigma(i)})$ is conjugate to y_i, and in the end we have indeed shown that $\varphi_i^{-1} \cdot [x_{\sigma(i)}, y_{\sigma(i)}]_c$ is in the $G \times G$-orbit of $[x_i, y_i]_c$.

4.3 The Group $\mathscr{S}(G)$

Let us use the notation θ and δ for the permutations induced on $\mathscr{P}_c$ by the automorphisms denoted by the same symbols in $Aut(\overline{G})$. They generate a subgroup $\langle \theta, \delta \rangle$

in $S(\mathscr{P}_c)$, the symmetric group of the set $\mathscr{P}_c$. One can check the identities $\theta^2 = 1$, $\delta^2 = 1$, $\delta\theta\delta = \theta\delta\theta$, and it follows that $\langle\theta, \delta\rangle$ is a homomorphic image of S_3.

We define $\mathscr{S}(G)$ to be the subgroup of $S(\mathscr{P}_c)$ of those permutations that commute with the action of $Out(G) \times \langle\theta, \delta\rangle$, and preserve the $G \times G$-orbits (bearing Remark 3 in mind). Thus $\mathscr{S}(G)$ is the intersection of the centralizer of a certain subgroup on the one hand, and a Young subgroup of $S(\mathscr{P}_c)$ on the other hand. (By "Young subgroup" we mean the product of symmetric groups associated to a partition of a set, here corresponding to the $G \times G$-orbits.)

For any group G, we have a map $\mathrm{GT}_1(G) \longrightarrow \mathscr{S}(G)$, by the lemmas just established. The rest of this section is dedicated to the proof of:

Theorem 5 *When G is simple and non-abelian, the map*

$$\mathrm{GT}_1(G) \longrightarrow \mathscr{S}(G)$$

is an isomorphism.

Recall that we have a model of $\overline{G}$ as the subgroup of the cartesian product G^r generated by $x = (x_1, x_2, \dots, x_r)$ and $y = (y_1, y_2, \dots, y_r)$. The map p_i is then just the projection onto the ith factor. In [4] one finds a proof of the following

Lemma 8 *Let G be a nonabelian, simple finite group. Then*

1. *The group $\overline{G}$ is all of G^r.*
2. *The normal subgroups of G^r are those of the form $\prod_I G_i$ for some $I \subset \{1, \dots, r\}$ (where G_i is the ith embedded copy of G in G^r.)*
3. *As a result, the maximal, proper normal subgroups of G^r are those of the form $\prod_{i\neq j} G_i$ for some j. This is precisely K_j.*

In the rest of this section G will always be nonabelian and simple, as well as finite. Let us add:

Lemma 9 *The automorphisms of G^r are as follows:*

1. *$Aut(G^r) \cong Aut(G) \wr S_r$.*
2. *$Out(G^r) \cong Out(G) \wr S_r$.*
3. *The action of $Out(\overline{G})$ on $\mathscr{P}_c$ is faithful.*

Proof Considering the action on the K_j's, we obtain a map $Aut(G^r) \to S_r$ which is clearly split surjective. Now suppose φ is an automorphism of G^r preserving all the K_j's. By taking intersections, we see that φ preserves all the normal subgroups of G^r, including G_1 and $K_1 \cong G^{r-1}$, these two satisfying $G_1 \times K_1 = G^r$. By induction, it is immediate that φ is of the form $\alpha_1 \times \cdots \times \alpha_r$. This proves (1).

An inner automorphism of G^r is the direct product of inner automorphisms of each G_i, so we have also (2).

Now suppose $\varphi \in Aut(\overline{G})$ acts trivially on $\mathscr{P}_c$. Then it must also act trivially on $\mathscr{P}/Aut(G)$ (the map $\mathscr{P}_c \to \mathscr{P}/Aut(G)$ is equivariant for this action). It follows that $\sigma(\varphi)$ is the trivial permutation of S_r. The map considered in (2) then sends φ to $(\varphi_1, \varphi_2, \ldots \varphi_r) \in Out(G)^r$, in a notation which is consistent with our earlier use of φ_i. As the action on $\mathscr{P}_c$ is trivial, we see that φ_i must be inner (that is, it represents the trivial element in $Out(G)$), so (2) implies that φ is itself inner. This concludes the proof of the lemma.

If we combine (3) of the lemma with Proposition 4, we see that $Out(\overline{G})$ injects into $C_S(Out(G))$, the centralizer of $Out(G)$ in $S = S(\mathscr{P}_c)$. However, since the action of $Out(G)$ on $\mathscr{P}_c$ is free with r orbits, its centralizer is itself a wreath product $Out(G) \wr S_r$. Comparing orders, we conclude that $Out(\overline{G})$ maps isomorphically onto $C_S(Out(G))$ via the action on $\mathscr{P}_c$.

It remains to check that the conditions defining $\mathrm{GT}_1(G)$ as a subgroup of $Out(\overline{G})$ correspond to what is stated in the theorem. This is immediate for the commutation with θ and δ. Lemma 7 does half the remaining work, by showing that the elements of $\mathrm{GT}_1(G)$ must preserve the $G \times G$-orbits in $\mathscr{P}_c$. The proof will be concluded by establishing the converse.

Indeed, if the action of $\varphi \in Aut(\overline{G})$ is such that $[\varphi_i^{-1}(x_{\sigma(i)}), \varphi_i^{-1}(y_{\sigma(i)})]_c$ is in the $G \times G$-orbits of $[x_i, y_i]_c$ for all $1 \le i \le r$ then $x_{\sigma(i)}$ and $\varphi_i(x_i)$ are conjugate; in other words $p_{\sigma(i)}(x)$ and $p_{\sigma(i)}(\varphi(x))$ are conjugate. If we recall that $\overline{G} = G^r$ and each p_j is just the projection onto the jth factor, then we see immediately that x and $\varphi(x)$ are conjugate in G^r, so in $\overline{G}$. Likewise for y. This concludes the proof of Theorem 5.

4.4 Properties of $\mathscr{S}(G)$

We write $S = S(\mathscr{P}_c)$ and $H = Out(G) \times \langle \theta, \delta \rangle$, while Y is the Young subgroup of those permutations in S which preserve the $G \times G$-orbits on $\mathscr{P}_c$. We have $\mathscr{S}(G) = C_S(H) \cap Y$.

Proposition 5 *The group $\mathscr{S}(G)$ is a product of wreath products $E_k \wr S_{r_k}$ where $\sum_k r_k = r$ and E_k is a subquotient of H.*

Moreover, each integer r_k satisfies

$$r_k \le \frac{|Z|m^2}{|G|}$$

where m is the size of the largest conjugacy class in G, and Z is the centre of G.

Of course we will mostly use this proposition when G is simple and non-abelian, so that $|Z| = 1$.

Proof We number arbitrarily the "orbits" $P_1, P_2, \ldots$ of $G \times G$ on $\mathscr{P}_c$. The use of quotes here refers to Remark 3. Other orbits in this proof are genuine.

Every orbit X of H has an ordered "partition" into the subsets $X^{(1)} = X \cap P_1$, $X^{(2)} = X \cap P_2, \ldots$ some of which may be empty. Call two of these H-orbits X_1

and X_2 *equivalent* when there is an H-equivariant bijection $X_1 \to X_2$ mapping $X_1^{(i)}$ onto $X_2^{(i)}$ for each i. Finally, a *block* is a subset of $\mathscr{P}_c$ obtained as the union of the H-orbits inside one equivalence class.

The image of an H-orbit under an element of $\mathscr{S}(G)$ is another H-orbit which is equivalent to the original one. It follows that $\mathscr{S}(G)$ preserves the blocks, as does H. Moreover, this allows for a decomposition of $\mathscr{S}(G)$ as a direct product of groups, one for each block: namely, if B is a block, define $\mathscr{S}(G)_B = C_{S(B)}(H) \cap Y_B$ where Y_B is the Young subgroup corresponding to the partition of B by the subsets $B \cap P_i$; then $\mathscr{S}(G)$ is the direct product of the various groups $\mathscr{S}(G)_B$.

Suppose that the H-orbits in the block B are $X_1, \ldots, X_s$. These are permuted by $\mathscr{S}(G)$, or $\mathscr{S}(G)_B$, yielding a homomorphism $\mathscr{S}(G)_B \to S_s$ which is easily seen to be split surjective. The kernel of this homomorphism is a direct product of s copies of the group E of self-equivalences of X_1 (in the above sense). The latter is a subgroup of the automorphism group of X_1 as an H-set; if $X_1 \cong H/K$ for some subgroup K of H, then this automorphism group is $N_H(K)/K$. This completes the description of $\mathscr{S}(G)_B$ as a wreath product $E \wr S_s$ where E is a subquotient of H.

There remains to prove the bound on r_k. If X_1 is an H-orbit, then an H-equivariant bijection $X_1 \to X_2$ is entirely determined by the image of a single point $p \in X_1$; if this bijection is to afford an equivalence between X_1 and X_2, then this image must be taken in the $G \times G$-"orbit" of p. As a result, there are no more orbits equivalent to X_1 than elements in the largest $G \times G$-"orbit". A $G \times G$-"orbit" on $\mathscr{P}$ has size $\leq m^2$; on the other hand as G/Z acts freely on $\mathscr{P}$, the fibres of the map $\mathscr{P} \to \mathscr{P}_c$ have size $|G|/|Z|$. Thus we see that a $G \times G$-"orbit" on $\mathscr{P}_c$ has size $\leq |Z|m^2/|G|$.

Keeping in mind that $\langle \theta, \delta \rangle$ is a homomorphic image of S_3, we draw:

Corollary 2 *A simple factor occurring in $\mathscr{S}(G)$ must be isomorphic to either:*

- C_2,
- C_3,
- *a subquotient of $Out(G)$,*
- *an alternating group A_s where $s \leq \frac{|Z|m^2}{|G|}$.*

Combining this with Theorem 5 yields a description of the possible simple factors in $\mathrm{GT}_1(G)$ when G is non-abelian and simple (namely, those in the corollary). Also using that $\mathrm{GT}(G)/\mathrm{GT}_1(G)$ is abelian (Lemma 2), we draw:

Corollary 3 *Let G be non-abelian and simple. Then a simple factor occurring in* $\mathrm{GT}(G)$ *must be isomorphic to either:*

- *a cyclic group,*
- *a subquotient of $Out(G)$,*
- *an alternating group A_s where $s \leq \frac{m^2}{|G|}$.*

Note that the classification of finite simple groups implies by inspection that the group $Out(G)$ is always solvable, so if one accepts this result then we conclude that the list of simple factors reduces to cyclic and alternating groups.

In any case, we can consider those non-abelian simple groups such that $Out(G)$ has order 1, 2 or 4: this includes the alternating groups for $n \geq 5$, almost all Chevalley groups over fields of prime order, and all 26 sporadic groups. The results above show that $\mathrm{GT}_1(G)$ can only have, as simple factors, the groups C_2 and C_3 as well as some alternating groups. In $\mathrm{GT}(G)$ one may encounter further cyclic groups.

4.5 A Complete Example

Take $G = PSL_3(\mathbb{F}_2)$, a simple group of order 168 with $Out(G) = C_2 = \langle\alpha\rangle$. The following information is obtained with the help of GAP.

There are 114 elements in $\mathscr{P}_c$; in the following some arbitrary numbering is used for them. Looking at the action of $G \times G$ we obtain the following partition of $\{1, \ldots, 114\}$:

$$\{1, 10, 27, 28, 96, 106\}, \{2, 11, 52, 57, 82, 86\}, \{3, 12\}, \{62, 63\}, \{107, 109\}$$
$$\{4, 13, 29, 49, 51, 54, 56, 70, 101, 102\}, \{5, 6, 31, 32, 71, 72\}, \{7, 34, 75\}, \{8, 84, 97\},$$
$$\{9, 36, 53, 73, 83, 85\}, \{14, 16, 38, 60, 91, 114\}, \{15, 61, 76\}, \{17, 98, 110\},$$
$$\{18, 40, 58, 59, 78, 99\}, \{19, 21\}, \{20, 41, 43\}, \{22, 42, 64\}, \{39, 77, 90\},$$
$$\{23, 25, 44, 48, 67, 79, 87, 103, 108, 111\}, \{24, 45, 68, 94, 105, 112\}, \{33, 35, 74\},$$
$$\{26, 46, 69, 89, 93, 104\}, \{30, 37, 50, 55, 100, 113\}, \{47, 65, 66, 80, 81, 88, 92, 95\}.$$

There are 4 subsets of size 2; 8 of size 3; 9 of size 6; one of size 8 and two of size 10. Thus $Y \cong C_2^4 \times S_3^8 \times S_6^9 \times S_8 \times S_{10}^2$.

Then we compute

- the permutation induced by α: (1,11) (2,10) (3,12) (4,101) (5,16) (6,14) (7,15) (8,17) (9,18) (13,102) (19,21) (20, 22) (23,48) (24,26) (25,103) (27,52) (28,57) (29,70) (30,100) (31,60) (32,38) (33, 77) (34,61) (35,39) (36,59) (37,113) (40,85) (41,42) (43,64) (44,87) (45,104) (46, 94) (47,88) (49,56) (50,55) (51,54) (53,58) (62,63) (65,80) (66,95) (67,111) (68, 93) (69,105) (71,114) (72,91) (73,78) (74,90) (75,76) (79,108) (81,92) (82,96) (83, 99) (84,110) (86,106) (89,112) (97,98) (107,109).
- the permutation induced by θ: (1,5) (2,14) (3,19) (4,23) (6,10) (7,15) (8,20) (9,24) (11,16) (12,21) (13,25) (17, 22) (18,26) (27,31) (28,72) (29,67) (30,100) (32,96) (34,61) (36,68) (38,82) (40, 69) (41,97) (42,98) (43,84) (44,49) (45,53) (46,99) (47,88) (48,101) (50,55) (51, 79) (52,60) (54,108) (56,87) (57,91) (58,104) (59,93) (62,107) (63,109) (64,110) (66, 95) (70,111) (71,106) (73,112) (75,76) (78,89) (81,92) (83,94) (85,105) (86,114) (102, 103).
- the permutation induced by δ: (2,3) (4,106) (5,35) (6,8) (7,85) (10,12) (13,100) (14,17) (15,40) (16,39) (19,22) (20,21) (23,68) (24,65) (25,103) (26,80) (27,52) (28,70) (29,57) (30,102) (31,75) (34, 73) (36,71) (37,96) (41,63) (42,62) (44,87) (45,111) (46,107) (47,88) (48,93) (49, 55) (50,56) (51,54) (53,72) (58,91) (59,114)

(60,76) (61,78) (66,108) (67,104) (69, 105) (74,84) (79,95) (81,92) (82,113) (83,97) (86,101) (89,112) (90,110) (94,109) (98, 99).

We can then ask GAP to compute $\mathrm{GT}_1(G)$ as the intersection of Y and the centralizer of the three permutations above. We find that $\mathrm{GT}_1(G)$ has order 512; its centre Z is elementary abelian of order 32; and $\mathrm{GT}_1(G)/Z$ is elementary abelian of order 16. In fact, finer use of GAP as described below allows us to improve this very last step of the computation, showing that $\mathrm{GT}_1(G) \cong C_2^3 \times D_8^2$, where D_8 is the dihedral group of order 8.

5 Computing Explicitly

In this Section we provide details on the use of the computer algebra system GAP in order to apply our results about $\mathscr{S}(G)$. No doubt many readers who are not computer enclined will wish to skip most of this, and we encourage them to browse the results themselves in Sects. 5.3 and 5.4.

We have chosen to give the explanation in a mathematical discourse interspersed with GAP commands. We feel that the readers having little familiarity with computational group theory will be able to understand what follows, while an opportinity is given to get a sense of "what is feasible with just one command" (and by contrast, what requires more effort). On the other hand, we find it useful to indicate some relevant GAP commands to those readers who will wish to implement their own calculations.

5.1 Computing $\mathrm{GT}_1(G)$

The first task is to construct $\overline{G}$ given G, and it is straightforward. We provide some details solely with the purpose of indicating some GAP functions to the reader. One builds the automorphism group of G using `AutomorphismGroup(G)`, then converts its generators into automorphisms of $G \times G$, see `DirectProduct(G, G)` and also the function `GroupHomomorphismByImages`.

Having thus constructed the group `A` of these automorphisms of $G \times G$, one appeals to `OrbitsDomain(A, GG)` (where `GG` is $G \times G$) to find the orbits. For each orbit `orb`, pick a representative `orb[1]` (which is an element of $G \times G$), extract the two elements `x` and `y` comprising the pair, and check whether we have the equality `Subgroup(G, [x, y]) = G`. If not, discard the orbit.

Picking representatives in the r remaining orbits, one constructs $\overline{G}$ as a subgroup of G^r, using `DirectProduct` again.

As for $\mathrm{GT}_1(G)$, we will rely on the next lemma.

Lemma 10 *Let C_x and C_y be the centralizers of the canonical generators x and y of $\overline{G}$, respectively. To each element $\varphi \in \mathrm{GT}_1(G)$ we may associate a unique double* *coset $D \in C_x\backslash\overline{G}/C_y$. In fact, if φ is induced by the automorphism $\tilde{\varphi}$ of $\overline{G}$ satisfying $x \mapsto x^f$ and $y \mapsto y$, then D is the double coset of f.*

Moreover, in the same notation, we have $1 \in C_x f\theta(f)\, C_y^{\theta(f)}$. Conversely if $\tilde{\varphi} \in Aut(\overline{G})$ satisfies $x \mapsto x^f$ and $y \mapsto y$ for some f such that $1 \in C_x f\theta(f)\, C_y^{\theta(f)}$, then the induced $\varphi \in Out(\overline{G})$ commutes with θ.

Proof By definition φ can be induced by such an automorphism. If f_1 and f_2 are both possible choices for f, yielding $\tilde{\varphi}_1$ and $\tilde{\varphi}_2$ both inducing φ, then we see that $\tilde{\varphi}_2 = c_t \circ \tilde{\varphi}_1$ where c_t is conjugation by t. It follows that $y^t = y$ so $t \in C_y$, and that $x^{f_2} = x^{f_1 t}$ so $s := f_1 t f_2^{-1} \in C_x$. In the end $f_1 = s f_2 t^{-1}$ so f_1 and f_2 are in the same double coset. The converse is obvious.

We turn to the last statement, which follows from the fact that $\tilde{\varphi} \circ \theta$ and $\theta \circ \tilde{\varphi}$ must differ by an inner automorphism, by definition of $\mathrm{GT}_1(G)$. So there must exist an element t such that $y^t = y^{\theta(f)}$ and $x^{ft} = x$. We see that $s = ft \in C_x$ and $u = t\theta(f)^{-1} \in C_y$, so that $1 = s^{-1} f u \theta(f) \in C_x f C_y \theta(f) = C_x f \theta(f)\, C_y^{\theta(f)}$. Again the converse is left to the reader.

This suggest the following method to compute $\mathrm{GT}_1(G)$. First, compute the centralizers `Cx:= Centralizer(GB, x)` and `Cy:= Centralizer(GB, y)`, where `GB` is $\overline{G}$. Then compute `DoubleCosets(GB, Cx, Cy)` (which is an optimized process in GAP). Now filter the double cosets, by excluding $C_x f C_y$ if the unit of $\overline{G}$ is not in `DoubleCoset(Cx, f*fth, Cy^fth)`, where `fth` is $\theta(f)$. Also exclude f if x^f and y generate a proper subgroup of $\overline{G}$.

The next step is to go through the remaining double cosets, and with each representative f define $\tilde{\varphi}$ with `GroupHomomorphismByImages(GB,GB, [x, y], [x^f, y])` (this is automatically well-defined by (3) of Lemma 1; however, the construction of $\tilde{\varphi}$ by GAP is surprisingly time-consuming, which is the reason for filtering out as many candidates for f as possible before reaching this stage.)

Finally, keep only those homomorphisms commuting with δ in $Out(\overline{G})$, which may be checked with `IsInnerAutomorphism(phi*delta*phi^(-1) *delta^(-1))`.

At this point, we have a list of automorphisms of $\overline{G}$ representing the elements of $\mathrm{GT}_1(G)$ with no repetition; the number of these automorphisms is the order of $\mathrm{GT}_1(G)$. Finding the group structure of $\mathrm{GT}_1(G)$ can be achieved, rather slowly, with the help of the commands

```
A:= AutomorphismGroup(GB);
int:= InnerAutomorphismsAutomorphismGroup(A);
quo:= NaturalHomomorphismByNormalSubgroup(A, int);
```

One can then create the list of all the elements `quo(phi)` where `phi` is taken from our list of automorphisms, and ask GAP to describe the group they generate.

5.2 Computing $\mathcal{S}(G)$

This is several orders of magnitude faster than computing $GT_1(G)$.

The first step is to construct $\mathcal{P}_c$. For this, one builds $G \times G$ and the embedded diagonal copy of G in $G \times G$, and one appeals to `OrbitsDomain(diagG, GG)`. As above, one filters out an orbit if a representative pair (x, y) fails to generate all of G. We obtain $\mathcal{P}_c$ as a GAP list, say `pairsconj`. Its length is $\ell = r\,|Out(G)|$.

Then one must build the orbits of $G \times G$ on $\mathcal{P}_c$, say stored as a list of lists of indices, those indices refering to `pairsconj`. It is impossible to rely on `OrbitsDomain` for the reasons given in Remark 3, since we are not dealing with genuine orbits; instead, for each `pairorbit` taken in `pairsconj`, we take the representative `pair:= pairorbit[1]`, then get its `C:= ConjugacyClass(GG, pair)` where `GG` is $G \times G$. Then we run through all the indices, and those corresponding to elements of $\mathcal{P}_c$ which happen to be in `C` we group together. After this has been done for each `pairorbit`, we have a list of lists of indices partitioning the set $\{1, \ldots, \ell\}$; for example in Sect. 4.5 we gave the corresponding partition of $\{1 \ldots 114\}$ when $G = PSL_3(\mathbb{F}_2)$. The corresponding Young subgroup `Y` may be constructed easily, but we will argue that it is not the best way to go.

Before turning to this though, we mention that we must construct two further permutations of $\{1, \ldots, \ell\}$ corresponding to θ and δ, and for this we follow Lemmas 5 and 6. We then do the same for each generator of $Out(G)$ which is not inner, and compute the corresponding permutations. We let GAP know that we call `H` the subgroup of S_ℓ generated by all these elements.

It is possible at this stage to ask GAP to compute

```
Intersection(Centralizer(SymmetricGroup(ell), H), Y);
```

but except in very small examples, the calculation will simply take too long (and will likely exhaust the available memory.)

Instead, we partially implement the ideas of Proposition 5 and its proof. Let us use the notation introduced in that proof. First we compute the orbits of H with `OrbitsDomain(H, [1..ell])`. Then we group these orbits according to a looser equivalence relation than the one used in the proof of Proposition 5: we call two orbits X_1 and X_2 equivalent if (i) they are isomorphic as H-sets, which in practice is checked by verifying whether the corresponding stabilizers are conjugate in H, *cf* `Stabilizer(H, orbit[1])` and `ConjugacyClassesSubgroups(H)` which together allow to associate to each orbit the position of the conjugacy class of the stabilizer in some numbering; and (ii) the *cardinality* of $X_1 \cap P_i$ is equal to the cardinality of $X_2 \cap P_i$, for each index i. The union of the orbits in one equivalence class we call a *packet*, and each packet is a union of the "blocks" defined in the aforementioned proof.

Now each packet is H-invariant, and $\mathcal{S}(G)$ splits as a direct product corresponding to the packets, which we see by arguing as we did in the proof of Proposition 5 with blocks. We can compute the image H' of H in the symmetric group of each packet by applying `RestrictedPermNC(g, packet)` to each generator `g` of H. Also, the Young subgroup Y' corresponding to the intersections of the P_i's with the packet is readily created in GAP.

It is now possible to ask directly for the computation of the centralizer of H', and its intersection with Y'. The product of all the resulting groups, for all packets, is $\mathscr{S}(G)$. For each factor in this product, we can prompt GAP for the composition series, see for example `DisplayCompositionSeries(factor)`. (Finer information, such as `StructureDescription(factor)`, can still take a very long time.)

5.3 Simple Groups of Small Order

We shall give information on $\mathrm{GT}_1(G)$ for 12 simple groups of small size. We start with the PSL_2 family (recall that $PSL_2(\mathbb{F}_4) \cong PSL_2(\mathbb{F}_5) \cong A_5$, $PSL_2(\mathbb{F}_7) \cong PSL_3(\mathbb{F}_2)$ and $PSL_2(\mathbb{F}_9) \cong A_6$). Write D_8 for the dihedral group of order 8.

Theorem 6 *We have:*

- *(Order 60)* $\mathrm{GT}_1(PSL_2(\mathbb{F}_4))$ *is trivial.*
- *(Order 168)* $\mathrm{GT}_1(PSL_2(\mathbb{F}_7)) \cong C_2^3 \times D_8^2$.
- *(Order 360)* $\mathrm{GT}_1(PSL_2(\mathbb{F}_9)) \cong C_2^{12} \times D_8$.
- *(Order 504)* $\mathrm{GT}_1(PSL_2(\mathbb{F}_8))$ *is trivial.*
- *(Order 660)* $\mathrm{GT}_1(PSL_2(\mathbb{F}_{11})) \cong C_2^{27} \times D_8^7$.
- *(Order 1092)* $\mathrm{GT}_1(PSL_2(\mathbb{F}_{13})) \cong C_2^{54} \times D_8^{17}$.
- *(Order 2448)* $\mathrm{GT}_1(PSL_2(\mathbb{F}_{17})) \cong C_2^{104} \times D_8^{50}$.
- *(Order 3420)* $\mathrm{GT}_1(PSL_2(\mathbb{F}_{19})) \cong C_2^{133} \times D_8^{74}$.
- *(Order 4080)* $\mathrm{GT}_1(PSL_2(\mathbb{F}_{16}))$ *is trivial.*

It seems tempting to conjecture that $\mathrm{GT}_1(PSL_2(\mathbb{F}_q)) \cong C_2^a \times D_8^b$, with $a = b = 0$ when q is a power of 2[1]. Let us now turn to the simple group of order 5616:

Theorem 7 *The group* $\mathrm{GT}_1(PSL_3(\mathbb{F}_3))$ *is isomorphic to*

$$C_2^{26} \times D_8^6 \times S_3^4 \times S_4^{21} \times S_6^{12} \times S_7^6 \times S_8^3 \times S_9^{11} \times A^3 \times B^8 \times C^5$$

where

$$A = (((C_2 \times C_2 \times C_2 \times C_2 \times C_2) \rtimes A_6) \rtimes C_2) \rtimes C_2,$$
$$B = ((((C_2 \times D_8) \rtimes C_2) \rtimes C_3) \rtimes C_2) \rtimes C_2,$$

and

$$C = (((C_2 \times C_2 \times C_2 \times C_2) \rtimes A_5) \rtimes C_2).$$

---

[1] Added in proof: the author now has a proof of this fact in general. Details to appear.

Two more simple groups from our intention remain. It has not been possible to obtain a complete description of $\mathrm{GT}_1(G)$ for these (in a reasonable amount of time), no doubt because of the appearance of simple factors of the form A_s for s large (18 and above). At least we have been able to find the corresponding simple factors.

Theorem 8 *The direct product of the simple factors of* $\mathrm{GT}_1(A_7)$ *is*

$$C_2^{152} \times C_3^{15} \times A_5^3 \times A_6^3 \times A_7 \times A_8^2 \times A_{10} \times A_{18}.$$

Theorem 9 *The direct product of the simple factors of* $\mathrm{GT}_1(M_{11})$ *is*

$$C_2^{465} \times C_3^{46} \times A_5^{10} \times A_6^9 \times A_7^{10} \times A_8^4 \times A_9^4 \times A_{10}^5 \times A_{11}^5 \times A_{12} \times A_{14}^2 \times A_{15}^4 \times A_{16} \times A_{17}^3 \times A_{18}^{12} \times A_{19} \times A_{20}^2 \times A_{23} \times A_{28} \times A_{31} \times A_{33}^2.$$

5.4 p-Groups

We are going to give information on the orders of $\mathrm{GT}_1(G)$ and $\mathscr{S}(G)$ for a few 2-groups. We offer the following table:

n	Number of groups	max $\lvert\mathrm{GT}_1(G)\rvert$	max $\lvert\mathscr{S}(G)\rvert$	Same order (1 or 2)
3	2	1	1	2 (2)
4	6	1	1	6 (6)
5	17	2	1	14 (14)
6	50	4	2	40 (40)
7	159	4	16	111 (109)

The first column contains a number n, indicating that the row is about groups of order 2^n. Among all these, we select those which can be generated by 2 elements, and which are non-abelian (otherwise $\mathrm{GT}_1(G)$ is not interesting). The next column gives the number of groups which we have kept. The largest order for $\mathrm{GT}_1(G)$, when G runs among those groups, is recorded in the next column, followed by the maximum for $|\mathscr{S}(G)|$. Finally, we give the number of groups for which $|\mathrm{GT}_1(G)| = |\mathscr{S}(G)|$ followed in parenthesis by the number of groups for which this common order is 1 or 2 (so that $\mathrm{GT}_1(G)$ and $\mathscr{S}(G)$ are at least abstractly isomorphic in these cases.)

It may be difficult to comment sensibly on this small-scale analysis. We venture to say that $\mathscr{S}(G)$ is "quite often" abstractly isomorphic to $\mathrm{GT}_1(G)$ (though not always), and that the order of $\mathrm{GT}_1(G)$ seem to grow very slowly with that of G (attempts to prove theoretical bounds on the order of $\mathrm{GT}_1(G)$ have produced very bad results, and there seems to be a phenomenon to understand here.)

6 Dessins d'enfants

Keeping the notation introduced before, we have seen that there is a natural action of $\mathrm{GT}(G)$ on the set $\mathscr{P}/Aut(G)$. However, in the process of constructing a map from $\mathrm{GT}_1(G)$ to $\mathscr{S}(G)$, we have also defined an action of $\mathrm{GT}(G)$ (in fact, of $Out(\overline{G})$) on $\mathscr{P}_c$, the set of pairs in $\mathscr{P}$ up to conjugation. This seems a little *ad hoc*. In this section we provide a partial "explanation", showing that at least for $\mathrm{Gal}(\overline{\mathbb{Q}}/\mathbb{Q})$ (a close cousin of $\mathrm{GT}(G)$) there are very good reasons for an action to exist on $\mathscr{P}_c$.

6.1 The Category of Dessins

Dessins d'enfants are essentially bipartite graphs drawn on compact, oriented surfaces, forming the 1-skeleton of a CW-complex structure (in particular, the complement of the graph is homeomorphic to a disjoint union of open discs). For precise definitions, and for all statements about dessins to follow, we refer to [2].

Dessins form a category $\mathfrak{Dessins}$, and the first striking results of the theory are equivalences between $\mathfrak{Dessins}$ and various other categories. To name but the most important one: algebraic curves over $\mathbb{C}$ with appropriate ramification, étale algebras over $\mathbb{C}(x)$ with a ramification condition, the other two categories obtained by replacing $\mathbb{C}$ by $\overline{\mathbb{Q}}$, and the category of finite sets with a right action of F_2, to be denoted simply by $\mathfrak{Sets}_{xy}$ (our usual notation for the generators of F_2 being x, y.)

Some dessins are called *regular*. Many definitions are possible; in $\mathfrak{Sets}_{xy}$, an object is regular if and only it is isomorphic to one of the following particular form: take a finite group G with two distinguished generators x and y, and let F_2 act on G on the right via the canonical homomorphism $F_2 \to G$, using right multiplication. The regular dessins (G, x, y) and (G', x', y') (as we will denote them) are isomorphic in $\mathfrak{Sets}_{xy}$ if and only if there is a group isomorphism $G \to G'$ with $x \mapsto x'$, $y \mapsto y'$, as the reader may check.

The automorphism group of the regular dessin (G, x, y) is then G itself, acting on itself by left multiplication. As a result of this discussion, we see that the set of isomorphism classes of regular dessins with automorphism group isomorphic to a fixed group G is in bijection with $\mathscr{P}/Aut(G)$, the set of pairs of elements generating G under the action of $Aut(G)$.

Thus we can rephrase the fact that $\mathrm{GT}(G)$ acts on $\mathscr{P}/Aut(G)$ by saying that $\mathrm{GT}(G)$ acts on the isomorphism classes of regular dessins with automorphism group G. What is more, one can define an action of $\mathrm{GT} = \lim_G \mathrm{GT}(G)$ on the isomorphism classes of all dessins, regular or not (*loc cit*).

The other fundamental result is the fact that there is also a natural action of $\mathrm{Gal}(\overline{\mathbb{Q}}/\mathbb{Q})$ on isomorphism classes of dessins. This action factorizes through the map $\varphi\colon \mathrm{Gal}(\overline{\mathbb{Q}}/\mathbb{Q}) \longrightarrow \mathrm{GT}$ (and is indeed instrumental in the very definition of φ); the classical result due to Belyi and Grothendieck that the action of $\mathrm{Gal}(\overline{\mathbb{Q}}/\mathbb{Q})$

is faithful implies then the injectivity of this homomorphism. In turn one can show that GT also acts faithfully, and even that the action on regular dessins is already faithful (Theorem 5.7 in [2]).

6.2 *Γ-Dessins*

In [2] we have defined a certain category $\mathfrak{Etale}(\overline{\mathbb{Q}}(x))$, whose objects are étale algebras over $\overline{\mathbb{Q}}(x)$ satisfying a certain ramification property, and we have built an equivalence of categories between $\mathfrak{Dessins}$ and $\mathfrak{Etale}(\overline{\mathbb{Q}}(x))$. The category $\mathfrak{Etale}(\overline{\mathbb{Q}}(x))$ is the one to use in order to define the action of $\mathrm{Gal}(\overline{\mathbb{Q}}/\mathbb{Q})$ on isomorphism classes of dessins.

What is more, we proved in *loc cit* that each $\lambda \in \mathrm{Gal}(\overline{\mathbb{Q}}/\mathbb{Q})$ actually defines a *functor* $F_\lambda\colon \mathfrak{Etale}(\overline{\mathbb{Q}}(x)) \longrightarrow \mathfrak{Etale}(\overline{\mathbb{Q}}(x))$, inducing the action. This gives more precise information, as we proceed to show. We shall work in a very general context, not for the sheer pleasure of writing abstract nonsense, but out of necessity: the equivalence between $\mathfrak{Dessins}$ (or the very practical $\mathfrak{Sets}_{xy}$) and $\mathfrak{Etale}(\overline{\mathbb{Q}}(x))$ is given by a zig-zag of explicit functors, whose inverses we know very little about beyond the fact that their existence is garanteed by the axiom of choice. This prevents us from being more direct and concrete.

So let $\mathfrak{C}$ be any category at all. Given a group Γ, we can define the category $\Gamma\mathfrak{C}$ whose objects are the pairs (X, ρ) where X is an object of $\mathfrak{C}$ and $\rho\colon \Gamma \to Aut_{\mathfrak{C}}(X)$ is a group homomorphism. (In the sequel we shall write $Aut(X)$ rather than $Aut_{\mathfrak{C}}(X)$ when no confusion can arise). The morphisms $(X, \rho) \to (Y, \rho')$ in $\Gamma\mathfrak{C}$ are those morphisms $f\colon X \to Y$ in $\mathfrak{C}$ which are equivariant in the sense that the following diagram commutes, for any $g \in \Gamma$:

$$\begin{array}{ccc} X & \xrightarrow{\rho(g)} & X \\ {\scriptstyle f}\downarrow & & \downarrow{\scriptstyle f} \\ Y & \xrightarrow{\rho'(g)} & Y \end{array}$$

Now let F be a self-equivalence of $\mathfrak{C}$. The operation $[X] \mapsto [F(X)]$ gives a permutation of the set of equivalence classes of objects in $\mathfrak{C}$, where we have written $[X]$ for the class of X.

However, more is true. Simply assuming that F is a functor from $\mathfrak{C}$ to itself, there is an induced homomorphism $a_X\colon Aut(X) \to Aut(F(X))$, and we can employ it to construct a self-functor $\tilde{F}$ of $\Gamma\mathfrak{C}$. On objects this is defined as $\tilde{F}(X, \rho) = (F(X), a_X \circ \rho)$, while on morphisms $\tilde{F}$ is simply the restriction of F. One checks readily that $\tilde{F}$ is indeed a functor, that $\widetilde{F \circ G} = \tilde{F} \circ \tilde{G}$, and that $\tilde{F}$ is the identity of $\Gamma\mathfrak{C}$ if F is the identity of $\mathfrak{C}$. In particular, if F is a self-equivalence of $\mathfrak{C}$, then $\tilde{F}$ is a self-equivalence of $\Gamma\mathfrak{C}$.

The comments we have made on F then apply to $\tilde{F}$: there is an induced permutation of the set of isomorphism classes of objects in $\Gamma\mathfrak{C}$. Since this discussion was conducted purely in the language of categories, it is clear that $\mathfrak{C}$ can be replaced by any category equivalent to it.

We also point out that the group $Aut(\Gamma)$ acts on the isomorphism classes of objects, by the rule $\alpha \cdot (X, \rho) \equiv (X, \rho \circ \alpha)$ (for $\alpha \in Aut(\gamma)$). When α is inner, we see readily that this action is trivial (in fact if α is conjugation by t, then $\rho(t) \colon X \to X$ is an isomorphism in $\Gamma\mathfrak{C}$ between (X, ρ) and $(X, \rho \circ \alpha)$). Thus we have an induced action of $Out(\Gamma)$, and it commutes visibly with the permutation induced by $\tilde{F}$.

Coming back to $\mathfrak{Etale}(\overline{\mathbb{Q}}(x))$, where we know that the action of $\lambda \in \mathrm{Gal}(\overline{\mathbb{Q}}/\mathbb{Q})$ is *via* some functor F_λ, we conclude:

Proposition 6 *There is an action of* $\mathrm{Gal}(\overline{\mathbb{Q}}/\mathbb{Q})$ *on the set of isomorphism classes of objects in* $\Gamma\mathfrak{Dessins}$, *for any group* Γ. *The same holds with* $\mathfrak{Dessins}$ *replaced by any equivalent category, such as* $\mathfrak{Sets}_{xy}$.

There is also an action of $Out(\Gamma)$ *on the same classes of objects, and the two actions commute.*

Moreover, the forgetful functor $\Gamma\mathfrak{Dessins} \to \mathfrak{Dessins}$ *induces a* $\mathrm{Gal}(\overline{\mathbb{Q}}/\mathbb{Q})$-*equivariant map between the sets of isomorphism classes.*

Finally, these actions restrict to the set of isomorphism classes of faithful, equivariant dessins.

The last sentence mentions *faithful* equivariant dessins, that is, those objects (X, ρ) in $\Gamma\mathfrak{Dessins}$ such that ρ is injective. The statement holds obviously.

We point out that the group Γ need not be finite here. This is one reason for not calling it G, which in this paper usually denotes a finite group. Still, the reader may complain that in the Abstract and Introduction, we mentioned G-dessins rather than Γ-dessins. The point is that we wanted to develop the properties of Γ-dessins in general, then consider a regular dessin X with automorphism group denoted G as in the rest of the paper, and *then* regard it as a G-dessin, that is, putting $\Gamma = G$ ultimately.

The reader may wish to skip ahead to Sect. 6.4 where two concrete examples of Γ-dessins are presented (with Γ a finite, cyclic group). However, a little result is required in order to prove that they are not isomorphic to one another, and we turn to this easy point now.

6.3 Γ-Objects in $\mathfrak{Sets}_{xy}$

As happens with many other concepts, the category $\mathfrak{Sets}_{xy}$ provides the most clean-cut statements about Γ-dessins. Recall from the definitions that an object in $\Gamma\mathfrak{Sets}_{xy}$ is a finite set with an action of F_2 on the right, and a commuting action of Γ on the left (since we usually define the composition in $Aut(X)$ such that it acts on the left on X, in general.)

Let us call an object in $\Gamma\mathfrak{Sets}_{xy}$ *regular* when it is regular as an object of $\mathfrak{Sets}_{xy}$ (that is, the concept ignores the Γ-action.)

Proposition 7 *Let G and Γ be groups. Consider the objects in $\Gamma\mathfrak{Sets}_{xy}$ which are regular and have automorphism group isomorphic to G. Then the set of isomorphism classes of such objects in $\Gamma\mathfrak{Sets}_{xy}$ is in bijection with the set of triples (g, h, φ) where $\langle g, h\rangle = G$ and $\varphi\colon \Gamma \to G$ is a homomorphism, modulo the relation*

$$(g, h, \varphi_1) \sim (g', h', \varphi_2)$$

which holds (by definition) if and only if there is an automorphism $\alpha \in Aut(G)$ and $t \in G$ such that $\alpha(g) = g'$, $\alpha(h) = h'$, and $\alpha(\varphi_1(\gamma)) = t^{-1}\varphi_2(\gamma)t$, for all $\gamma \in \Gamma$.

Proof It is clear from the definitions that a triple (g, h, φ) does define a regular object in $\Gamma\mathfrak{Sets}_{xy}$, and that each such object can be obtained in this way. What needs to be checked is the condition expressing that (g, h, φ_1) and (g', h', φ_2) yield isomorphic objects in $\Gamma\mathfrak{Sets}_{xy}$.

So assume that there is $\alpha_1 : G \to G$ giving such an isomorphism; that is, α is a map of sets which is equivariant with respect to both the F_2-actions on the right, and the Γ-actions on the left. Let $t = \alpha_1(1)$ and define $\alpha_2 \colon G \to G$ by $\alpha_2(g) = t^{-1}g$. Finally, put $\alpha = \alpha_2 \circ \alpha_1$. We have $\alpha(1) = 1$, and as we know that α commutes with the F_2-actions on the right, implying, in particular, that $\alpha(x) = \alpha(1 \cdot x) = \alpha(1) \cdot x' = x'$ and $\alpha(y) = y'$, it follows easily that α is in fact a homomorphism. Moreover, for $\gamma \in \Gamma$ we have $\alpha(\varphi_1(\gamma)) = \alpha_2(\alpha_1(\varphi_1(\gamma) \cdot 1)) = \alpha_2(\varphi_2(\gamma) \cdot \alpha_1(1)) = t^{-1}\varphi_2(\gamma)t$.

It is straightforward to reverse this argument and prove, conversely, that whenever α and t exist, the objects defined by (g, h, φ_1) and (g', h', φ_2) are indeed isomorphic in $\Gamma\mathfrak{Sets}_{xy}$.

For $\Gamma = 1$ we find, as we already know, that the set of isomorphism classes of regular dessins with G as automorphism group is in bijection with $\mathscr{P}/Aut(G)$. Of more interest is the following:

Corollary 4 *For $\Gamma = G$, the set of isomorphism classes in $G\mathfrak{Sets}_{xy}$ of faithful, regular objects, with G as automorphism group, is in bijection with $\mathscr{P}_c$.*

In particular $\mathrm{Gal}(\overline{\mathbb{Q}}/\mathbb{Q})$ *acts on $\mathscr{P}_c$ as well as $\mathscr{P}/Aut(G)$, and the map*

$$\mathscr{P}_c \longrightarrow \mathscr{P}/Aut(G)$$

is $\mathrm{Gal}(\overline{\mathbb{Q}}/\mathbb{Q})$*-equivariant.*

Proof Start with an object (g, h, φ). Since we consider only faithful objects, and since $\Gamma = G$ here, the map φ is an automorphism. Taking $\alpha = \varphi^{-1}$ in the proposition, we see that the same object is represented by (g', h', id) where $g' = \varphi^{-1}(g)$, $h' = \varphi^{-1}(h)$ (and id is the identity map).

We conclude that the objects under consideration can be represented by triples of the form (g, h, id). By the proposition, (g_1, h_1, id) and (g_2, h_2, id) represent isomorphic objects precisely when there is an automorphism α of the form $\alpha(\gamma) = t^{-1}\gamma t$, that is an inner automorphism, taking g_1 to g_2 and h_1 to h_2. In the end the set of isomorphism classes is precisely $\mathscr{P}_c$.

Of course this falls short of a proof that GT, let alone $\mathrm{GT}(G)$ or $Out(\overline{G})$, acts on $\mathscr{P}_c$, but this corollary makes the result much less surprising and much more natural. Also note that our *ad hoc* arguments to the effect that $Out(\overline{G})$ does act on this set do not guarantee any compatibility with the action of the image of $\mathrm{Gal}(\overline{\mathbb{Q}}/\mathbb{Q}) \to Out(\overline{G})$.

6.4 A Complete Example; Cyclic Dessins

We follow the usual workflow for dessins. We start with an informal picture of a dessin on the sphere:

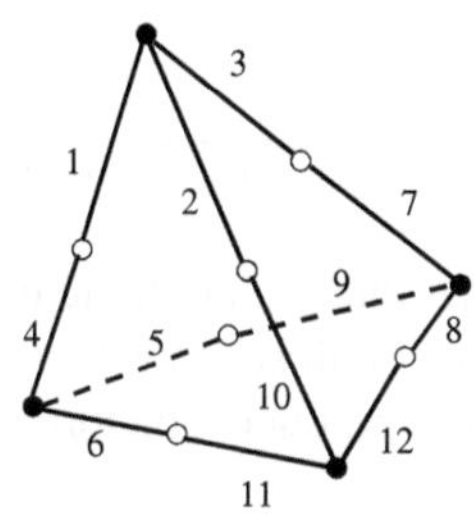

The theory guarantees that enough information is conveyed in the picture to define an object unambiguously. To proceed with this, we move to $\mathfrak{Sets}_{xy}$. Having numbered the "darts" (=edge between a black vertex and a white vertex), we write down the two permutations x and y of the set $\{1, \dots, 12\}$ which take each dart to the next one in the positive rotation around the incident black, resp. white, vertex. These are

$$x = (123)(456)(789)(10, 11, 12) \quad \text{and} \quad y = (14)(2, 10)(37)(59)(6, 11)(8, 12).$$

Then $X = (\{1, \dots, 12\}, x, y)$ is our object in $\mathfrak{Sets}_{xy}$. The subgroup G of S_{12} generated by x and y has order 12, and is in fact isomorphic to A_4; it acts freely and transitively, and it follows that X is regular. For the rest of the discussion, we identify G with the set $\{1, \dots, 12\}$, by identifying $g \in G$ with the image of 1 under g, and the natural action of G on this set is by right multiplication.

The automorphism group of X is given by all the multiplications by elements of G *on the left* on $\{1, \dots, 12\} = G$. Thus this group is (isomorphic to) G itself. For example the permutation $\tilde{x}$ of the set $\{1, \dots, 12\}$ corresponding to the action of x by left multiplication is the unique automorphism of X taking 1 to 2; from the picture we know that this must be the rotation around the black vertex incident with the dart 1, that is

$$\tilde{x} = (123)(4, 10, 7)(6, 12, 9)(11, 8, 5).$$

(This can also be checked by computation). A similar reasoning gives

$$\tilde{y} = (14)(8,12)(2,5)(3,6)(10,9)(11,7).$$

See Example 3.8 in [2] for more details.

Let $C_3 = \langle r \rangle$ be the cyclic group of order 3, and let us define C_3-dessins with X as the underlying dessin. Define two homomorphisms $\varphi_1, \varphi_2 \colon C_3 \to G$ by $\varphi_1(r) = x$ and $\varphi_2(r) = x^{-1}$. Then $X_1 = (G, x, y, \varphi_1)$ and (G, x, y, φ_2) are C_3-dessins.

Let us show that they are not isomorphic. If they were, by Proposition 7 there would exist $\alpha \in Aut(G)$ and $t \in G$ such that $\alpha(x) = x$, $\alpha(y) = y$ and $\alpha(\varphi_1(r)) = t^{-1}\varphi_2(r)t$. The first two conditions impose $\alpha = Id$ of course, and the last one reads $x = t^{-1}x^{-1}t$. However x and x^{-1} are not conjugate in G, so t cannot exist.

To explore the Galois action, we continue with the usual workflow, and compute a Belyi map. A possible choice is

$$f(z) = -64\frac{(z^3+1)^3}{(z^3-8)^3 z^3}$$

(taken from [5]). The simple fact that the coefficients of f are in $\mathbb{Q}$ means that X is fixed by $\mathrm{Gal}(\overline{\mathbb{Q}}/\mathbb{Q})$. However, we shall see that X_1 and X_2 are not.

First we can ask a computer to produce a picture of $f^{-1}([0,1])$.

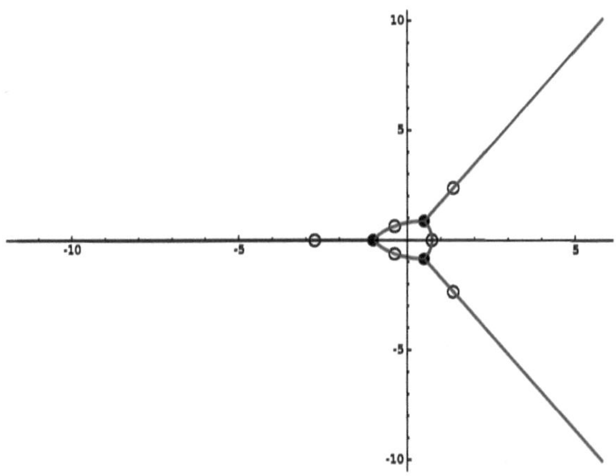

Here the white vertex on the real axis is $w = -(1+\sqrt{3})$, a root of $w^2+2w-2 = 0$. The dart from ∞ to w we number as 1, and we number all the others so that x and y are as above.

We know that there must exist Moebius transformations inducing the actions of $\tilde{x}$ and $\tilde{y}$ as above, and we find easily that they are

$$\mu_x \colon z \mapsto j^2 z \quad \text{and} \quad \mu_y \colon z \mapsto \frac{-z+2}{z+1}$$

respectively, where $j = e^{\frac{2i\pi}{3}}$. (The use of j^2 rather than j mirrors the fact that we consider the positive rotation around ∞, which is also the clockwise rotation around 0). We can now think of X as the Belyi pair $(\mathbb{P}^1, f)$ and of $Aut(X)$ as the group generated by μ_x and μ_y. The C_3-dessins X_1 and X_2 are obtained from X by throwing in the homomorphism $C_3 \to Aut(X)$ mapping r to μ_x or μ_x^{-1}.

If $\lambda \in \mathrm{Gal}(\overline{\mathbb{Q}}/\mathbb{Q})$ satisfies $\lambda(j) = j^2$ (and there are such elements!), then the action of λ exchanges X_1 and X_2. [In fact, since the dessin X is fixed by $\mathrm{Gal}(\overline{\mathbb{Q}}/\mathbb{Q})$, we have a homomorphism $\mathrm{Gal}(\overline{\mathbb{Q}}/\mathbb{Q}) \to Out(Aut(X))$. The non-trivial element of $Out(A_4) \cong C_2$ sends x to x^{-1} and y to $y^{-1} = y$.]

In a nutshell: *the tetrahedron can be made into a C_3-dessin in two non-isomorphic ways, by picking a rotation of order 3 whose axis carries a black vertex and the centre of the opposite face; the two choices are enabled by the two possibilities for the orientation; these are interchanged by the action of* $\mathrm{Gal}(\overline{\mathbb{Q}}/\mathbb{Q})$.

More generally, let us call a dessin *cyclic* when it is a C_n-equivariant dessin for some n. Starting with a regular dessin (G, x, y), a cyclic structure on it is simply given by an element of G of order dividing n; the non-isomorphic cyclic structures correspond to the conjugacy classes of such elements. If the dessin is fixed by the Galois group, then $\mathrm{Gal}(\overline{\mathbb{Q}}/\mathbb{Q})$ permutes these conjugacy classes.

7 Concluding Comments

In this paper we have explored the properties of the groups $\mathrm{GT}(G)$ and $\mathscr{S}(G)$, from first principles. Motivation for this is twofold: on the one hand, each of these acts on the set of (isomorphism classes of) regular dessins with G as automorphism group, extending the action of $\mathrm{Gal}(\overline{\mathbb{Q}}/\mathbb{Q})$, and this is our finest information about the latter action; on the other hand, $\mathrm{Gal}(\overline{\mathbb{Q}}/\mathbb{Q})$ injects in the inverse limit GT of all the groups $\mathrm{GT}(G)$. By contrast one cannot form an inverse limit with the groups $\mathscr{S}(G)$, but they are much easier to compute with, and in good cases $\mathscr{S}(G)$ agrees with $\mathrm{GT}(G)$, e.g. when G is simple and non-abelian. We have also tried, using the apparatus of G-dessins, to give a conceptual explanation for the existence of $\mathscr{S}(G)$.

Many people have attempted to classify the generating pairs up to conjugacy or up to isomorphism, within their favorite finite group G. This may or may not have been motivated by the enumeration of dessins. We believe that, once this is done, trying to compute $\mathrm{GT}(G)$ is a natural idea, that will lead to a wealth of new information.

We shall conclude with a few open problems for the reader. Some of these have been implicitly touched upon in the text. There are many other questions, which are too vague to be mentioned here.

Obviously one may ask for computations of $\mathrm{GT}(G)$ or $\mathrm{GT}_1(G)$ for various groups G, and obtaining results "by hand", that is, without computers, would certainly have value. For example, in a subsequent publication we shall describe $\mathrm{GT}_1(PSL_2(\mathbb{F}_q))$, confirming the pattern emerging in Theorem 6. It would be interesting to have many other such results available, so as to understand better what it is about G that makes $\mathrm{GT}(G)$ large or small. The question of *intuition* is wide open.

Intuition is much needed for the next problem: find an entire (infinite) family of groups $(G_i)_{i\in I}$ such that you can compute the inverse limit

$$\lim_{i\in I} \mathrm{GT}(G_i);$$

then, give an interpretation of the map $\mathrm{Gal}(\overline{\mathbb{Q}}/\mathbb{Q}) \to \mathrm{GT} \to \lim_i \mathrm{GT}(G_i)$. This would be a direct generalization of the cyclotomic character (obtained for cyclic groups, see the Introduction). In this paper, in a sense, we have done this for the dihedral groups, unfortunately obtaining a trivial answer (see Proposition 3 and the subsequent discussion).

Perhaps less ambitious, but already quite hard, is the question of finding a single explicit element in $\lim_i \mathrm{GT}(G_i)$. Recall that, beyond the identity and complex conjugation, exhibiting elements of $\mathrm{Gal}(\overline{\mathbb{Q}}/\mathbb{Q})$ is no simple matter at all. With any luck, the problem at hand might be less difficult.

In Sect. 5.4 we have mentioned that there is no known bound on the order of the group $\mathrm{GT}(G)$. It is an exciting problem to find one, since it would provide a bound for the "order" of the profinite group $\mathrm{Gal}(\overline{\mathbb{Q}}/\mathbb{Q})$, in some sense.

Finally, if GT is similar to $\mathrm{Gal}(\overline{\mathbb{Q}}/\mathbb{Q})$ in any reasonable sense, then it should be interesting to have a look at its closed subgroups of index 2, or equivalently at the (continuous) homomorphisms $\mathrm{GT} \to \mathbb{F}_2$; they form the group which would be written $H^1(\mathrm{GT}, \mathbb{F}_2)$ in cohomological notation. Indeed, the cohomology ring $H^*(\mathrm{Gal}(\overline{\mathbb{Q}}/\mathbb{Q}), \mathbb{F}_2)$ is generated by its elements of degree 1 (part of Milnor's conjecture, now a theorem by Voevodsky, states that this is true for any field replacing $\mathbb{Q}$, showing the depth of this result). It follows from the fact that the abelianization map $\mathrm{Gal}(\overline{\mathbb{Q}}/\mathbb{Q}) \to \hat{\mathbb{Z}}$ can be lifted to a map $\mathrm{GT} \to \hat{\mathbb{Z}}$, as we have seen, that any map $\mathrm{Gal}(\overline{\mathbb{Q}}/\mathbb{Q}) \to \mathbb{F}_2$ can also be lifted to GT; in other words, the homomorphism

$$H^1(\mathrm{GT}, \mathbb{F}_2) \longrightarrow H^1(\mathrm{Gal}(\overline{\mathbb{Q}}/\mathbb{Q}), \mathbb{F}_2)$$

is surjective. One may ask whether it is injective as well, that is: does GT have homomorphisms onto $\mathbb{F}_2$ which are identically trivial on $\mathrm{Gal}(\overline{\mathbb{Q}}/\mathbb{Q})$? A probably harder question being: what is the abelianization of GT?

References

1. Drinfeld, V. G.: On quasitriangular quasi-Hopf algebras and on a group that is closely connected with $\mathrm{Gal}(\overline{\mathbf{Q}}/\mathbf{Q})$. Algebra i Analiz **2**, 149–181 (1990).
2. Guillot, P.: An elementary approach to dessins d'enfants and the Grothendieck-Teichmüller group. L'enseignement mathématique **60**, 293–375 (2014).
3. Isaacs, I. M.: Finite group theory. Graduate Studies in Mathematics vol 92, American Mathematical Society, Providence, RI (2008).

4. Jones, G.: Regular dessins with a given automorphism group. Contemporary Mathematics **629**, 245–260 (2014).
5. Magot, N., Zvonkin, A.: Belyi functions for Archimedean solids. Discrete Math. **217**, 249-271 (2000).
6. Neukirch, J.: Algebraic number theory. Grundlehren der Mathematischen Wissenschaften vol 322, Springer-Verlag, Berlin (1999).

[illegible]

[illegible]

[illegible]

[illegible]

Discrete Groups and Surface Automorphisms: A Theorem of A.M. Macbeath

W.J. Harvey

Abstract This short article re-examines the interaction between group actions in hyperbolic geometry and low-dimensional topology, focussing in particular on some contributions of Murray Macbeath to the study of Riemann surface automorphisms. A brief account is included of a potential extension to hyperbolic 3-manifolds.

1 Introduction

The classical results of Klein, Hurwitz and others on automorphisms of Riemann surfaces were based on the theory of projective algebraic curves. This contrasts somewhat with the approach used today, which makes essential use of the uniformisation theorem, covering spaces and the geometry of non-Euclidean crystallographic groups. Behind all this stands the rigorous theory of uniformisation which was worked out in the years before 1910 via Dirichlet's Principle by Hilbert and Courant and completed by Koebe and (using other methods) by Poincaré, thus establishing a firm basis for a systematic geometric account of surface topology. Group actions in the hyperbolic plane were analysed by Dehn and Nielsen, while the 2-volume book of Fricke and Klein [5] explored at length the immense range of discrete hyperbolic plane groups involved in this theory, formulating a classification of Fuchsian groups into distinct parameter spaces associated with each *signature* (or geometric type). At the same time, the formulation of an abstract notion of manifold, signalled by Weyl's ground-breaking book on Riemann surfaces [19], now just over a hundred years old, heralded an upsurge of interest in geometric topology generally and low dimensional manifolds in particular.

It is worth noting that something of a hiatus in the systematic development of discrete group actions began in the late 1920s. Thus, after Fricke's construction of parameter spaces for Fuchsian groups and the work of Dehn and Nielsen on surface topology, the problem of moduli for Riemann surfaces remained unresolved until

W.J. Harvey (✉)
King's College London, London, UK
e-mail: bill.harvey@kcl.ac.uk

J. Širáň and R. Jajcay (eds.), *Symmetries in Graphs, Maps, and Polytopes*,
Springer Proceedings in Mathematics & Statistics 159,
DOI 10.1007/978-3-319-30451-9_9

the theory of complex analytic deformations was established, first in outline by Teichmüller from 1938 to 1943, and then in rigorous detail by the school of Lars Ahlfors and Lipman Bers in the late 1950s. The latter developments will not concern us here; that material can now be found in many sources, including the collected works of these two authors, [1] and [4].

2 Hurwitz's Theorem Revisited

A brief paper of Siegel from 1945 [17] led Murray Macbeath to formulate a systematic new approach to the study of Riemann surface automorphisms in the late 1950s. In 1893, A. Hurwitz showed that, for values of the genus $g \geq 2$, the maximum number of automorphisms of a surface is $84(g-1)$, a bound attained by the famous Klein quartic curve of genus 3, with automorphism group the simple group $PSL(2, \mathbb{F}_7)$ of order 168. This finite group action had been discovered by Klein (1879) in the appropriate setting of non-Euclidean plane geometry; for more details of that fascinating story and some contemporary developments, see [9]. Soon after, Poincaré began his own study of the discrete subgroups of the Lie group $G = PSL(2, \mathbb{R})$ (which he called Fuchsian, much to Klein's annoyance), motivated by his sudden realisation that the groups emerging from his study of uniformisation by differential equations are these same groups of isometries of the hyperbolic plane,

$$\mathscr{H}^2 = \{z = x + iy \in \mathbb{C} : y > 0\}, \quad \text{with the Poincaré metric} \quad ds_h = \frac{|ds|}{2y}.$$

The (sense-preserving) isometry group of $\mathscr{H}^2$ is isomorphic to the group $G = PSL(2, \mathbb{R})$ acting transitively by fractional linear transformations: if A is a 2×2 real matrix with $\det A \neq 0$, the corresponding mapping is

$$T_A : z \mapsto \frac{az + b}{cz + d}, \quad \text{when} \quad A = \begin{bmatrix} a & b \\ c & d \end{bmatrix}.$$

The invariant Haar measure of G induces an invariant notion of area μ in the homogeneous space $\mathscr{H}^2 \cong G/PSO(2)$, known as the *Gauss-Bonnet area measure*; it coincides (up to a multiplicative constant) with the hyperbolic area element induced by the Poincaré metric ds_h.

For a Fuchsian group, a discrete subgroup Γ of G, the action on $\mathscr{H}^2$ is properly discontinuous and there is a *fundamental domain*, which we take here to mean a closed (Borel-measurable) subset F, with interior F^0, satisfying two characteristic properties:-

(i) $F^0 \cap \gamma F^0 = \emptyset$ for $\gamma \in \Gamma$ with $\gamma \neq \mathrm{Id}$;
(ii) $\bigcup_{\gamma \in \Gamma} \gamma F = \mathscr{H}^2$.

The Dirichlet region with a chosen centre point p_0 not fixed by any group element is a convenient construction, giving a (convex hyperbolic) polygonal fundamental domain for each Fuchsian group: one takes the subset of those points of $\mathcal{H}^2$ for which the distance $d_h(p, p_0)$ from p to p_0 is minimal among the points in the orbit $\Gamma \cdot p$. For full details of these basic concepts of hyperbolic geometry, see for instance Beardon's book [2, 12].

For Γ co-compact and torsion-free, i.e. such that the quotient orbit space $S = \mathcal{H}^2/\Gamma$ is a compact Riemann surface, the hyperbolic area $\mu(F)$ of any fundamental domain for Γ is a positive number, independent of the choice of fundamental set. In particular, by the Gauss-Bonnet Theorem, the area of a fundamental set for a group which has quotient a genus g closed surface is $4\pi(g-1) = (-2\pi\chi(S)$. Siegel showed in [17] that, within the range of all possible Fuchsian groups in G, there is a unique group (up to conjugacy) with the smallest positive value for the area.

Theorem 1 *For all co-compact Fuchsian groups, the minimum value of the invariant area* μ *is* $\pi/21$.

This value corresponds to the triangle group $\Gamma_0 = \langle x, y : x^2 = y^3 = (xy)^7 = 1\rangle$.

Now we let $K < \Gamma$ be a subgroup of finite index in a Fuchsian group with compact quotient space. Choosing a finite set of coset representatives, $\gamma_1, \ldots, \gamma_n$, we see that the union of these translates $\biguplus_{j=1}^{n} \gamma_j F$ of a given fundamental domain F for Γ forms a fundamental domain F_K for the subgroup K, and invariance of the measure implies that

Theorem 2 *The index of the subgroup K,* $[\Gamma : K]$*, is equal to the quotient* $\mu(F_K)/\mu(F)$.

This is the *Gauss-Bonnet Index Theorem* for Fuchsian groups. A simple consequence of this and the result of Siegel is the *Hurwitz Theorem*: choose a Fuchsian cocompact group $K \cong \pi_1(S)$ and let Γ denote the group of all possible lifts to the universal covering of S of the automorphisms of S. Then Γ contains K as a normal subgroup, and since $|\mathrm{Aut}(S)| = [\Gamma : K]$, we obtain at once the following result.

Corollary 3 *The order of an automorphism group acting on a genus g Riemann surface is at most* $84(g-1)$.

Of course, this is not the end of the story: the same line of reasoning produces the Riemann-Hurwitz branching formula, involving the genus of $\mathcal{H}^2/\Gamma$ and the orders of the maximal periodic generators of Γ, and there is a natural extension to subgroups of finite index in arbitrary non-Euclidean crystallographic groups, leading to a vast catalogue of results analysing the patterns of conformal and anti-conformal group actions on hyperbolic surfaces.

Macbeath's paper [10] fills in the details of the above proof and unveils a method (the *'Macbeath trick'* mentioned by Marston Conder in his conference talk at Malvern) for constructing, from a single finite index torsion-free normal subgroup $K \lhd \Gamma$, an infinite sequence K_n, $n \in \mathbb{N}$ of finite index characteristic normal subgroups of Γ. For each of these subgroups, the corresponding quotient surface S_n is of course a smooth covering of $S_0 = \mathscr{H}^2/K$ and the finite groups Γ/K_n are automorphism groups of the surfaces, with orders determined by multiplying by the index just like the Euler characteristic. This proves the following result, *Macbeath's Theorem.*

Theorem 4 *If there is a Riemann surface S of genus $g \geq 2$ with a group of $h(g-1)$ automorphisms, where h is a rational number with denominator dividing $g-1$, then for infinitely many values of the integer k there is a k-sheeted covering surface S_k of genus g_k with $h(g_k - 1)$ automorphisms.*

The sequence of characteristic subgroups employed in [10] is defined as the set of product groups $K_n = [K, K].\{K^n\}$, where $\{K^n\}$ denotes the Burnside n-kernel generated by all n-th powers in K and $[K, K]$ is the commutator subgroup. It is then an easy exercise to show that the index of K_n in the surface group K is n^{2g}, so that we know from Euler characteristic considerations that S_n has genus $g_n = n^{2g}(g-1)+1$.

If the over-group concerned is Γ_0, then the induced finite groups are all (by Siegel's result) *Hurwitz groups* acting on the surfaces S_n, that is, we have produced an infinite family of surfaces with automorphism groups for which the Hurwitz bound on their order is attained.

The original exposition of this approach to Fuchsian groups and surface automorphisms was presented in a widely circulated set of lecture notes, [11] from the Summer School in Topology at Dundee in 1961; they can be obtained in pdf format by email request to this author (bill.harvey@kcl.ac.uk). A very pleasant account by Macbeath of the whole story, in the context of Klein's study of his quartic and the genus 3 action of the simple group of order 168, can be found in [13].

3 Automorphisms and Geometry in Three Dimensional Hyperbolic Space

Poincaré also initiated the study of hyperbolic 3-space with its analogous metric structure closely linked to the matrix group $PSL(2, \mathbb{C})$ and conformal geometry on the boundary 2-sphere, but progress in understanding the topological structure of 3-manifolds was slow. After the revolutionary geometric ideas, results and conjectures worked out by W.P. Thurston in the mid-1970s, a furious concentration of research effort ensued which has swept away most of the topological and group-theoretic difficulties which confronted 3-manifold topology at that time. In the process, two crucial facts emerged, the first largely due to the efforts of G. Perelman.

- (Geometrisation.) All compact 3-manifolds possess a natural geometric structure, modelled on one of the eight geometries that Thurston described.
- (Hyperbolic structure predominates.) By far the majority of compact 3-manifolds are hyperbolic.

We can now formulate a very simple topological characterisation of the class of compact hyperbolic 3-manifolds, thanks to recent breakthrough work by Jeremy Kahn and Vlad Markovic with some essential further topological and group-theoretic input from I. Agol and from D. Wise.

Theorem 5 *A compact 3-manifold has a hyperbolic structure if and only if it has a finite covering which fibers over the circle with fibre a compact surface of genus at least 2 and with pseudo-Anosov holonomy.*

An accessible summary treatment of these developments which brings out well the range of ideas and work involved can be found in a recent Bourbaki Seminar report [3]. The key result which drives them appears in [7]. It confirms a remarkable string of conjectures made by Thurston [18], following his proof that both Haken manifolds and pseudo-Anosov surface bundles over the circle carry hyperbolic structures.

In the present context, it is natural to look for a parallel approach to study automorphism groups of (compact) hyperbolic 3 manifolds via the structure of 3D hyperbolic orbifolds. This turns out to be possible in principle, but the combinatorial patterns which exist have not yet been completely understood. However, for the privileged class of surface bundles, we can reason as follows.

A compact hyperbolic surface bundle induced by a pseudo-Anosov map $\varphi : S \to S$ is, by definition, a 3-manifold obtained from the product of a surface S of genus at least 2 with a closed interval by identifying the two end surfaces using φ:

$$M_\varphi = S \times \{0 \le t \le 1\}/\{x \times 0 \sim \varphi(x) \times 1 \quad \text{for each} \quad x \in S\}.$$

Now any automorphism of M_φ must preserve the fibration structure since, by Macbeath's method of lifting automorphisms to the universal cover, it is induced by conjugation with a hyperbolic isometry—equally this follows by Mostow rigidity. But such an isometry must induce an automorphism of a typical fibre surface S preserving the holomorphic quadratic form (Teichmüller differential) on S which determines the hyperbolic axis in $\mathscr{H}^3$ up to conjugacy. Note that any hyperbolic isometry f of $\mathscr{H}^3$ preserving an axis must be contained in the stabiliser of the axis and, furthermore, lies in the normaliser of $\pi_1(M_\varphi)$, a discrete subgroup of $PSL(2, \mathbb{C})$ which means that f lies in a discrete subgroup of that stabiliser. We can assume for convenience that this is the vertical axis I at O, the origin in the horizontal plane $\mathbb{C}$, and it follows that the group of automorphisms of a hyperbolic surface bundle is very restricted.

Theorem 6 *Let $M = M_\varphi$ be a smooth hyperbolic surface bundle over the circle induced by a pseudo-Anosov homeomorphism $\varphi : S \to S$ of a genus g surface. The automorphism group of the fibered hyperbolic 3-manifold M is either cyclic or dihedral, with order bounded above by a linear function $(2g - 1)$ of the genus of S.*

Proof (Sketch.) The stabiliser in $PSL(2, \mathbb{C})$ of an axis A is isomorphic to $G(A) = \mathbb{Z}/2 \times \mathbb{C}^*$ and the intersection of this stabiliser with $K = \pi_1(M)$ is cyclic, generated by some loxodromic element. Note that any discrete subgroup of $G(A)$ is dihedral or cyclic. The automorphisms of M, when lifted to the universal covering, generate a discrete overgroup $\Gamma > K$, the normaliser of $\pi_1(M)$, just as in the Riemann surface case. But in contrast to the case of surfaces, where the $(2, 3, 7)$—triangle group Γ_0 gives the Hurwitz upper bound, the orbifold fibre surface must in this case be *sufficiently large*, in the sense that it contains an essential closed loop and admits a pseudo-Anosov automorphism. This implies that the fibre subgroup Γ, which must contain the surface subgroup $\pi_1(S)$ with cyclic or dihedral quotient automorphism group, has at least 4 generators, and restrictions on the orders of torsion generators coming from the so-called lcm condition (see [6]) imply that the smallest area Γ (in terms of Euler-Poincaré characteristic) is a $(2, 2, n, m)$—group for suitable periods n, m dividing the index. Hence the index, if the quotient is cyclic, is at most $2g - 1$ by a short argument using Theorem 2.2. The argument in the dihedral case is similar but a little more complicated.

A more detailed discussion and complete proof will be published elsewhere. Notice that this result does not hold if the fibration is not smooth: an example of a hyperbolic orbifold fibering in three mutually orthogonal ways, which goes back to Sullivan (and probably Thurston), is described briefly in Otal's text [16].

We note, finally, that Macbeath's result on a sequence of characteristic subgroups applies here for any given example as in the theorem, to produce an infinite family of fibered 3-manifolds which cover it and enjoy the same symmetry property.

More general results are known about automorphisms of hyperbolic manifolds in dimension 3 which chime with dimension 2; for instance a finite volume or compact hyperbolic 3-manifold may have any finite group as automorphism group. See, for instance, [8] or [14]. At present, however, a normal form for crystallographic groups in hyperbolic 3-space is unknown and no general direct analysis of automorphism groups analogous to the 2D case seems possible. A good account of the basic facts about 3D hyperbolic volumes can be found in Milnor's paper [15], and more recently the smallest volume manifolds and orbifolds have been determined, both compact and cusped. Clearly, much remains to be done in analysing the combinatorial structure of hyperbolic 3-manifolds and their automorphisms.

References

1. L.V. Ahlfors, Collected Papers (2 vols.) Birkhauser, Boston. 1982.
2. A.F. Beardon, Geometry of Discrete Groups, Graduate Texts in Math. vol 91, Springer Verlag, 1983.
3. N. Bergeron, *La conjecture des sous-groupes de surfaces [d'apres J. Kahn & V. Markovic].* Séminaire Bourbaki, 64éme année, Juin 2012, no. 1055.
4. L. Bers, Papers on Complex Analysis (2vols), Editors, I. Kra & B. Maskit. AMS, Providence, R.I. 1998.
5. R. Fricke & F. Klein, Vorlesungenüber die Theorie der automorphen Functionen. B. Teubner, Leipzig. 2 vols. (1898, 1912).
6. W.J. Harvey, *Cyclic groups of automorphisms of a compact Riemann surface* Quart. J. Math. (Oxford), **17** (1966), 86–97.

7. J. Kahn & V. Markovic, *Immersing almost-geometric surfaces in a closed hyperbolic 3-manifold*, Ann. of Math. (2) **175** (2012), 1127–1190.
8. S. Kojima, *Isometry transformations of hyperbolic 3-manifolds*, Topology & Appls. **37** (1988), 297–307.
9. Silvio Levy (editor), The Eightfold Way. MSRI Publications **35**, Cambridge Univ. Press, 1999.
10. A.M. Macbeath, *On a theorem of Hurwitz*, Proc. Glasgow Math. Assoc. **5**(1961), 90–96.
11. A.M. Macbeath, *Discontinuous Groups and Birational Transformations, 'The Dundee Notes'*, in Proceedings of Summer School at Queen's College, Dundee, July, 1961. Reissued with corrections, Birmingham University, 1979.
12. A.M. Macbeath, *Generic Dirichlet polygons and the modular group*, Glasgow Math. J. **27** (1985), 129–141.
13. A.M. Macbeath, *Hurwitz groups and surfaces,* in Levy The Eightfold Way, 103–113.
14. A. D. Mednykh, *Automorphisms of hyperbolic manifolds*, A.M.S. Transl. (2) vol. **151** (1992), 107–119.
15. J. Milnor, *Hyperbolic geometry: the first 150 years*, Bull. Amer. Math. Soc. (New Series), **6** (1982), 9–24.
16. J-P. Otal, The Hyperbolisation Theorem for fibered 3-manifolds. SMF-AMS Texts & Monographs **7**, AMS (Providence RI), 2001.
17. C.L. Siegel, *Some remarks on discontinuous groups,* Ann. of Math. (2) **46** (1945), 708–718.
18. W.P. Thurston, *Three-dimensional manifolds, Kleinian groups and hyperbolic geometry*, Bull. A.M.S. (N. S.) **6** (1982), 357–381.
19. Hermann Weyl, Die Idee der Riemannschen Fläche. B.Teubner, 1913.

Isometric Point-Circle Configurations on Surfaces from Uniform Maps

Milagros Izquierdo and Klara Stokes

Abstract We embed neighborhood geometries of graphs on surfaces as point-circle configurations. We give examples coming from regular maps on surfaces with a maximum number of automorphisms for their genus, and survey geometric realization of pentagonal geometries coming from Moore graphs. An infinite family of point-circle v_4 configurations on p-gonal surfaces with two p-gonal morphisms is given. The image of these configurations on the sphere under the two p-gonal morphisms is also described.

1 Introduction

Consider the (rank two) set system of points and blocks where the points are the vertices and the blocks are the neighborhoods of the vertices of an r-regular graph on v vertices. Such set systems are called *neighborhood geometries* of graphs, and were first defined in [1], within a more general context. If any two vertices have distinct neighborhoods, then the system has the following two properties: (1) each vertex appears in r blocks and (2) each block contains r vertices. A set system, or a geometry, with properties (1) and (2) with v points and v blocks is classically known as a (balanced) v_r *configuration*. We say that a rank two set system is *connected* if there is a sequence of subsequently incident points and blocks between each pair of points. If the intersection of each pair of blocks contains at most d elements, then we will say that it is of *combinatorial linear dimension* d, since two distinct linear spaces of dimension d can intersect in d linearly independent points, but not in $d+1$. The blocks of a (combinatorial) geometry of linear dimension $d = 1, 2, 3$ will sometimes

M. Izquierdo (✉)
Department of Mathematics, Linköping University, 58183 Linköping, Sweden
e-mail: milagros.izquierdo@liu.se

K. Stokes
School of Engineering Science, University of Skövde, 54128 Skövde, Sweden
e-mail: klara.stokes@his.se

J. Širáň and R. Jajcay (eds.), *Symmetries in Graphs, Maps, and Polytopes*,
Springer Proceedings in Mathematics & Statistics 159,
DOI 10.1007/978-3-319-30451-9_10

be called lines, planes and 3-spaces, and so on. A combinatorial configuration is *linear* if it is of linear dimension 1. Most configurations in the literature are linear.

A classical example of a configuration of combinatorial linear dimension 2 is the Möbius 8_4 configuration, which is also a geometric configuration of 8 planes intersecting in quadruples on 8 points. The literature also contains many examples of geometric point-circle configurations. A classical example is the Miquel $(8_3, 6_4)$ configuration with 6 circles intersecting in triples on 8 points. Two circles in the real plane intersect in at most 2 points, so combinatorially non-linear point-circle configurations in the plane correspond to configurations of combinatorial linear dimension 2. Note that if the point-circle configuration is embedded on a surface of genus $g > 0$, then two circles may intersect in more than 2 points.

The neighborhood geometry of a graph is a combinatorial geometry with the following properties, described in [1]: It is linear exactly when the graph does not contain any cycle of length 4. There is a combinatorial polarity (that is, a duality of order two) in a neighborhood geometry defined by mapping each vertex to its neighborhood. If the graph is bipartite, then one obtains two disconnected neighborhood geometries. The polarity then maps a point in the first connected component to its neighborhood, which is then a block in the second component. Therefore, the two components are duals of each other, that is, one is obtained from the other by interchanging the roles of the points and the blocks. This is true also if the graph is not r-regular. If the graph is not bipartite then the set system it defines consists of a single connected component, and so it is self-dual. The two disjoint geometries defined by a bipartite r-regular graph are dual but are not necessarily isomorphic and then not self-dual, although they have the same parameters. This is the case for example if the graph is the incidence graph of a configuration which is not self-dual. However, the union of the connected components of a neighborhood geometry is always self-dual.

Combinatorially, the incidence graph of the neighborhood geometry of a graph is the Kronecker double cover of the graph [2]. Any combinatorial property of neighborhood geometries of graphs is therefore a property of the Kronecker cover of graphs.

The neighborhood geometry of a graph has been applied to 1-skeletons of regular polytopes in real Euclidean d-space in order to construct *geometric* point-hyperplane realizations (that is, of linear dimension $d-1$) of self-polar symmetric configurations (symmetric as in *with the maximal number of symmetries*) [3]. The construction was also generalized there by using the t-neighborhoods of the vertices of the polytope, that is, the vertices at distance t from a given vertex.

It was observed in [3] that the neighborhood geometry of the 1-skeleton of a spherical polytope in real Euclidean 3-space, which is a point-plane configuration, also defines a point-circle configuration in the real Euclidean plane through stereographic projection, whenever the points in each plane are concyclic. Indeed, the circle-preserving property of the stereographic projection implies that any point-circle configuration drawn on the sphere can also be drawn in the real Euclidean plane.

In [2], point-circle configurations were realized in the plane as neighborhood geometries of unit-distance graphs. For the neighborhood geometry of a graph embedded in the plane to be realized in terms of the circles passing through all the vertices in each neighborhood of the graph, it is necessary to ensure that these vertices are concyclic. Since three points define a circle, if the graph is 3-regular this condition is always satisfied. If the valencies of the vertices is larger than three, then embeddings with concyclic neighborhoods are special. A unit-distance embedding of the graph in the plane is an example of an embedding with this property. The 1-skeleton of a quasi-regular polyhedron in Euclidean 3-space is an example of an embedding with the same property in three dimensions.

In [4], the construction of point-circle configurations on spherical polyhedron was generalized to surfaces in general. The motivation was there to give geometric realizations as point-circle configurations of certain pentagonal geometries coming from Moore graphs.

For a more general overview of lineal and point-circle configurations, see the books [5, 6].

In this article we will use the construction in [4] to construct point-circle realizations of neighborhood geometries on several classical surfaces, as well as on an infinite family of surfaces for an infinite number of genera. We also give two realizations in terms of points and isometric circles of a certain 9_4 configuration of linear dimension 2, one in the real plane and one on an orientable surface of genus 4, the latter realizing all the combinatorial automorphisms of the configuration.

2 Constructing Configurations of Points and Isometric Circles on Surfaces

Let U be either the Riemann sphere, the complex Euclidean plane or the hyperbolic plane. A uniform tiling of U is a collection of congruent polygons that partitions and fills up the entire space. If this tiling has p q-gons meeting in each vertex, then the stabilizer of the tiling is a cocompact subgroup G of a triangle group $\Gamma(p, 2, q)$ or, if we allow orientation-reversing elements, a discrete torsion-free group of automorphisms of U in which $\Gamma(p, 2, q)$ has index 2. Since the polygons are congruent, the neighbors of each vertex are concyclic on isometric circles. The distance is the spherical, the Euclidean or the hyperbolic distance respectively.

An (compact) orientable Riemann surface is a closed topological surface S with analytic structure. A non-orientable Riemann surface is a closed topological surface S with dianalytic structure where the conjugation $z \to \bar{z}$ is allowed. The quotient of U under the action of G is a Riemann surface $S = U/G$. The group is called the surface group of S and U is its universal covering space. By the Poincaré uniformization theorem any Riemann surface is the quotient $S = U/G$, where U is either the Riemann sphere, the complex Euclidean plane or the hyperbolic plane and G

is a discrete torsion-free group of automorphisms of U, possibly with orientation-reversing elements. The quotient of the polygonal tiling by the action of G is a uniform map of type $\{p, q\}$ on the surface [7–10]. In a uniform map of type $\{p, q\}$ the vertices have valency p, the edges have valency 2 and the faces have valency q. A map is regular if its automorphism group acts transitively on the triples of incident vertices, edges and faces, so a regular map is always uniform.

The image of the isometric circles through the neighborhoods of the vertices of a uniform tiling of U under the quotient by G are isometric circles through the neighbors of each vertex of the corresponding uniform map of U/G. The result is a configuration of a finite number of points and circles on the surface. Each circle contains p points and p circles go through each point. We have proved the following result.

Theorem 1 *[4] A uniform map on a surface produces a configuration of points and isometric circles on the same surface.*

Since the map completely determines the geometric point-circle configuration, the automorphism group of the configuration coincides with the automorphism group of the map. Therefore, applying Theorem 1 to regular maps will give configurations with many geometric symmetries. This motivates the study of point-circle configurations defined by regular maps in general.

Two non-isomorphic graphs can define the same neighborhood geometry. For example, both the Petersen graph and the Desargues graph has the Desargues configuration as neighborhood geometry. Since the Desargues graph is bipartite (it is the incidence graph of the Desargues configuration), it defines the Desargues configuration as a point-circle configuration twice. For other examples, see [11]. Also, the same r-regular graph can be embedded in a Riemann surface as a uniform map in several ways. Consequently, there may be many ways to realize the same configuration in terms of points and circles on some surface.

In a paper from 1949, Coxeter explored the relation between self-dual configurations and arc-transitive graphs (he used the term *regular graph*) [12]. He embedded the incidence graph of the configurations as regular maps on surfaces. For bipartite graphs, this is exactly what we also do. However, our approach goes further. We obtain a geometric configuration defined by the incidences of elements from two classes of distinct geometric objects, points and circles, on the surface. As an immediate consequence of this, we see that all configurations represented in [12] as regular maps of incidence graphs, are actually point-circle configurations on the same surfaces.

For graphs that are not bipartite, our construction is essentially different from Coxeter's approach. However, his general principle *"interesting configurations are represented by interesting graphs"* [12] may still be applied.

3 Point-Circle Configurations from Some Classical Regular Maps

In this section we apply Theorem 1 to some classic regular maps, giving point-circle configurations on surfaces of genus two, three and four.

3.1 The Bolza Curve

The Bolza curve is a complex projective curve of genus 2 with automorphism group $GL(2,3)$ of order 48. It is the curve of genus 2 with maximum number of automorphisms. Its surface group (as Riemann surface) is a normal subgroup of the triangle group $\Gamma(2,3,8)$. It has a regular map of type $\{3,8\}$ with 16 vertices, 24 edges and 6 octagonal faces. The neighborhood geometry of the underlying bipartite graph is the unique linear 8_3 configuration known as the Möbius-Kantor configuration, which is therefore realized as a point-circle configuration on the Bolza curve. This configuration has the antipodal property, mutually non-collinear points occur in pairs. The dual map has 6 vertices, 24 edges and 16 triangular faces. The neighborhood geometry of the underlying graph is a degenerated 6_4 configuration of 3 circles through 6 points where each circle appears twice. Each pair of distinct circles meet in 2 points.

3.2 Klein's Quartic

Klein's quartic projective curve, given by the equation $x^3y + y^3z + z^3x = 0$ over the complex field is the curve of smallest genus that attains the Hurwitz bound. Its genus is $g = 3$ and its automorphism group is $PSL(2,7)$ of order $84(g-1) = 168$, so it is the curve of genus 3 with maximum number of automorphisms. There is an epimorphism from the triangle group $\Gamma(2,3,7)$ to $PSL(2,7)$ and its kernel is the surface group uniformizing Klein's quartic, as Riemann surface. This results in the classical regular heptagonal map on the surface of type $\{3,7\}$ with 56 vertices of valency 3, 84 edges and 24 heptagonal faces. The neighborhood geometry of the underlying non-bipartite graph is a self-polar 56_3 configuration, which is linear, since the graph has no 4-cycles.

The dual map (obtained by interchanging the roles of vertices and faces) has 24 vertices of valency 7, 84 edges and 56 triangular faces. The 24 vertices correspond to the 24 Weierstrass points of the surface. The neighborhood geometry of the underlying non-bipartite graph is a self-polar 24_7 configuration of combinatorial linear dimension 2. Given any point p, there are exactly two points which are not concylic with p. All other points are concyclic with p exactly twice.

Note that it is possible to embed the Fano plane (the projective plane over $\mathbb{F}_2$), which is a self-polar linear 7_3 configuration, in Klein's quartic in terms of the incidences of a hypermap related to the 24 Weierstrass points of the surface [13]. These 24 points are also exactly the inflection points of the surface. There are 8 triangles with these points as vertices and inflection tangents as sides, and one of these triangle is the coordinate triangle [14]. The other 7 triangles are then the blocks of the Fano plane, with the incidences defined like the incidences of the hyperpoints and hyperedges described in [13]. Therefore this embedding of the Fano plane in Klein's quartic is actually an embedding of the Fano plane as a configuration of planes, where the planes are defined by the triangles.

3.3 Bring's Curve

Bring's curve is a complex projective curve of genus 4 given by the equations $\sum_{i=1}^{5} x_i = \sum_{i=1}^{5} x_i^2 = \sum_{i=1}^{5} x_i^3 = 0$. Its automorphism group is the symmetric group acting on 5 elements. It is the curve of genus 4 with maximum number of automorphisms.

Consider the triangle group $\Gamma(2, 4, 5)$. There is an epimorphism from $\Gamma(2, 4, 5)$ to the symmetric group S_5 and the kernel is the surface group of Bring's curve (as Riemann surface), which is normal in $\Gamma(2, 4, 5)$. The surface allows a regular map of type $\{4, 5\}$. This map has 30 vertices of valency 4, 60 edges and 24 pentagonal faces. The neighborhood geometry of the underlying graph is a self-polar 30_4 configuration. Since the graph has no 4-cycles, the configuration is linear.

The dual map has 24 vertices, 60 edges and 30 quadratic faces. The underlying graph is bipartite, and is therefore the incidence graph of its neighborhood geometry, a self-polar 12_5 configuration of combinatorial linear dimension 2. Given any point p, there is exactly one point which is not concylic with p. All other points are concyclic with p exactly twice. This gives the configuration an antipodal property.

4 Pentagonal Geometries as Point-Circle Configurations from Moore Graphs

A pentagonal geometry is a (linear) combinatorial configuration with the property that, for any point p, all points that are not collinear with p are on a single line, which is called the opposite line of p [15]. The lines in a pentagonal geometry are of two types, lines that are the opposite line of some point, and lines that are not.

A pentagonal geometry in which all lines are opposite lines is self-polar by the polarity that makes correspond each point to its opposite line. The reduced Levi graph (in the sense of [11]) defined by the polarity of a self-polar pentagonal geometry is the graph in which the vertices are pairs of one point and its polar line and two vertices are joined by an edge if the point of one vertex is incident with the line of the other vertex. Therefore this graph is exactly the deficiency graph of the geometry, that is, the graph in which the vertices are the points and two vertices are joined by an edge if the points are not collinear. The pentagonal geometry can be recovered from the reduced Levi graph as its neighborhood geometry. More generally, the neighborhood geometry of the reduced Levi graph of a self-polar configuration is always equal to the original configuration. This construction of pentagonal geometries was first described in [15], where it also was proved that pentagonal geometries with $r = k$ are exactly the ones with a Moore graph of diameter 2 as reduced Levi graph.

There are only three known Moore graphs of diameter 2; the cycle graph of length 5, the Petersen graph and the Hoffman-Singleton graph. These graphs have degree 2, 3 and 7, respectively. The existence of a Moore graph of degree 57 is still an open question. The pentagonal geometries obtained from these graphs are, respectively, the ordinary pentagon, the Desargues' configuration and a pentagonal geometry with parameters $(7, 7)$ and with 50 points and 50 lines. In [15], it was also proved that all pentagonal configurations of order $(k, k + 1)$ can be constructed from pentagonal geometries of order $(k + 1, k + 1)$ through the removal of one point and its opposite line. There are therefore at most three such pentagonal geometries, with $k = 2, 6$ and maybe 56.

Regular embeddings of the two smallest Moore graphs are well-known. The cycle graph can be embedded with full automorphism group as a regular map on the Riemann sphere with two pentagonal faces. The Petersen graph has an embedding in the real projective plane as a regular map with six pentagonal faces, obtained from the classical spherical map $\{3, 5\}$ by identifying antipodal points. By Theorem 1, this implies that both the pentagon and the Desargues configuration allow realizations in terms of points and isometric circles with full automorphism group on surfaces of orientable genus 0 and non-orientable genus 1, respectively.

There is no regular embedding of the Hoffman-Singleton graph, but there are uniform pentagonal embeddings of type $\{7, 5\}$ on non-orientable surfaces of genus 57 with automorphism group of the map either trivial, of order 5 or of order 7 [16]. By Theorem 1 this implies that the pentagonal geometry $(7, 7)$ on 50 points and 50 blocks can be realized as a point-circle configuration on a surface of non-orientable genus 57 with any of these three automorphism groups [4]. It is also possible to realize this pentagonal geometry as a point-hypersphere configuration with full automorphism group in Euclidean space of 24 dimensions as the geometric neighborhood geometry of the well-known embedding of the Hoffman-Singleton graph in the Leech lattice [4].

5 Point-Circle Configurations on p-gonal Surfaces with Two Cyclic p-gonal Morphisms

In this section we give an infinite family of Riemann surfaces, one for each genus, admitting point-circle configurations. For this purpose we use Theorem 1 and the so called p-gonal surfaces. Before the p-gonal surfaces can be properly introduced some notation is needed.

The universal covering space U of an (orientable) Riemann surface U/G covers it with infinitely many sheets. Each point of U/G is the representative of exactly one orbit (fiber) of the points in U under the action of the surface group G, which is a torsion-free discrete group of automorphisms of U, possibly with orientation-reversing elements.

By considering also cocompact discrete groups H of automorphisms of U with elliptic elements, one obtains a surface U/H with singular points, a geometric orbifold. An orbifold is a more general concept than a Riemann surface. When H has no elliptic elements, then U/H is a Riemann surface.

Let G be a finite index n subgroup of a discrete subgroup H of automorphisms of U (a Fuchsian group). The inclusion $G \hookrightarrow H$ induces a (possibly ramified) covering $f : U/G \to U/H$ of degree n. The covering f is determined by the action of H on the G-cosets $\theta : H \to \Sigma_{|H:G|}$. If G is normal in H, then the covering $f : U/G \to U/H$ is a regular covering given by the monodromy $\theta : H \to H/G$. Assume that G has no elliptic elements, so that G is a surface group uniformizing a Riemann surface U/G. Assume also that the covering morphism $U/G \to U/H$ is of degree a prime number p. Then there is an automorphism $\phi : S \to S$ such that the deck-transformation group of the covering is generated by ϕ, that is, $\langle \phi \rangle \sim H/G = C_p$. If the genus of the underlying surface of U/H is 0, then we say that the surface U/G is a p-gonal surface and $f : U/G \to U/H$ is a p-gonal morphism [17, 18].

According to Castelnuovo-Severi [19] a compact Riemann surface which allows more than one p-gonal morphism has genus g satisfying $g \leq (p-1)^2$. Additionally, it is known that if S_g has several p-gonal morphisms, then these morphisms are all conjugate [18].

For every prime $p \geq 3$ there is a family of surfaces with two distinct cyclic p-gonal morphisms [17] of genus $(p-1)^2$, implying that the Castelnuovo-Severi inequality is sharp. For each p, one of the surfaces in this family, which we call Y_p, is quasi-platonic, that is, its surface group is normal in the triangle group $\Gamma(4, 2, 2p)$. The automorphism group of this surface[1] is $(C_p \times C_p) \rtimes D_4$, where C_n and D_n are the cyclic and the dihedral groups of order n and $2n$, respectively. Since the surface group is a normal subgroup in $\Gamma(4, 2, 2p)$, the surface allows a regular map of type $\{4, 2p\}$ of genus $(p-1)^2$. The underlying graph of this map is a bipartite 4-regular symmetric graph on $2p^2$ vertices of girth 4. The neighborhood geometry of this graph has p^2 points and p^2 blocks and it is self-polar, because the graph is symmetric. Each pair of points appears in exactly 0, 1 or 2 blocks, so it is not linear, but (combinatorially) planar. By Theorem 1, the geometry is realized as a point-circle configuration on the surface Y_p. There are 4 points on each circle and 4 circles through each point.

The neighborhood geometry of the underlying bipartite graph of the dual map is less interesting, since it is a degenerate $2p_{2p}$ configuration of one circle through $2p$ points, for the $2p$ circles are all the same.

[1]In [20] an erroneous presentation of this group for $p = 3$ was given. A presentation is $\langle a, b, s, t / a^3 = b^3 = s^2 = t^4 = (st)^2 = (sa)^2 = sbsb^2 = t^3atb^2 = t^3bta = 1 \rangle$, as correctly stated in [21]. The two p-gonal automorphisms are then ab and ab^{-1}.

The automorphism group of Y_p contains two conjugate p-gonal morphisms such that the orbit space of their action on Y_p is the Riemann sphere with $2p$ singular points, all of degree p. The action of each one of the two p-gonal morphism on the regular map of type $\{4, 2p\}$ on Y_p divides the vertices in $2p$ orbits of p vertices each. The result is a 2-regular bipartite graph on $2p$ vertices embedded as a map on the sphere. By Theorem 1, its neighborhood geometry is a point-circle configuration with p points and p circles on the sphere, 2 points on each circle and 2 circles through each point. We have proved the following.

Theorem 2 *There is an infinite family of combinatorially planar self-polar configurations p_4^2, realizable as point-circle configurations on orientable p-gonal surfaces of genus $(p-1)^2$. The orbits of such a configuration under the action of the two p-gonal morphisms defines a p_2 configuration of p circles and p points on the sphere which is isomorphic to the cycle graph of order p.*

The smallest member of this family of point-circle configurations on p-gonal surfaces, with $p = 3$, is perhaps the most interesting one. Note that this configuration occurs in [22] as one of two new interesting small v_4 configurations with the property that the last incidence in a geometric realization is implied by the other incidences. The graph of the regular map on the 3-gonal surface, resulting from Theorem 2 with $p = 3$, is bipartite on 18 vertices. Its neighborhood geometry has 9 points and 9 blocks. Each pair of points are in at least one block, therefore either in 1 or 2 blocks. Interestingly, given a point x, the points that are not in the same block twice (and therefore once) with x form a single block. This property can be seen as a generalization of the property defining a pentagonal (linear) geometry: given a point x, the points that are not collineal with x are collinear on a single line of the geometry, and they are the only points on that line (see Sect. 4). We define the property in its general form for a configuration of combinatorial linear dimension d as follows.

Definition 1 A configuration of combinatorial linear dimension d has the generalized pentagonal property if, given any point x, there are points which are in the same block as x exactly d or $d-1$ times and the points that are in the same block as x exactly $d-1$ times form a single block.

For $d = 1$ this is the defining property of the pentagonal geometries. For $d = 2$, there is the 9_4 configuration we just described. Its incidence graph, the 4-regular symmetric graph on 18 vertices is listed as the second symmetric graph of order 18 in [23]. In the same list one also finds an example for $d = 3$, the neighborhood geometry defined by the third symmetric graph of order 26, a 6-regular symmetric graph on 26 vertices with automorphism group of order 156. These examples are all self-polar configurations by construction and somehow they generalize the pentagon. A d-dimensional self-polar geometry with the generalized pentagonal property must have $1 + r^2/d$ points and $1+r^2/d$ blocks. In particular d must divide the number of points per block r. The pentagonal geometries are in general not self-polar linear configurations and may be non-balanced, that is, they may have more lines through a point than points on a line.

It would be interesting to know if there are examples of non-balanced configurations of combinatorial linear dimension d with the generalized pentagonal property.

The neighborhood geometry of the underlying symmetric graph of the regular $\{4, 6\}$ map on Y_3 is also the smallest member of the family of planar point-circle configurations defined as the neighborhood geometries of the generalized cuboctahedron graph in [2]. Apart from this example, the family of point-circle configurations from generalized cuboctahedron graphs and the family defined by the regular maps of type $\{4, 2p\}$ on the surfaces Y_p are disjoint. In [2] it is stated as an open question whether there is an isometric realization in terms of points and circles of any of the configurations in the family coming from the generalized cuboctahedron graphs. In this article we have given a realization of the first member of that family in terms of points and isometric circles on an orientable surface of genus 4. The automorphism group of this realization equals the automorphism group of the combinatorial configuration.

This special 9_4 configuration can also be realized as a point-circle configuration in the complex plane as the neighborhood geometry of the $3\{4\}2$ regular and complex polygon. The points and edges of this polygon are the points and lines of the generalized quadrangle with parameters (2, 1) (a grid on 9 points with 3 points on each line and 2 lines through each point). Our 9_4 configuration is the neighborhood geometry of the symmetric, distance-regular collinearity graph of this generalized quadrangle.

It is described in [24] how a complex polygon can be represented in the real plane, and a real representation of this particular $3\{4\}2$ polygon can be found on page 108. The regularity of the polygon and the properties of this representation imply that the graph that has as vertices the vertices of the real representation and as edges the sides of the triangles representing the complex edges of the polygon in the real plane, is a unit-distance graph in the real plane. This implies that this particular neighborhood geometry can be realized as an isometric point-circle configuration in the Euclidean real plane. This gives also a planar answer to the open question mentioned before from [2]. Note that the realization in the real plane has a smaller automorphism group than the one on a surface of genus 4.

Acknowledgments The authors want to thank Marston Conder for helpful discussions. Calculations were mainly done with Magma. The second author acknowledges partial financial support from the Spanish MEC project ICWT (TIN2012-32757).

References

1. C. Lefèvre-Percsy, N. Percsy and D. Leemans, New geometries for finite groups and polytopes. Bull. Belg. Math. Soc. 7, 583–610 (2000).
2. G. Gévay, T. Pisanski, Kronecker covers, V-construction, unit-distance graphs and isometric point-circle configurations, Ars Mathematica Contemporanea, 7:2, 317–336 (2014).
3. G. Gévay, Symmetric configurations and the different levels of their symmetry, Symmetry Cult. Sci. 20, 309–329 (2009).
4. K. Stokes and M. Izquierdo, Geometric point-circle pentagonal geometries from Moore graphs. Ars Mathematica Contemporanea 11, 215–229 (2016).

5. B. Grünbaum, Configurations of Points and Lines. American Mathematical Society, Providence, RI, 2009.
6. T. Pisanski and B. Servatius, Configurations from a Graphical Viewpoint. Birkäuser Advanced Texts, Basler Lehrbücher, Springer, 2013.
7. M. Conder and B. Everitt, Regular maps on non-orientable surfaces. Geometriae Dedicata 56, 209–219 (1995).
8. G. González-Diez, Variations on Belyi's theorem. Q. J. Math. 57:3, 339–354 (2006).
9. G.A. Jones and D. Singerman, Theory of maps on orientable surfaces, Proc. London Math. Soc. 3:37, 273–307 (1978).
10. D. Singerman and R.I. Syddal, The Riemann Surface of a Uniform Dessin. Contributions to Algebra and Geometry, 44:2, 413–430 (2003).
11. R. Artzy, Self-dual configurations and their Levi graphs. Proc. Amer. Math. Soc. 7, 299–303 (1956).
12. H.S.M. Coxeter, Self-dual configurations and regular graphs. Bull. Amer. Math. Soc. 56, 413–455 (1950).
13. P. Martin and D. Singerman, The geometry behind Galois' final theorem, European J. Combin. 33:7, 1619–1630 (2012).
14. I.V. Dolgachev, Classical Algebraic Geometry: A Modern View. Cambridge University Press, Cambridge, UK, 2012.
15. S. Ball, J. Bamberg, A. Devillers and K. Stokes, An Alternative Way to Generalize the Pentagon. Journal of Combinatorial Designs, 21:4, 163–179 (2013).
16. M. Conder and K. Stokes, Minimum genus embeddings of the Hoffman-Singleton graph. Manuscript.
17. A.F. Costa, M. Izquierdo and D. Ying, On cyclic p-gonal Riemann surfaces with several p-gonal morphisms. Geo. Dedicata 147, 139–147 (2009).
18. G. González-Diez, On prime Galois covering of the Riemann sphere. Ann. Mat. Pure Appl. 168, 1–15 (1995).
19. R.D.M. Accola, On cyclic trigonal Riemann surfaces, I. Trans. Am. Math. Soc. 283, 423–449 (1984).
20. A.F. Costa, M. Izquierdo and D. Ying, On Riemann surfaces with non-unique cyclic trigonal morphism. Manuscripta Math., 118 (4), 443–453 (2005).
21. D. Ying, Cyclic Trigonal Riemann Surfaces of Genus 4. Linköping Studies in Science and Technology. Thesis No. 1125 (URN: urn:nbn:se:liu:diva-5678), Linköping, 2004.
22. D. Glynn, Theorems of points and planes in three-dimensional projective space, Journal of the Australian Mathematical Society 88, 75–92 (2010).
23. M. Conder, List of symmetric graphs of order 2 to 30, www.math.auckland.ac.nz/~conder.
24. H.S.M. Coxeter, Regular complex polytopes. Cambridge University Press, Cambridge, UK, (1974).

Dessins, Their Delta-Matroids and Partial Duals

Goran Malić

Abstract Given a map $\mathcal{M}$ on a connected and closed orientable surface, the delta-matroid of $\mathcal{M}$ is a combinatorial object associated to $\mathcal{M}$ which captures some topological information of the embedding. We explore how delta-matroids associated to dessins behave under the action of the absolute Galois group. Twists of delta-matroids are considered as well; they correspond to the recently introduced operation of partial duality of maps. Furthermore, we prove that every map has a partial dual defined over its field of moduli. A relationship between dessins, partial duals and tropical curves arising from the cartography groups of dessins is observed as well.

1 Introduction

A *map* on a connected and orientable closed surface X is a cellular embedding of a connected graph G (loops and multiple edges are allowed). By this we mean that the vertices of G are distinguished points of the surface, and the edges are open 1-cells drawn on the surface so that their closures meet only at the vertices; furthermore, the removal of the all the vertices and all the edges from the surface decomposes the surface into a union of open 2-cells, which are called the *faces* of the map (Fig. 1).

To every map on X a *clean dessin d'enfant* corresponds. A clean dessin d'enfant is a pair (X, f) where X is a compact Riemann surface (or, equivalently, an algebraic curve) defined over $\mathbb{C}$ and $f: X \to \mathbb{CP}^1$ is a holomorphic ramified covering of the Riemann sphere, ramified at most over a subset of $\{0, 1, \infty\}$, with ramification orders over 1 all equal to 2. Vertices of the map correspond to the points in the fiber above 0, whilst the preimages $f^{-1}(\langle 0, 1\rangle)$ of the open unit interval, glued together at the fiber above 1, form the edges.

The following theorem of Belyĭ [3, 4] is considered as the starting point of the theory of dessins d'enfants.

G. Malić (✉)
School of Mathematics,University of Manchester,
Oxford Road, Manchester M13 9PL, UK
e-mail: goran.malic@manchester.ac.uk

J. Širáň and R. Jajcay (eds.), *Symmetries in Graphs, Maps, and Polytopes*,
Springer Proceedings in Mathematics & Statistics 159,
DOI 10.1007/978-3-319-30451-9_11

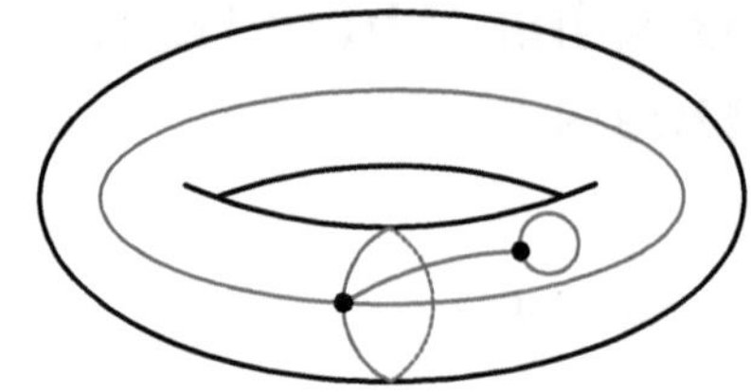

Fig. 1 A map with 2 vertices, 4 edges, and 2 faces on a genus 1 surface

Theorem 1 (Belyĭ) *Let X be an algebraic curve defined over* $\mathbb{C}$. *Then X is defined over the field* $\overline{\mathbb{Q}}$ *of algebraic numbers if, and only if there is a holomorphic ramified covering* $f : X \to \mathbb{CP}^1$ *of the Riemann sphere, ramified at most over a subset of* $\{0, 1, \infty\}$.

As a direct consequence, given any dessin (X, f), both the algebraic curve X and the covering map f are defined over $\overline{\mathbb{Q}}$ and therefore the absolute Galois group $\mathrm{Gal}(\overline{\mathbb{Q}}|\mathbb{Q})$ acts naturally on both. One of the major themes of the theory of dessin d'enfants is the identification of combinatorial, topological or geometric properties of dessins which remain invariant under the aforementioned action. We will call such invariants *Galois invariants*. A number of Galois invariants have been documented and an incomplete list can be found in Sect. 3.2 of this paper or in [25, Sect. 2.4.2.2].

A *delta-matroid* is a combinatorial object associated to a map $\mathcal{M}$ on a surface X which records a certain independence structure. It is completely determined by the spanning *quasi-trees* of $\mathcal{M}$, that is the spanning sub-graphs of the underlying graph of $\mathcal{M}$ which can be embedded as a map with precisely one face in some surface, not necessarily the same one as X. We will study the behaviour of the delta-matroid of a clean dessin under the action of $\mathrm{Gal}(\overline{\mathbb{Q}}|\mathbb{Q})$; the main conclusion is that the delta-matroid itself is not Galois invariant, however further consideration suggests that the self-dual property of delta-matroids might be, and in some cases is, preserved by the action.

A *partial dual* of a map with respect to some subset of its edges is an operation which generalises the geometric dual of a map. It was recently introduced in [10] and generalised to hypermaps in [11]. It was shown in [12] that the delta-matroids of partial duals of a map $\mathcal{M}$ correspond to the *twists* of the delta-matroid of $\mathcal{M}$. We give a proof of this correspondence without invoking the machinery of ribbon graphs used in [12] and use it to show that a map always has a partial dual defined over its field of moduli.

Towards the end of the paper we discuss the connection between maps, partial duals, and tropical curves. An *abstract tropical curve* is a connected graph without vertices of degree 2 and with edges decorated by the set of positive reals and ∞. We associate a tropical curve to a map via the *monodromy graph* of a map. The vertices of these graphs correspond to the partial duals of the map and the tropical curves obtained in this way show some similarities with maps when considering the action of $\mathrm{Gal}(\overline{\mathbb{Q}}|\mathbb{Q})$ on them. For example, the number of vertices, edges and the genus of tropical curves remains invariant under the action of $\mathrm{Gal}(\overline{\mathbb{Q}}|\mathbb{Q})$.

The paper is structured as follows. In Sect. 2 we define (not just clean) dessins d'enfants, describe the correspondence between dessins and bipartite maps and give a permutation representation.

In Sect. 3 we revisit Belyĭ's theorem and go into more detail about the action of $\mathrm{Gal}(\overline{\mathbb{Q}}|\mathbb{Q})$ on dessins. Some Galois invariants are described in Sect. 3.2 as well.

In Sect. 4 we introduce matroids and delta-matroids and describe how they arise from maps on surfaces.

In Sect. 5 we discuss the behaviour of delta-matroids of maps when the maps are acted upon by $\mathrm{Gal}(\overline{\mathbb{Q}}|\mathbb{Q})$. Special consideration is given to maps with self-dual delta-matroids in Sect. 5.2.

In Sect. 6 partial duals of maps are introduced, with remarks on the partial duals of hypermaps. We discuss both the combinatorial and geometric interpretation. In Sect. 6.1 we give a link from [12] between partial duals and delta-matroids and use it to show that a map always has a partial dual defined over its field of moduli.

In Sect. 7 we present a relationship between maps, their partial duals and tropical curves and note some similarities between the tropical curves associated to dessins that are in the same orbit of $\mathrm{Gal}(\overline{\mathbb{Q}}|\mathbb{Q})$.

2 Dessins and Bipartite Maps

Throughout this paper X shall denote a compact Riemann surface or its underlying connected and closed orientable topological surface. Furthermore, since compact Riemann surfaces are algebraic, X shall denote an algebraic curve as well. We consider X to be oriented, with positive orientation. Permutations shall be multiplied from left to right.

Definition 1 A *dessin d'enfant*, or just *dessin* for short, is a pair (X, f) where X is a compact Riemann surface (or, equivalently, an algebraic curve) defined over $\overline{\mathbb{Q}}$ and $f\colon X \to \mathbb{CP}^1$ is a holomorphic ramified covering of the Riemann sphere, ramified at most over a subset of $\{0, 1, \infty\}$.

The pair (X, f) is called a *Belyĭ pair* as well, whilst the map f is called a *Belyĭ map* or a *Belyĭ function*. Sometimes we will denote a dessin by $D = (X, f)$ to emphasise both the curve and the Belyĭ map. A dessin is of *genus* g if X is of genus g.

Two dessins (X_1, f_1) and (X_2, f_2) are isomorphic if they are isomorphic as coverings, that is if there is an orientation preserving homeomorphism $h\colon X_1 \to X_2$ such that $f_2 \circ h = f_1$.

Under the terminology of Grothendieck and Schneps [30, 31], a dessin is called *pre-clean* if the ramification orders above 1 are at most 2, and *clean* if they all are precisely equal to 2. The associated Belyĭ maps are called *pre-clean* and *clean Belyĭ maps*, respectively.

Definition 2 A bipartite map on X is a map on a topological surface X with bipartite structure, that is the set of vertices can be decomposed into a disjoint union $B \cup W$ such that every edge is incident with precisely one vertex from B and one vertex from W. Vertices from B and W are called black and white, respectively.

Two bipartite maps $\mathcal{M}_1$ on X_1 and $\mathcal{M}_2$ on X_2 are isomorphic if there is an orientation preserving homeomorphism $X_1 \to X_2$ which restricts to a bipartite graph isomorphism. When working with bipartite maps we shall adopt the following.

Convention 1 The segments incident with precisely one black and one white vertex in a bipartite map shall be called *darts*. Since every map can be thought of as a bipartite map by considering the edge midpoints as white vertices (see Fig. 3), we shall reserve the term *edge* for maps only. To summarise, a bipartite map has darts, not edges, whilst an edge of a map has precisely two darts.

To every bipartite map on a topological surface X a dessin corresponds, and vice-versa. This correspondence is realised in the following way: given a dessin (X, f), the preimage $f^{-1}([0, 1])$ of the closed unit interval will produce a bipartite map on the underlying surface of the curve X such that the vertices of the map correspond to the points in the preimages of 0 and 1, and the darts correspond to the preimages of the open unit interval. The bipartite structure is obtained by colouring the preimages of 0 in black and the preimages of 1 in white.

On the other hand, given a bipartite map on a topological surface X, colour the vertices in black and white so that the bipartite structure is respected. To the interior of each face add a single new vertex and represent it with a diamond $\diamond$, so that it is distinguished from the black and white vertices. Now triangulate X by connecting the diamonds with the black and white vertices that are on the boundaries of the corresponding faces. Following the orientation of X, call the triangles with vertices oriented as $\bullet$-$\circ$-$\diamond$-$\bullet$ positive, and call other triangles negative (see Fig. 2). Now map the positive and negative triangles to the upper and lower half-plane of $\mathbb{C}$, respectively, and map the sides of the triangles to the real line so that the black, white

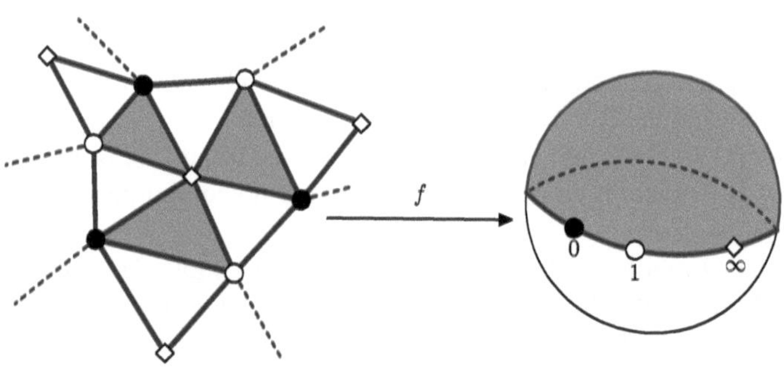

Fig. 2 The positive (*shaded*) and negative triangles are mapped to the upper and lower-half plane, respectively. The sides of the triangles are mapped to $\mathbb{R} \cup \{\infty\}$ so that the *black* and *white vertices* map to 0 and 1, respectively, and the face centres map to ∞

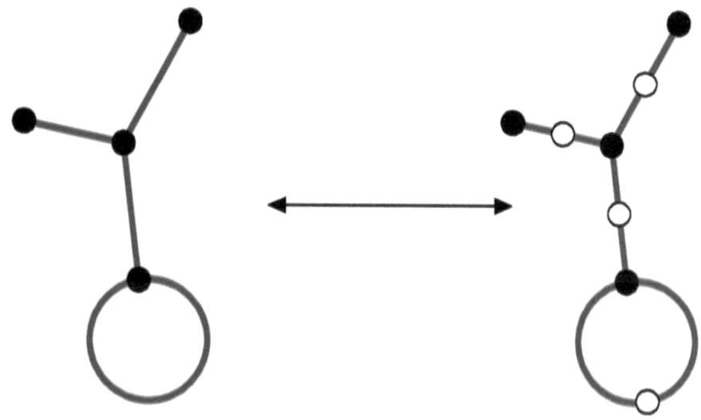

Fig. 3 A map (*left*) is transformed into a clean dessin (*right*) by adding edge midpoints as *white vertices*. In the other way, from a clean dessin we obtain a map by ignoring the *white vertices*

and diamond vertices are mapped to 0, 1 and ∞, respectively. As a result, a ramified cover $f : X \to \mathbb{CP}^1$, ramified only over a subset of $\{0, 1, \infty\}$ will be produced. We now impose on X the unique Riemann surface structure which makes f holomorphic. For a detailed description of this correspondence *see* [16, Sects. 4.2 and 4.3].

Remark 1 In the introduction we stated that maps correspond to clean dessins. Here we explain why this is the case: a given map with n edges can be refined into a bipartite map $2n$ darts by adding the edge midpoints of the map as white vertices. The corresponding Belyĭ function will obviously have ramification orders at the white vertices equal to 2. In the other way, given a clean dessin, we first obtain a bipartite map with $2n$ darts in which every white vertex is incident to precisely two darts, since all the ramification orders above 1 are equal to 2. By ignoring the white vertices we obtain a map with n edges. See Fig. 3 for an example.

From now on we shall think of dessins both as bipartite maps, and as Belyĭ pairs. Consequently, clean dessins are synonymous with maps.

2.1 A Permutation Representation of Dessins

Throughout this section let (X, f) be a dessin with n darts (or, equivalently, such that f is a degree n ramified covering). The goal of this section is to describe how each such dessin can be represented by a triple $(\sigma, \alpha, \varphi)$ of permutations in S_n. However, let us first introduce the following labelling convention to which we will conform throughout the rest of this paper.

Convention 2 We label the darts of a dessin with the elements of the set $\{1, \dots, n\}$ so that, when standing at a black vertex, and looking towards an adjacent white vertex, the label is placed on the 'left side' of the dart. See Fig. 4 for an example.

Following the previous convention, label the darts of a dessin arbitrarily. Now let σ and α denote the permutations which record the cyclic orderings of the labels

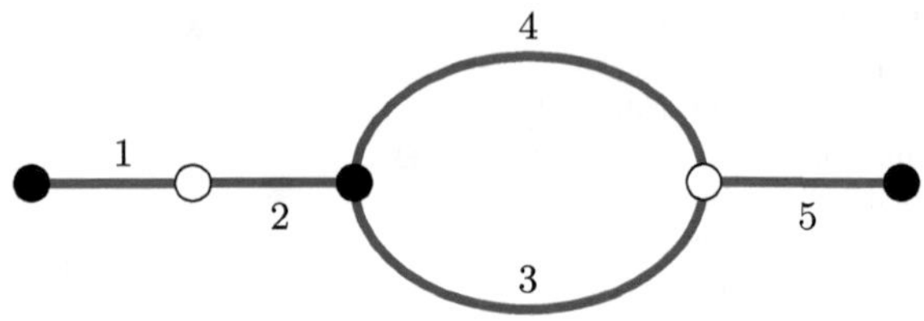

Fig. 4 Labelling of darts. The labels are always on the left when looking from a *black vertex* to its adjacent *white vertices*

around the black and white vertices, respectively, and let φ denote the permutation which records the counter-clockwise ordering of the labels within each face.

Example 1 For the dessin in Fig. 4 we have $\sigma = (1)(2\,3\,4)(5)$, $\alpha = (1\,2)(3\,5\,4)$ and $\varphi = (1\,4\,5\,2)(3)$. The cycles of length 1 are usually dropped from the notation. Note that the cycle corresponding to the 'outer face' is, from the reader's perspective, recorder clockwise. This does not violate our convention since that face should be viewed from the opposite side of the sphere [25, Remark 1.3.18(3)].

Since the labelling was arbitrary, a change of labels corresponds to simultaneous conjugation of σ, α and φ by some element in S_n. Therefore, any dessin can be represented, up to conjugation, as a triple of permutations.

Definition 3 The length of a cycle in σ or α corresponding to a black or a white vertex, respectively, is called the *degree* of the vertex. The length of a cycle in φ corresponding to a face is called the *degree* of the face. Thus, the degree of a vertex is the number of darts incident to it, while the degree of a face is half the number of darts on its boundary.

A triple $(\sigma, \alpha, \varphi)$ representing a dessin $D = (X, f)$ satisfies the following properties:

- the group $\langle \sigma, \alpha, \varphi \rangle$ acts transitively on the set $\{1, \ldots, n\}$ and
- $\sigma\alpha\varphi = 1$.

The first property above is due to the fact that dessins are connected while the second is due to the following: consider three non-trivial simple loops γ_0, γ_1 and γ_∞ on $\mathbb{CP}^1 \backslash \{0, 1, \infty\}$ based at $1/2$ and going around 0, 1 and ∞ once, respectively. The lifts of these loops under f correspond to paths on X that start at some and end at another (possibly the same) point in $f^{-1}(1/2)$. We observe the following.

- Every dart of D contains precisely one element of $f^{-1}(1/2)$ since f is unramified at $1/2$.
- The cardinality of $f^{-1}(1/2)$ is precisely n. Hence there is a bijection between $f^{-1}(1/2)$ and $\{1, \ldots, n\}$.
- With respect to this bijection, σ, α and φ can be thought of as permutations of the set $f^{-1}(1/2)$.

Therefore the loops γ_0, γ_1 and γ_∞ induce σ, α and φ. Since the product $\gamma_0\gamma_1\gamma_\infty$ is trivial, the corresponding permutation $\sigma\alpha\varphi$ must be trivial as well.

We have now seen that to every dessin with n darts we can assign a triple of permutations in S_n such that their product is trivial and the group that they generate acts transitively on the set $\{1, \ldots, n\}$. In a similar fashion we can show that this assignment works in the opposite direction: given three permutations σ, α and φ in S_n such that $\sigma\alpha\varphi = 1$ and the group that they generate acts transitively on $\{1, \ldots, n\}$, we can construct a dessin with n darts so that the cyclic orderings of labels around vertices correspond to the cycles of σ, α and φ, up to simultaneous conjugation. Therefore, up to simultaneous conjugation, a dessin is uniquely represented by a transitive triple $(\sigma, \alpha, \varphi)$ with $\sigma\alpha\varphi = 1$, and such a triple recovers a unique dessin (up to isomorphism).

Remark 2 Obviously, dessins correspond to 2-generated transitive permutation groups since we can set $\varphi = (\sigma\alpha)^{-1}$. However, we prefer to emphasise all three permutations.

We shall use the notation $D = (\sigma, \alpha, \varphi)$ to denote that a dessin D is represented by the triple $(\sigma, \alpha, \varphi)$.

Definition 4 The subgroup of S_n generated by σ, α and φ is called *the monodromy group* of $D = (\sigma, \alpha, \varphi)$ and denoted by $\mathrm{Mon}(D)$.

The monodromy group is actually defined up to conjugation in order to account for all the possible ways in which a dessin can be labelled.

Example 2 The monodromy group of the dessin in Fig. 3 is (isomorphic to) $\mathrm{PSL}_3(2)$. The monodromy group of the dessin in Fig. 4 is S_5.

3 Belyĭ's Theorem and the Galois Action on Dessins

One of the most mysterious objects in mathematics is the absolute Galois group $\mathrm{Gal}(\overline{\mathbb{Q}}|\mathbb{Q})$, the group of automorphisms of $\overline{\mathbb{Q}}$ that fix $\mathbb{Q}$ point-wise, and the study of its structure is one of the goals of the Langlands program. Grothendieck, in his remarkable *Esquisse d'un Programme* [20], envisioned an approach towards understanding $\mathrm{Gal}(\overline{\mathbb{Q}}|\mathbb{Q})$ as an automorphism group of a certain topological object; the starting point of his approach is Belyĭ's theorem, which we restate here.

Theorem 2 (Belyĭ) *Let X be an algebraic curve defined over $\mathbb{C}$. Then X is defined over $\overline{\mathbb{Q}}$ if, and only if there is a holomorphic ramified covering $f: X \to \mathbb{CP}^1$, ramified at most over a subset of* $\{0, 1, \infty\}$.

Aside from Belyĭ's own papers [3, 4], various other proofs can be found in, for example, [33, Theorem 4.7.6] or [16, Chap. 3] or the recent new proof in [17]. Belyĭ himself concluded that the above theorem implies that $\mathrm{Gal}(\overline{\mathbb{Q}}|\mathbb{Q})$ embeds into the outer automorphism group of the profinite completion of the fundamental group of $\mathbb{CP}^1\backslash\{0, 1, \infty\}$, however it was Grothendieck who observed that $\mathrm{Gal}(\overline{\mathbb{Q}}|\mathbb{Q})$ must therefore act faithfully on the set of dessins as well. This interplay between algebraic, combinatorial and topological objects is what prompted Grothendieck to develop his *Esquisse*. For more detail, see [31] or [33].

3.1 Galois Action on Dessins

Let $D = (X, f)$ be a dessin. If X is of genus 0, then necessarily $X = \mathbb{CP}^1$ and $f: \mathbb{CP}^1 \to \mathbb{CP}^1$ is a rational map with critical values in the set $\{0, 1, \infty\}$. If $f = p/q$, where $p, q \in \mathbb{C}[z]$, then Belyĭ's theorem implies that $p, q \in \overline{\mathbb{Q}}[z]$. Moreover, the coefficients of both p and q generate a finite Galois extension K of $\mathbb{Q}$. Therefore $p, q \in K[z]$, and $\mathrm{Gal}(K|\mathbb{Q})$ acts on f by acting on the coefficients of p and q, that is if $\theta \in \mathrm{Gal}(K|\mathbb{Q})$ and

$$f(z) = \frac{a_0 + a_1 z + \cdots + a_m z^m}{b_0 + b_1 z + \cdots b_n z^n},$$

$$\text{then } f^\theta(z) = \frac{\theta(a_0) + \theta(a_1) z + \cdots + \theta(a_m) z^m}{\theta(b_0) + \theta(b_1) z + \cdots \theta(b_n) z^n}.$$

If X is of genus 1 or 2, then as an hyperelliptic algebraic curve it is defined by the zero-set of an irreducible polynomial F in $\mathbb{C}[x, y]$. This time we must take into consideration the coefficients of both F and f which, due to Belyĭ's theorem again, generate a finite Galois extension K of $\mathbb{Q}$. Similarly as in the genus 0 case, $\mathrm{Gal}(K|\mathbb{Q})$ acts on D by acting on the coefficients of both F and f simultaneously. When the genus of X is at least 3, the action is exhibited similarly.

It is not immediately clear that the action of some automorphism in $\mathrm{Gal}(K|\mathbb{Q})$ on a Belyĭ map f will produce a Belyĭ map. This indeed is the case and we refer the reader to the discussion in [25, Sect. 2.4.2].

Since any $\mathbb{Q}$-automorphism of K extends to an $\mathbb{Q}$-automorphism of $\overline{\mathbb{Q}}$ [7, Chap. 3], we truly have an action of $\mathrm{Gal}(\overline{\mathbb{Q}}|\mathbb{Q})$ on the set of dessins.

We shall denote by $D^\theta = (X^\theta, f^\theta)$ the dessin that is the result of the action of $\theta \in \mathrm{Gal}(\overline{\mathbb{Q}}|\mathbb{Q})$ on $D = (X, f)$. We shall also say that D^θ is *conjugate* to D.

The following example is borrowed from [25, Example 2.3.3].

Example 3 Let $D = (X, f)$ be a dessin where X is the elliptic curve

$$y^2 = x(x-1)(x-(3+2\sqrt{3})),$$

and $f: X \to \mathbb{CP}^1$ is the composition $g \circ \pi_x$, where $\pi_x: X \to \mathbb{CP}^1$ is the projection to the first coordinate and $g: \mathbb{CP}^1 \to \mathbb{CP}^1$ is given by

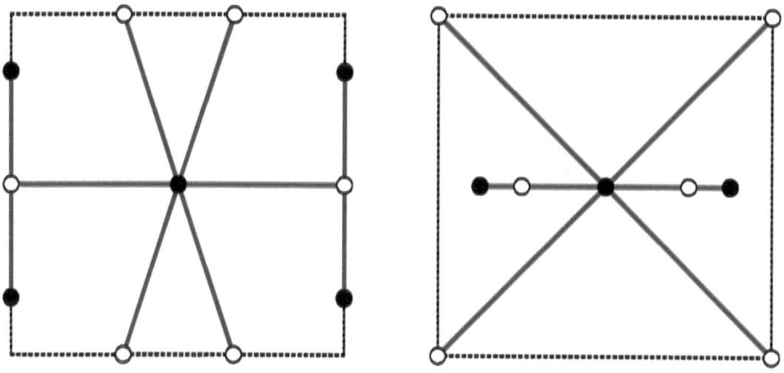

Fig. 5 The two dessins (X, f) and (X^θ, f^θ) from Example 3. The *dotted lines* indicate the boundary of the polygon representation of an orientable genus 1 surface with the usual identification of the *left*-and-*right* and *top*-and-*bottom* sides

$$g(z) = -\frac{(z-1)^3(z-9)}{64z}.$$

The corresponding bipartite map is depicted on the left in Fig. 5.

Note that we must consider $g \circ \pi_x$ and not just π_x since π_x is not a Belyĭ map; it is ramified over four points, namely $0, 1, 3+2\sqrt{3}$ and ∞. However, g maps these four points onto the set $\{0, 1, \infty\}$ and therefore $g \circ \pi_x$ is a true Belyĭ map.

The Galois extension that the coefficients of X and f generate is $K = \mathbb{Q}(\sqrt{3})$ and the corresponding Galois group has only one non-trivial automorphism θ given by $\theta: \sqrt{3} \mapsto -\sqrt{3}$. Therefore X^θ is the elliptic curve $y^2 = x(x-1)(x-(3-2\sqrt{3}))$. The curve X^θ is non-isomorphic to X, which can easily be seen by computing the j-invariants of both.

What about f^θ? In this case, $\pi_x : X^\theta \to \mathbb{CP}^1$ is unramified over $3+2\sqrt{3}$ and ramified over $3-2\sqrt{3}$. However, g maps $3-2\sqrt{3}$ to 0 as well, and since g is defined over $\mathbb{Q}$, the Belyĭ functions f and f^θ coincide. The bipartite map corresponding to (X^θ, f^θ) is depicted on the right in Fig. 5.

This action of $\mathrm{Gal}(\overline{\mathbb{Q}}|\mathbb{Q})$ on dessins is faithful already on the set of *trees*, i.e. the genus 0 dessins with precisely one face and with polynomials as Belyĭ functions. However, this is not straight-forward (proofs can be found in [16, 30]) and, surprisingly, it is much easier to show faithfulness in genus 1 [16, Sect. 4.5.2]. Moreover, the action is faithful in every genus [16, Sect. 4.5.2].

3.2 Galois Invariants

Here we shall list a number of properties of dessins which, up to various notions of equivalence, remain invariant under the action of $\mathrm{Gal}(\overline{\mathbb{Q}}|\mathbb{Q})$. Such properties are called *Galois invariants* of dessins. We shall use the notation $D \simeq D'$ to indicate that two dessins D and D' are conjugate.

Invariant 1 (Passport) Let $D = (\sigma, \alpha, \varphi)$ be a dessin with n darts. The cycle types of σ, α and φ define three partitions λ_σ, λ_α and λ_φ of n. The *passport* of D is the sequence $[\lambda_\sigma, \lambda_\alpha, \lambda_\varphi]$. If $D' = (\sigma', \alpha', \varphi')$ and $D \simeq D'$, then $[\lambda_\sigma, \lambda_\alpha, \lambda_\varphi] = [\lambda_{\sigma'}, \lambda_{\alpha'}, \lambda_{\varphi'}]$. In other words, conjugate dessins have the same passport.

We compactly record a partition of, for example, $n = 17 = 3+3+3+3+2+1+1+1$ as $3^4 21^3$. If a double-digit number appears in the partition, for example $23 = 11 + 11 + 1$, then we record it as $(11)^2 1$.

Example 4 The dessin in Fig. 3 has the sequence $[3^2 1^2, 2^4, 71]$ as its passport. The dessin in Fig. 4 has $[31^2, 32, 41]$ as its passport. The two dessins in Fig. 5 both have $[61^2, 42^2, 62]$ as their passport.

The passport is a very crude invariant, however much useful information can be extracted from it. For example, the number of black vertices, white vertices, darts and faces is invariant and hence the genus of the surface must also be invariant. Moreover, we can conclude that every orbit of the action is finite since there are only finitely many dessins with a given passport.

Invariant 2 (Monodromy group) If $D \simeq D'$, then $\mathrm{Mon}(D) \cong \mathrm{Mon}(D')$. In other words, conjugate dessins have isomorphic monodromy groups.

Example 5 The monodromy group of the dessin D on the left side in Fig. 5 is the nilpotent group given by the external wreath product of $\mathbb{Z}_2$ by the alternating group A_4. Since the dessin on the right side of the same figure is conjugate to D, its monodromy group is isomorphic to $\mathrm{Mon}(D)$.

The monodromy group is a much finer invariant than the passport since dessins with the same passport may have non-isomorphic monodromy groups.

Invariant 3 (Automorphism group) Let $D = (\sigma, \alpha, \varphi)$. The centre of $\mathrm{Mon}(D)$ in S_n is the *automorphism group* of D, denoted by $\mathrm{Aut}(D)$. If $D \simeq D'$, then $\mathrm{Aut}(D) \cong \mathrm{Aut}(D')$.

If the automorphism group of a dessin D acts transitively on the set $\{1, \ldots, n\}$ or, equivalently, if $|\mathrm{Aut}(D)| = n$, then we say that the dessin is *regular*. It has been shown in [18, 21] that $\mathrm{Gal}(\overline{\mathbb{Q}}|\mathbb{Q})$ acts faithfully on the set of regular dessins as well.

Invariant 4 (Cartography group) The *cartography group* $\mathrm{Cart}(D)$ of a dessin D is the monodromy group of the map obtained from D by colouring all the white vertices black and adding new white vertices to the midpoints of edges. Therefore, for maps or clean dessins we have $\mathrm{Cart}(D) = \mathrm{Mon}(D)$. As it was the case with the monodromy group, conjugate dessins have isomorphic cartography groups.

Since the cartography groups are subgroups of S_{2n}, when n is large they are in general more difficult to compute than the monodromy groups. However, Jones and Streit have shown in [24] that the cartography group can be used to distinguish between the orbits of $\mathrm{Gal}(\overline{\mathbb{Q}}|\mathbb{Q})$ when the monodromy group does not suffice. That is, there are non-conjugate dessins with isomorphic monodromy groups but non-isomorphic cartography groups.

More Galois invariant groups that arise from the monodromy group in a similar fashion can be found in [27].

Invariant 5 (Duality) Given a dessin $D = (X, f)$ we define its *dual* dessin D^* to be the dessin corresponding to the Belyĭ pair $(X, 1/f)$. Clearly, if $D_1 \simeq D_2$, then $D_1^* \simeq D_2^*$.

In terms of permutation representations, if $D = (\sigma, \alpha, \varphi)$, then D^* will have the triple $(\varphi^{-1}, \alpha^{-1}, \sigma^{-1})$ as its permutation representation. Geometrically this means that the black vertices and the face centres of the dual are the face centres and the black vertices of D, respectively, while the white vertices remain unchanged, except for the orientation of the labels. The darts of D^* are the curved segments that connect the face centres and the white vertices of D. See Fig. 6 for an example.

Remark 3 If D is a map then D^* corresponds to the geometric dual of a map. If e is an edge of D, then the unique edge e^* in D^* which intersects e at the appropriate white vertex is called the *coedge* of e.

Invariant 6 (Self-duality) We say that a dessin is self-dual if it is isomorphic to its dual. If D is self dual and $D \simeq D'$, then D' is self-dual as well. We shall considered self-duality again in Sect. 5.2.

Invariant 7 (Field of moduli) Let D be a dessin and

$$\mathrm{Stab}(D) = \{\theta \in \mathrm{Gal}(\overline{\mathbb{Q}}|\mathbb{Q}) \mid D^\theta = D\}$$

the stabiliser of D in $\mathrm{Gal}(\overline{\mathbb{Q}}|\mathbb{Q})$. The *field of moduli* of D is the fixed field corresponding to $\mathrm{Stab}(D)$, that is the field

$$\{q \in \overline{\mathbb{Q}} \mid \theta(q) = q, \text{ for all } \theta \in \mathrm{Stab}(D)\}.$$

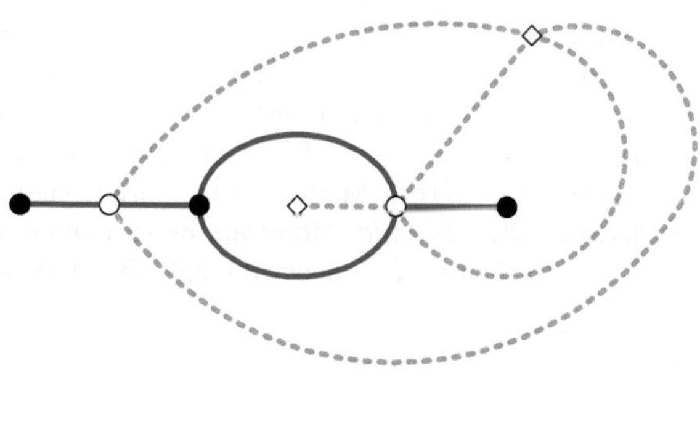

Fig. 6 The dessin (full) from Fig. 4 and its dual (*dashed*)

Alternatively, the field of moduli of D is the intersection of all *fields of definition* of D, i.e. all the fields in which we can write down a Belyĭ pair for D.

Fields of moduli are notoriously difficult to compute, and moreover, there are dessins whose Belyĭ pairs cannot be realised over their own fields of moduli! [25, e.g. 2.4.8 and 2.4.9]. Therefore, a natural question to ask is when can a dessin be defined over its field of moduli. Based on the work of Birch in [2] (see also [32]), a necessary, but not sufficient condition was given in [36].[1]

Theorem 3 *A dessin can be defined over its field of moduli if there exists a black vertex, or a white vertex, or a face center which is unique for its type and degree.*

4 Matroids and Delta-Matroids

It is often said that matroids are a combinatorial abstraction of linear independence. Formally we have

Definition 5 Given a non-empty finite set E, a *matroid on* E is a non-empty family $M(E)$ of subsets of E which is closed under taking subsets, i.e.

- if $J \in M(E)$ and $I \subseteq J$, then $I \in M(E)$,

and satisfies the following *augmentation axiom*:

- if $I, J \in M(E)$ with $|I| < |J|$, then there exists $x \in J \setminus I$ such that $I \cup \{x\} \in M(E)$.

The elements of $M(E)$ obviously mimic the properties of linearly independent sets of vectors and are hence called *independent sets*. Subsets of E which are not independent are called *dependent*. Maximal independent sets are called *bases*, and, as the reader might suspect, any two bases of $M(E)$ are of the same size [29, Lemma 1.2.1]. Two matroids $M(E)$ and $M(E')$ are isomorphic if there is a bijection $\psi : E \to E'$ such that $\psi(I)$ is independent if, and only if I is independent.

Matroids were introduced by Hassler Whitney [35] and, as the name suggests, arise naturally from matrices; the collection of linearly independent sets of columns in a matrix forms a matroid [29, Proposition 1.1.1]. Matroids which are isomorphic to matroids arising from matrices are called *representable*.

A multitude of examples of matroids arise from graphs as well. Given an abstract undirected graph $G = (V, E)$, the collection of its acyclic sets of edges forms a matroid $M(G)$ [19, Theorem 4]. The independent sets of this matroid are in fact subsets of E, however we denote it by $M(G)$ to emphasise that the matroid is arising from a graph. The spanning forests of G correspond to the bases of $M(G)$. If G is connected then the trees and the spanning trees correspond to the independent sets and the bases of $M(G)$. Matroids which are isomorphic to matroids arising from graphs are called *graphic*. Moreover, every graphic matroid is isomorphic to the graphic matroid of some connected graph [29, Proposition 1.2.8].

[1] See also Theorem 2.4.14 in [25].

Convention 3 It is customary in matroid theory to drop the braces and commas when specifying sets. For example, *abc* stands for the set $\{a, b, c\}$.

Given a matroid $M(E)$ we can completely recover the independent sets by describing only the collection $\mathcal{B}$ of its bases. On the other hand, if $\mathcal{B}$ is a non-empty collection of subsets of some non-empty set E, then $\mathcal{B}$ will be the collection of bases of a matroid if, and only if the following *exchange axiom* is satisfied [29, Corollary 1.2.5]:

- if $B_1, B_2 \in \mathcal{B}$ and $x \in B_1 \setminus B_2$, then there is $y \in B_2 \setminus B_1$ such that $(B_1 \setminus x) \cup y \in \mathcal{B}$.

Let us look at a simple example of a graphic matroid.

Example 6 Let G be the map obtained from the bipartite map in Fig. 4 by colouring all the white vertices into black vertices (see Fig. 7). The bases of $M(G)$ are the sets 1235 and 1245 and they correspond precisely to the spanning trees of the map.

Let $\mathcal{B}$ be the collection of bases of some matroid $M(E)$ and let

$$\mathcal{B}^* = \{E \setminus B \mid B \in \mathcal{B}\}$$

be the collection of the complements of its bases. This collection is clearly non-empty and it can be shown that it satisfies the exchange axiom [29, Chap. 2]. The matroid with $\mathcal{B}^*$ as its collection of bases is called the *dual matroid* of $M(E)$, and is denoted by $M^*(E)$.

Example 7 Let us go back to the map in Fig. 7. As we have seen in Example 6, the bases of this map are 1235 and 1245. Recall that the unique edge of the dual map which intersects an edge e of the map is labelled by e^*. Therefore the bases of the dual map should be the coedges 4^* and 3^*. In Fig. 7 we can see that this indeed is the case.

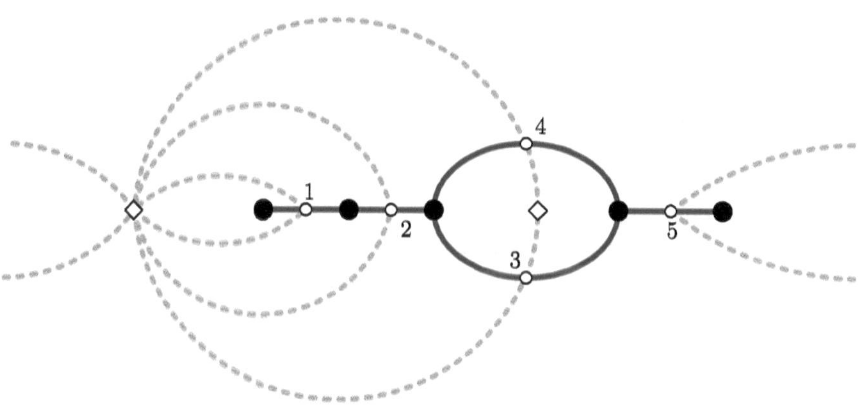

Fig. 7 A map obtained from the dessin in Fig. 4 by colouring the *white vertices* into *black* and adding new *white vertices* at the edge midpoints. The dual map (*dashed*) is formed by connecting the face centres to the (new) *white vertices*. The segments on the *left* and *right* go around the *sphere* and connect into a loop

We say that a matroid is *cographic* if it is isomorphic to the dual of some graphic matroid. The following theorem of Whitney [34] establishes a matroidal characterisation of planarity.

Theorem 4 (Whitney's planarity criterion) *Let G be a connected graph. Then G is planar if, and only if $M(G)$ is cographic. Moreover, if G is a plane map, then $M^*(G) = M(G^*)$, where G^* is the geometric dual of G.*

4.1 Delta-Matroids

As we have seen in Theorem 4, the dual matroid of a plane map is the matroid of the dual map. This correspondence does not hold for graphs that are not planar. However, we would like to extend this property to non-planar graphs and their cellular embeddings, that is to maps on surfaces of any genus. To that effect, we introduce the following.

Definition 6 A *delta-matroid $\Delta(E)$ on $E = \{1, \ldots, n\}$* is a non-empty collection $\mathcal{F}$ of subsets of E satisfying the following *symmetric axiom*:

- if $F_1, F_2 \in \mathcal{F}$ and $x \in F_1 \triangle F_2$, then there is $y \in F_2 \triangle F_1$ such that $F_1 \triangle \{x, y\} \in \mathcal{F}$.

Here $\triangle$ denotes the symmetric difference of sets. The elements of $\mathcal{F}$ are called *feasible sets*. Two delta-matroids $\Delta(E)$ and $\Delta(E')$ are isomorphic if there is a bijection $\psi : E \to E'$ preserving feasible sets. We shall use the notation $\Delta(E) \cong \Delta(E')$ to indicate that $\Delta(E)$ and $\Delta(E')$ are isomorphic delta-matroids.

It is straightforward to show that every matroid is a delta-matroid, however not every delta-matroid is a matroid, as we shall see.

Delta-matroids, also known as symmetric or Lagrangian matroids [8, Chap. 4], were first introduced by Bouchet [5] and later generalized to the so-called *Coxeter matroids* by Gelfand and Serganova [14, 15]. A systematic treatment of Coxeter matroid theory can be found in [8].

Delta-matroids arise from maps in a fashion similar to which graphic matroids arise from graphs. However, instead of spanning trees we shall consider *bases* of maps. To that effect, let $\mathcal{M}$ be a map on X with n edges labelled by the set $E = \{1, 2, \ldots, n\}$. Label the edges of the dual map $\mathcal{M}^*$ by the set $E^* = \{1^*, 2^*, \ldots, n^*\}$ so that j^* is the coedge corresponding to j. Call an n-subset B of $E \cup E^*$ admissible if precisely one of j or j^* appears in it.

Definition 7 An admissible n-subset B of $E \cup E^*$ is called a *base* if $X \backslash B$ is connected.

It was shown in [6, Proposition 2.1] that the bases of $\mathcal{M}$ are equicardinal and spanning, that is each base includes a spanning tree of the underlying graph of $\mathcal{M}$.

Definition 8 A *quasi-tree* is a map with precisely one face. A *spanning quasi-tree of a map* $\mathcal{M}$ is a quasi-tree obtained from a base B of $\mathcal{M}$ by ignoring the starred elements.

Remark 4 We are allowing the case of an empty spanning quasi-tree. This occurs precisely when there is a base $B = E^*$. In that case, $X \setminus E^*$ is connected and therefore $\mathcal{M}^*$ has precisely one face. Hence $\mathcal{M}$ has only one vertex and we think of the empty spanning quasi-tree as the degenerate map on the sphere with one vertex and no edges.

Let $\mathcal{B}$ denote the collection of bases of a map $\mathcal{M}$, and let $\mathcal{F}$ denote the collection of the spanning quasi-trees of $\mathcal{M}$, that is the collection

$$\mathcal{F} = \{E \cap B | B \in \mathcal{B}\}.$$

Analogously to matroids, the spanning quasi-trees of a map form a delta-matroid [8, Theorem 4.3.1].

Theorem 5 *If* $\mathcal{M}$ *is a map on* X*, then* $\mathcal{F}$ *is the collection of feasible sets of a delta-matroid.*

The delta-matroid arising from a map $\mathcal{M}$ shall be denoted by $\Delta(\mathcal{M})$ or $\Delta(D)$ when we are assuming that D is a clean dessin.

Example 8 Let $\mathcal{M}$ be a map on a genus 1 surface X with two vertices, three edges and one face, as shown and labelled in Fig. 8. Since the map itself has precisely one face, then $X \setminus \mathcal{M}$ must be connected. Therefore 123 is a base. It is easy to see that no 2-subset of 123, together with an appropriate coedge, is a base. The remaining admissible 3-sets are 12*3*, 1*23*, 1*2*3 and 1*2*3*. Out of those four, only 12*3* and 1*2*3 do not disconnect X. Therefore, the feasible sets are 123, 1, 3.

In general one does not need to go through all possible admissible n-subsets of $E \cup E^*$ and check which ones are bases. It is enough to find one base which can then be used to find the *representation* of the delta-matroid as an n by $2n$ matrix over $\mathbb{Q}^E \oplus \mathbb{Q}^{E^*}$. The linearly independent admissible n-sets of columns of the representation will correspond to the bases of the map [8, Theorem 4.3.5]. However, we shall not consider representations of delta-matroids in this paper.

We note that the Definition 6 can be modified so that a delta-matroid is specified by a collection of admissible n-sets [8, Sect. 4.1.2]. In that case we must replace F_1, $F_2, \mathcal{F}, x, y$ and $\{x, y\}$ with $B_1, B_2, \mathcal{B}, \{x, x^*\}, \{y, y^*\}$ and $\{x, x^*, y, y^*\}$, respectively. The reason that we chose our definition is due to the fact that if $\mathcal{M}$ is a map on the sphere, then its feasible sets correspond precisely to its spanning trees and therefore the delta-matroid in question is a matroid.

As in the case of matroids, there exists a notion of a dual delta-matroid.

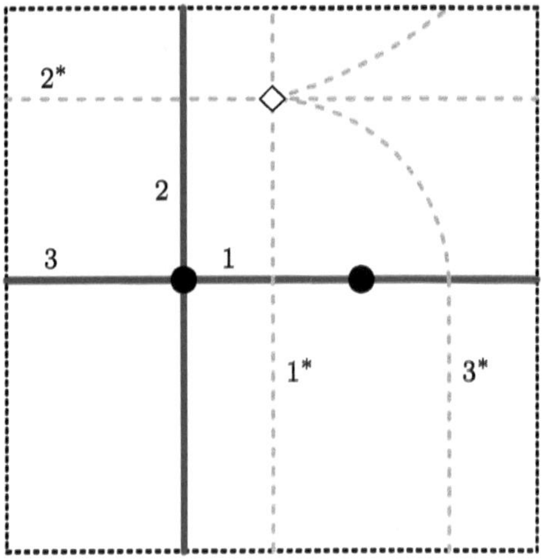

Fig. 8 The bases of the map are 123, 12*3* and 1*2*3. Hence $\Delta(\mathcal{M}) = \{123, 1, 3\}$. The edges 1 and 3 are the spanning quasi-trees of $\mathcal{M}$ which can be embedded as maps only on the *sphere*

Proposition 1 *Let $\Delta(E)$ be a delta-matroid with $\mathcal{F}$ as its collection of feasible sets. Then the collection*

$$\mathcal{F}^* = \{E \backslash F | F \in \mathcal{F}\}$$

is the collection of feasible sets of some delta-matroid on E.

This proposition is easily seen to be true by noting that

$$F_1 \triangle F_2 = (E \backslash F_1) \triangle (E \backslash F_2).$$

The delta-matroid on E with $\mathcal{F}^*$ as the collection of its feasible sets is called the *dual delta-matroid of* $\Delta(E)$ and is denoted by $\Delta^*(E)$.

Theorem 6 *Let $\mathcal{M}$ be a map and $\mathcal{B}$ the collection of its bases. Let $\mathcal{M}^*$ be its dual map and $\Delta(\mathcal{M}^*)$ the delta-matroid of $\mathcal{M}^*$. Then $\Delta^*(\mathcal{M}) \cong \Delta(\mathcal{M}^*)$.*

Proof The bases of $\mathcal{M}$ and $\mathcal{M}^*$ clearly coincide. Therefore, the collection of feasible sets of $\Delta(\mathcal{M}^*)$ is

$$\mathcal{F}' = \{E^* \cap B | B \in \mathcal{B}\}.$$

If F is a feasible set of $\Delta(\mathcal{M})$, then $E \backslash F$ is a feasible set of $\Delta^*(\mathcal{M})$, and we have

$$\begin{aligned} E \backslash F &= E \cap F^c = E \cap (B \cap E)^c \\ &= E \cap (B^c \cup E^*) = E \cap B^c \\ &= E \cap B^*, \end{aligned}$$

where B^* is the admissible n-subset obtained from B by starring and un-starring the un-starred and starred elements, respectively. Denote by $\psi : E \to E^*$ the bijection $\psi(i) = i^*$. From the computation above we have

$$\psi(E \backslash F) = \psi(E) \cap \psi(B^*) = E^* \cap B.$$

Hence $\Delta^*(\mathcal{M})$ and $\Delta(\mathcal{M}^*)$ are isomorphic. Moreover, by relabelling the edges of $\mathcal{M}^*$ with the elements of E we can even achieve equality between the two delta-matroids.

If we recall that for plane maps the feasible sets correspond to spanning trees, we immediately recover Theorem 4. In other words, a delta-matroid $\Delta(\mathcal{M})$ is a matroid if, and only if $\mathcal{M}$ is a plane map.

5 Galois Action on the Delta-Matroids of Maps

Since delta-matroids do not take into account the bipartite structure of dessins, throughout this section we shall consider maps only. Nevertheless, this restriction is not a significant one, as established by the following corollary [30, p. 50] to Theorem 2.

Corollary 1 *Let X be an algebraic curve defined over $\mathbb{C}$. Then X is defined over $\overline{\mathbb{Q}}$ if, and only if there is a clean Belyĭ map $f : X \to \mathbb{CP}^1$.*

This corollary is due to the fact that if $\vartheta : X \to \mathbb{CP}^1$ is a Belyĭ function, then $f = 4\vartheta(1 - \vartheta)$ is a clean Belyĭ function on the same curve X. The dessin to which it corresponds is a familiar one: it is the dessin obtained from (X, ϑ) by colouring all the white vertices black and adjoining the edge midpoints as the white vertices.

As we have seen, delta-matroids of maps are defined through a topological property, namely connectedness, and therefore we cannot expect that conjugate maps will have isomorphic delta-matroids. This indeed is the case, as we will see in the following examples.

Example 9 Let A, B_+ and B_- be the three genus 0 clean dessins depicted in Fig. 9 with Belyĭ functions

$$f(z) = 16\frac{(391 + 550\nu + 455\nu^2)(z + 2\nu)(z + 1)^2 z^5}{(16z - \nu + 7\nu^2 - 4)(-8z + 4\nu + 3\nu^2 - 4)^2},$$

where ν is a root of the irreducible polynomial

$$7\nu^3 + 2\nu^2 - \nu - 4.$$

Fig. 9 From *left* to *right*: dessins A, B_+ and B_-

The dessin A corresponds to its real root, while B_+ and B_- correspond to its imaginary roots with positive and negative real parts, respectively [1, Figs. 87–89]. Clearly, any two are conjugate.

Since these dessins are plane maps, their delta-matroids are matroids and the feasible sets are their spanning trees. Dessins B_+ and B_- clearly have isomorphic delta-matroids with two feasible sets, while A has only one feasible set.

Example 10 Let us look at some delta-matroids which are not matroids. Let A_+, A_- and B be the three genus 1 clean dessins as depicted and labelled in Fig. 10. The Belyĭ pairs of the three dessins have coefficients in the fixed field corresponding to the Galois group of the irreducible polynomial

$$256\nu^3 - 544\nu^2 + 1427\nu - 172,$$

and any two are conjugate. Similarly to the previous example, the dessin B corresponds to the Belyĭ pair defined over $\mathbb{R}$ while the Belyĭ pairs for A_+ and A_- are complex-conjugate. Due to the complicated expressions involved, we shall omit the equations for the Belyĭ pairs. However, the reader may look them up in [1, pp. 39–40].

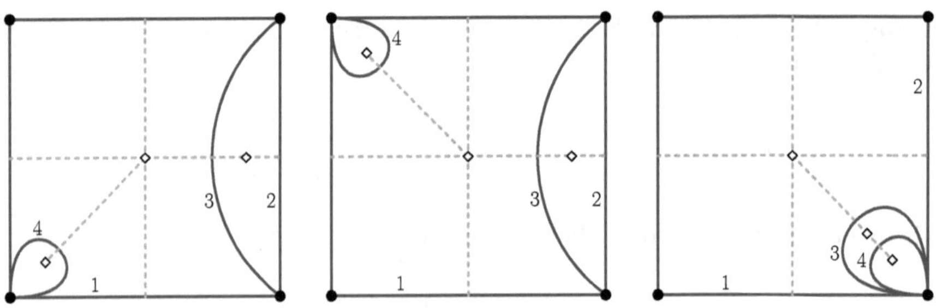

Fig. 10 From *left* to *right*: dessins A_+, A_- and B

The bases of A_+ and A_- are 123^*4^*, 12^*34^* and $1^*2^*3^*4^*$ hence the feasible sets are 12, 13 and $\emptyset$. However, B has only two bases, namely 123^*4^* and $1^*2^*3^*4^*$ and therefore has only two feasible sets: 12 and $\emptyset$. The reason why delta-matroids fail to be Galois invariant is illustrated clearly in this example: a delta-matroid takes into account the topology of edges and hence distinguishes between non-contractible and contractible loops on the surface whereas $\mathrm{Gal}(\overline{\mathbb{Q}}|\mathbb{Q})$ does not!

5.1 Trivial Delta-Matroidal Galois Invariants

The simplest dessins are the trees, that is genus 0 dessins with precisely one face. As we have already mentioned in the last paragraph before Sect. 3.2, the action of $\mathrm{Gal}(\overline{\mathbb{Q}}|\mathbb{Q})$ on the set of trees is very rich since it is faithful. However, delta-matroids associated to trees do not reveal much information as every tree has precisely one feasible set, the tree itself.

Similarly, $\mathrm{Gal}(\overline{\mathbb{Q}}|\mathbb{Q})$ will preserve the delta-matroid of a genus 0 dessin which has n faces of degree 1 and one face of arbitrary degree. Such a dessin is a tree with m loops attached to it. Again, every such dessin clearly has only one feasible set, namely the tree obtained by removing the m loops. Therefore, we have the following proposition.

Proposition 2 *Let D be a genus 0 clean dessin which is either*

(i) a tree,
(ii) a tree with m degree 1 faces attached, or
(iii) the dual dessin of a dessin of type (i) *or* (ii).

If D' *is a dessin conjugate to D, then* $\Delta(D') \cong \Delta(D)$.

Proof In the cases (i) and (ii) the proof is trivial if we recall that the passport of a dessin is a Galois invariant. Hence the conjugate dessin D' must be of the same type as D in both cases. Since the delta-matroids of those dessins are one and the same feasible set, namely the (underlying) tree, we must have $\Delta(D') \cong \Delta(D)$.

For (iii), recall from Invariant 5 that the duals of conjugate dessins are conjugate as well. Since D^* is of type (i) or (ii) we have $\Delta(D'^*) \cong \Delta(D^*)$. Combining with Theorem 6 we have

$$\Delta^*(D') = \Delta(D'^*) \cong \Delta(D^*) = \Delta^*(D).$$

Now by noting that $(\Delta^*)^* = \Delta$, we recover $\Delta(D') \cong \Delta(D)$.

As we have seen in Example 9, the case (ii) cannot be improved even to trees with only one degree 2 face attached. The following conjugate dessins found in [37] show that case (i) cannot be extended to quasi-trees.

Example 11 Let T_5 denote the fifth Chebyshev polynomial of the first kind and consider its square

$$T_5^2(x) = 25x^2 - 200x^4 + 560x^6 - 640x^8 + 256x^{10}.$$

This polynomial is a clean Belyĭ map with critical points in the set

$$\left\{0, \frac{1 \pm \sqrt{5}}{4}, \frac{-1 \pm \sqrt{5}}{4}, \sqrt{\frac{5 \pm \sqrt{5}}{8}}, \sqrt{\frac{-5 \pm \sqrt{5}}{8}}\right\}.$$

Therefore, if X is the algebraic curve

$$y^2 = (x-1)(x+1)\left(x - \sqrt{\frac{5+\sqrt{5}}{8}}\right),$$

then the composition $t = T_5^2 \circ \pi_x$, where $\pi_x\colon X \to \mathbb{CP}^1$ is the projection to the first coordinate, is a clean Belyĭ map. Clearly $D = (X, t)$ will have precisely one face since $t^{-1}(\infty) = \{\infty\}$, as we can see in Fig. 11.

Let D be labelled as in Fig. 11 and let B be a base of D. If the edge 10 is in B then no coedges can appear since cuts along the two edges 10 and e^*, for any $e \in \{1, \ldots, 9\}$, will clearly disconnect X. Therefore, $B = 12\ldots10$ is the only base containing the edge 10. On the other hand, if 10^* is in B then at least one coedge $e^* \in \{1^*, \ldots, 9^*\}$ must appear since $1\ldots9(10)^*$ disconnects D. But if two or more coedges in $\{1^*, \ldots, 9^*\}$ appear in B then D will again be disconnected. Therefore, $\Delta(D)$ has precisely 10 feasible sets, namely $12\ldots10$ and $1\ldots\hat{e}\ldots9$, where $\hat{e}$ denotes the omission of $e \in \{1, 2, \ldots 9\}$.

Now let θ be an automorphism in $\mathrm{Gal}(\overline{\mathbb{Q}}|\mathbb{Q})$ such that

$$\theta\colon \sqrt{\frac{5+\sqrt{5}}{8}} \mapsto \sqrt{\frac{5-\sqrt{5}}{8}}.$$

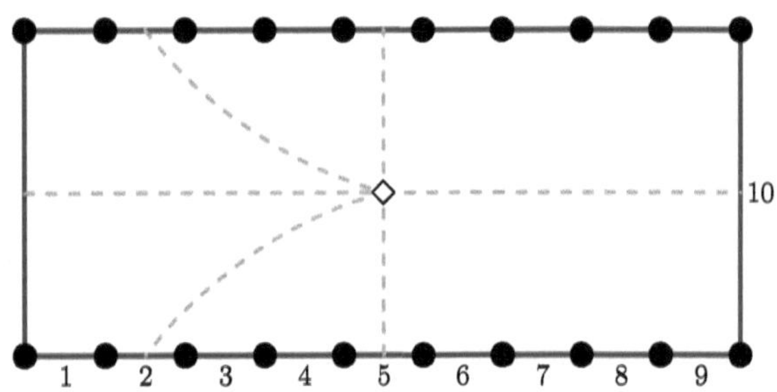

Fig. 11 The dessin (X, t). The only feasible set containing 10 is the entire dessin. Any two coedges $1 \leq e, f \leq 9$ disconnect X so other feasible sets must be of the form $1 \cdots \hat{e} \cdots 9$, where $\hat{e}$ is omitted

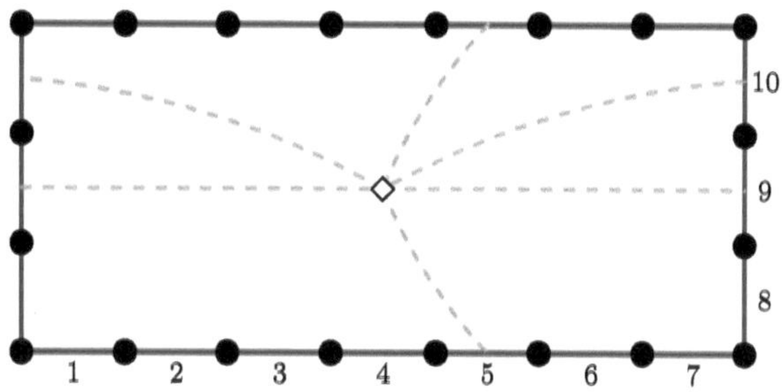

Fig. 12 The dessin $D^\theta = (X^\theta, t)$. There are at least 18 feasible sets obtained by adjoining $1 \cdots \hat{e} \cdots 7$, where $\hat{e}$ is omitted, to 89, 8(10) or 9(10)

Since T_5^2 is defined over the rationals, then $(T_5^2)^\theta$ coincides with T_5^2 and therefore t^θ and t coincide as well. However, X^θ, which is given by

$$y^2 = (x-1)(x+1)\left(x - \sqrt{\frac{5-\sqrt{5}}{8}}\right),$$

is a curve not isomorphic to X. Hence D^θ and D are non-isomorphic conjugate dessins. The corresponding map is shown in Fig. 12.

Let D^θ be labelled as in Fig. 12 and B a base of D^θ. If the edges 8, 9 and 10 are in B, then B must be the entire dessin. Now suppose that 8, 9 and 10* are in B. Then the rest of B must be of the form $1 \ldots \hat{e} \ldots 7$, where $\hat{e} \in \{1, \ldots 7\}$ is omitted. We can conclude the same for bases that contain 8, 9*, 10 or 8*, 9, 10. Therefore $\Delta(D^\theta)$ has at least 19 feasible sets and cannot be isomorphic to $\Delta(D)$.

Question 1 As we have seen, $\mathrm{Gal}(\overline{\mathbb{Q}}|\mathbb{Q})$ alters significantly the delta-matroids of conjugate dessins. In the cases where the delta-matroid is preserved, most information about the dessin is not captured. Is there an interesting family of dessins for which delta-matroids could provide some useful information?

5.2 *Self-Duality of Maps and Matroids*

Recall that a map is self-dual if it is isomorphic to its dual. As an example, any map in Fig. 9 is self-dual.

We say that a delta-matroid is *self-dual* if $\Delta(E) \cong \Delta^*(E)$. Combining with Theorem 6, the delta-matroid of a map $\mathcal{M}$ is self-dual if, and only if $\Delta(\mathcal{M}) \cong \Delta(\mathcal{M}^*)$.

Self-dual maps clearly have self-dual delta-matroids. The following example demonstrates that the converse need not be true.

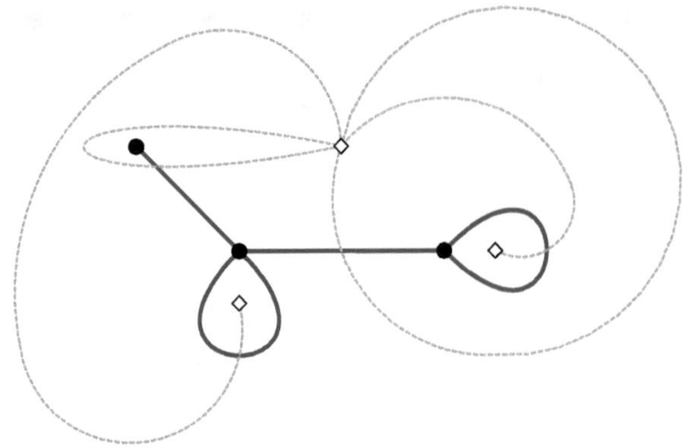

Fig. 13 A map which is not self-dual but has a self-dual delta-matroid

Example 12 Consider the map in Fig. 13. It is not self-dual since it has only one vertex of degree 1, while the dual map has two. However, both have precisely one feasible set corresponding to the unique spanning tree. Clearly their delta-matroids are isomorphic, as the two feasible sets are of the same size.

By a theorem of Steinitz[2] [28, p. 63], a 3-connected planar simple graph G has, up to isomorphism, a unique embedding on the sphere. Moreover, if the delta-matroid of G is self-dual, then G, as a planar map, is self-dual as well. Hence a 3-connected planar simple graph is self-dual as a map if, and only if its delta-matroid is self-dual. As we have mentioned in Sect. 3.2, the property of being self-dual is a Galois invariant, and therefore the conjugates of 3-connected plane simple maps with self-dual delta-matroids must have a self-dual delta-matroid. Can the same be said, at least in the genus 0 case, for all clean dessins with self-dual delta-matroids? It is easy to see by inspecting the catalogue [1] that this is the case for genus 0 dessins with 4 edges or less. However, this might be due to the simplicity of orbits involved; the largest orbit in the catalogue consists of only 3 dessins. Here we pose the following question.

Question 2 Given a genus 0 clean dessin D, if the delta-matroid of D is self-dual, does the same hold for any dessin conjugate to D?

Since in the genus 0 case the feasible sets of D correspond to spanning trees, and if v is the number of vertices, then any feasible set must have $v-1$ edges. Moreover, if F is a feasible set of D, then $E\backslash F$ is a feasible set of D^* and therefore D must have $2v-2$ edges. Euler's formula now implies that the number f of faces of D has to be $f = v$. Therefore, if a counterexample is to be found, its passport should be of the following form

$$[a_1^{\alpha_1} \cdots a_j^{\alpha_j}, 2^{2v-2}, b_1^{\beta_1} \cdots b_k^{\beta_k}],$$

with the following equalities satisfied:

[2]Also, see 8.2.16 in [29]. There the same theorem is attributed to Whitney.

$$\alpha_1 + \cdots + \alpha_j = \beta_1 + \cdots + \beta_k = v,$$
$$a_1\alpha_1 + \cdots + a_j\alpha_j = b_1\beta_1 + \cdots + b_k\beta_k = 4v - 4.$$

In higher genus feasible sets are not all of the same size and therefore there are less constraints on the passport. This would suggest that a question analogous to Question 2 is even less likely to have a positive answer.

Question 3 Are there some other properties of delta-matroids that are invariant under the action of $\mathrm{Gal}(\overline{\mathbb{Q}}|\mathbb{Q})$?

6 Partial Duals and Twists of Delta-Matroids

A *partial dual* of a map is a generalisation of the geometric dual of a map. It was first introduced in [10] and later generalised to hypermaps in [11], where a representation as a triple of permutations is given as well. In this paper we shall first define partial duals combinatorially and then explain the geometric counterpart, thus working in the opposite direction of [11]. We shall consider maps only but give some remarks on hypermaps as well. Throughout this section $D = (\sigma, \alpha, \varphi)$ will denote a clean dessin with n edges, hence α will be of the form $\alpha = c_1 \cdots c_n$, where $c_1, \ldots, c_n$ are n disjoint transpositions. We are identifying the edges of D with the cycles of α so that the j-th edge corresponds to the transposition c_j. The notation D/j stands for the map D with the edge j contracted, while $D \setminus j$ stands for the map D with the edge j deleted.

Definition 9 Let $D = (\sigma, \alpha, \varphi)$ be a map. The *partial dual with respect to an edge* j of D is the map

$$\partial_j D = (\sigma c_j, \alpha, c_j\varphi).$$

The following theorem shows that the partial dual with respect to an edge is well defined.

Theorem 7 *Let* $D = (\sigma, \alpha, \varphi)$ *be a map. Then* $\sigma c_j \alpha c_j \varphi = 1$ *and the group* $\langle \sigma c_j, \alpha, c_j\varphi \rangle$ *acts transitively on* $\{1, \ldots, 2n\}$.

Proof Since c_j commutes with α we clearly have $\sigma c_j \alpha c_j \varphi = 1$. If $n = 1$ we are done since in that case $\partial_j D$ corresponds to the geometric dual of D. Hence suppose that $n > 1$.

Without loss of generality set $c_j = (1\ 2)$ and let $a, b \in \{1, \ldots, 2n\}$. If $(a\ b)$ is a cycle in α, then $a^\alpha = b$ and we are done. Otherwise, let σ_1 and σ_2 (with possibly $\sigma_1 = \sigma_2$) be the cycles of σ corresponding to the (black) vertices of D incident to the darts 1 and 2, respectively. Since we are assuming $n > 1$, the two cycles σ_1 and σ_2 cannot both be trivial and neither can be equal to c_j.

We may assume that $a, b \notin \{1, 2\}$ as well since if, say, $a = 1$ and σ_1 is not trivial, then $a^{\sigma c_j} \notin \{1, 2\}$. If σ_1 is trivial, then

$$a^{(\sigma c_j)^2} = 2^{\sigma c_j}.$$

Since σ_2 is not trivial, we clearly must have $2^{\sigma c_j} \notin \{1, 2\}$.

Case (i). Suppose that σ_1 and σ_2 are disjoint. Consider the not necessarily connected map $D \setminus j = \hat{D} \cup \tilde{D}$ obtained from D by deleting the edge j. Let $\hat{\sigma}$, $\hat{\alpha}$ and $\tilde{\sigma}$, $\tilde{\alpha}$ be the restrictions of σ and α on $\hat{D}$ and $\tilde{D}$, respectively. Clearly $\hat{\sigma}$ coincides with the restriction of σc_j on $\hat{D}$, and similarly $\tilde{\sigma}$ coincides with the restriction of σc_j on $\tilde{D}$.

If a and b both belong to the same connected component, say $\hat{D}$, then there is $\hat{g} \in \langle \hat{\sigma}, \hat{\alpha} \rangle$ such that $a^{\hat{g}} = b$. If $\hat{g}$ is of the form

$$\hat{g} = \hat{\sigma}^{v_1} \hat{\alpha}^{w_1} \cdots \hat{\sigma}^{v_k} \hat{\alpha}^{w_k},$$

and since on $\hat{D}$ we have $\hat{\sigma} = \sigma c_j$ and $\hat{\alpha} = \alpha$, then for

$$g = (\sigma c_j)^{v_1} \alpha^{w_1} \cdots (\sigma c_j)^{v_k} \alpha^{w_k}$$

we must have $a^g = b$ as well.

If a belongs to $\hat{D}$ and b to $\tilde{D}$, then suppose that the vertex that corresponds to σ_1 in D is in $\hat{D}$. Let d be a dart in $\hat{D}$ such that in the map D we have $d^{\sigma} = 1$. By repeating the previous argument, there is $g \in \langle \sigma c_j, \alpha \rangle$ such that $a^g = d$. By acting with σc_j on d twice we first map d to 2 and then to some dart in $\tilde{D}$. Therefore, $a^{g(\sigma c_j)^2}$ and b are now both in $\tilde{D}$. By reusing the same argument as before we can find $h \in \langle \sigma c_j, \alpha \rangle$ such that

$$a^{g(\sigma c_j)^2 h} = b.$$

Case (ii). Suppose that σ_1 and σ_2 coincide, that is

$$\sigma_1 = \sigma_2 = (1\ p_1\ \cdots p_r\ 2\ q_1 \cdots\ q_s).$$

The product $\sigma_1 c_j$ will split σ_1 into two cycles σ_1' and σ_2' such that

$$\sigma_1' = (1\ p_1\ \cdots p_r),$$
$$\sigma_2' = (2\ q_1 \cdots\ q_s).$$

Let D' be the not necessarily connected map obtained from D by splitting the vertex corresponding to $\sigma_1 = \sigma_2$ so that the orderings of the darts around the two new vertices correspond to σ_1' and σ_2'. By connecting the new vertices with an edge with darts labeled by $\{2n + 1, 2n + 2\}$, a connected map with σ_1' and σ_2' disjoint is obtained. Now case *(ii)* follows from *(i)* by noting that D and D' with the new edge $(2n+1\ 2n+2)$ contracted are equivalent maps.

Remark 5 When D is a general dessin, i.e. a bipartite map (or equivalently, a hypermap), and c_j a cycle in α, then the partial dual with respect to the j-th white vertex (equivalently, j-th hyperedge) is the bipartite map

$$\partial_j D = (\sigma c_j, c_j^{-1}\hat{\alpha}, c_j\varphi),$$

where $\hat{\alpha}$ denotes the permutation obtained from α by omitting the cycle c_j.

The geometric interpretation of the partial dual $\partial_j D$ for $c_j = (1\,2)$ is the following. Suppose that $n > 1$ and c_j is not a loop. Let σ_1 and σ_2 be the two cycles of σ which contain 1 and 2, respectively. Draw the dual edge j^* of j by crossing j at the white vertex. The coedge j^* is incident to at most two face centers marked with $\diamond$ as before; draw a segment joining a face center to a black vertex of j if, and only if the black vertex is on the boundary of the corresponding face. As a result, four triangles are formed. Using the orientation of the underlying surface of D shade the two triangles with vertices oriented as $\bullet - \circ - \diamond - \bullet$. Exactly one of those triangles has the dart 1 as its side. Label the $\circ - \diamond$ segment of that triangle with 1^*, and proceed similarly with the other triangle. See Fig. 14.

Now contract j, and if j^* is not already a loop, glue the endpoints of j^* together and consider them as a single white vertex. If necessary, add a handle to the underlying surface of D so that $(D/j) \cup \{j^*\}$ is a map. Then $\partial_j D$ is obtained by relabeling j^*, 1^* and 2^* into j, 1 and 2, respectively. The cycle corresponding to the new vertex is given by $\sigma_1\sigma_2 c_j$. See Fig. 15.

If c_j is a loop we proceed in the reverse direction. That is, first we break the loop at its white vertex so that the two endpoints fall onto some, possibly the same, face centers. If need be, remove a handle from the underlying surface. Then we split $\sigma_1 = \sigma_2$ into two vertices and add an edge j^* between them so that the former loop j intersects it at its midpoint. Next we label the darts of j^* as before. Finally, the partial dual is completed by deleting j and relabeling j^* to j together with its darts. See Fig. 16.

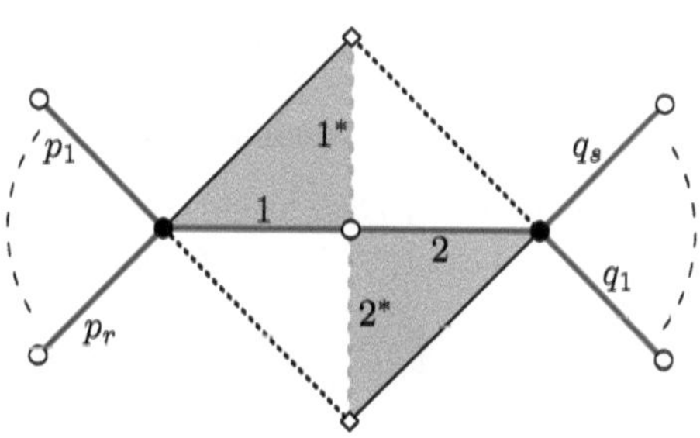

Fig. 14 The darts of the coedge are labeled so that i and i^* are sides of the same shaded *triangle*, for $i = 1, 2$. Here $\sigma_1\sigma_2 = (1\ p_1 \cdots p_r)\ (2\ q_1 \cdots q_s)$

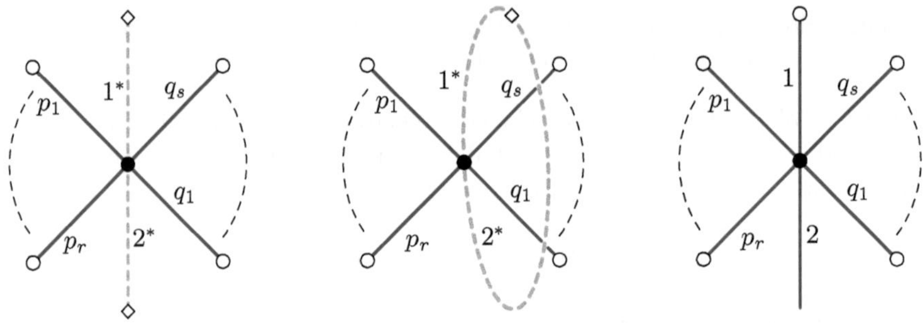

Fig. 15 From *left* to *right*: contraction, then gluing of the endpoints and relabelling. By comparing with Fig. 14 we see that $\sigma_1\sigma_2 c_j = (1\ p_1 \cdots\ p_r\ 2\ q_1 \cdots\ q_s)$

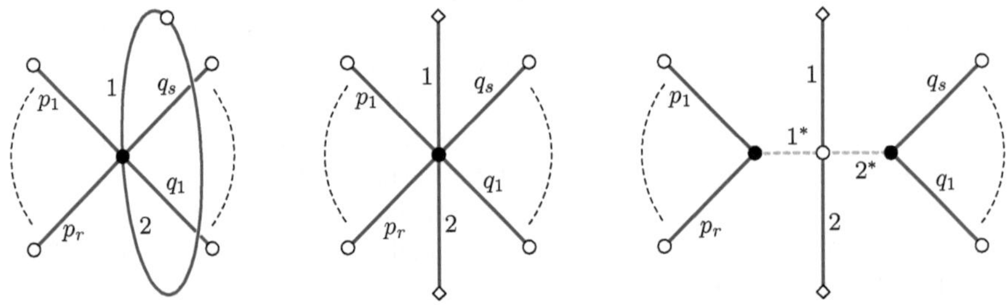

Fig. 16 From *left* to *right*: a map with a cycle $\sigma_1 = \sigma_2 = (1\ p_1 \cdots\ p_r\ 2\ q_1 \cdots\ q_s)$. The loop is then broken at its white vertex and the two endpoints fall onto face centers. We split the vertex and add a new edge j^*. The final step is obtained by deleting j and relabelling. By comparing with Fig. 15 we see that $\sigma_1 c_j = \sigma_2 c_j = (1\ p_1 \cdots\ p_r)(2\ q_1 \cdots\ q_s)$

Example 13 Let D be the genus 0 dessin given by the triple

$$D = \big((1\ 4)(2\ 3), (1\ 2)(3\ 4), (1\ 3)(2\ 4)\big).$$

Let $c_1 = (12)$. Then $\partial_1 D$ is the genus 1 dessin given by the triple

$$\partial_1 D = \big((1\ 4\ 2\ 3), (1\ 2)(3\ 4), (1\ 3\ 2\ 4)\big).$$

See Fig. 17 for the geometric counterparts.

Since the cycles of α commute, the following is well defined.

Definition 10 Let D be a map, E its set of edges and $S = \{i_1, \ldots, i_k\}$ some subset of E. Then the partial dual of D with respect to the set of edges S is the map

$$\partial_S D = \partial_{i_k} \cdots \partial_{i_1} D = (\sigma c_{i_1} \cdots c_{i_k}, \alpha, c_{i_k} \cdots c_{i_1} \varphi).$$

The geometric interpretation is immediately clear; the partial dual with respect to the set S is obtained by dualising the edges in S one at a time.

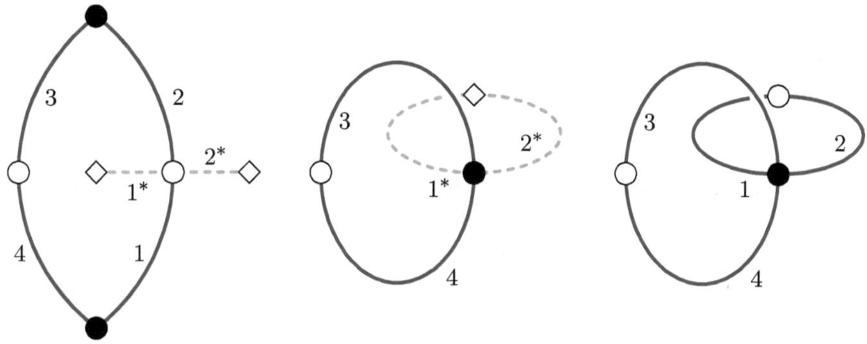

Fig. 17 From *left* to *right*: the map D from Example 13, an intermediate step, and its partial dual $\partial_1 D$

Remark 6 When D is a general dessin, the partial dual with respect to some subset of hyperedges is obtained analogously to Remark 5.

The following lemma, borrowed directly from [10, 11], lists some properties of the operation of partial duality.

Lemma 1 *Let D be a map, E its set of edges and S some subset of E. Then*

(a) $\partial_E D = D^*$
(b) $\partial_S \partial_S D = D$.
(c) If $j \in E \setminus S$, *then* $\partial_j \partial_S D = \partial_{S\cup\{j\}} D$.
(d) If S' is some other subset of E, then $\partial_{S'} \partial_S D = \partial_{S\triangle S'} D$.
(e) Partial duality preserves orientability of hypermaps.
(f) If X is the underlying surface of $\partial_S D$, then X is the underlying surface of $\partial_{E\setminus S} D$ as well.

We shall comment only on part (f) of the lemma as other properties follow directly from the definition. For the partial dual $\partial_{E\setminus S} D$ we have

$$\partial_{E\setminus S} D = \partial_{E\triangle S} D = \partial_E \partial_S D.$$

Therefore, $\partial_{E\setminus S} D$ and $\partial_S D$ are dual maps and hence they are embedded on homeomorphic surfaces. Moreover, if f is the clean Belyĭ function of $\partial_{E\setminus S} D$, then the two corresponding Belyĭ pairs are (X, f) and $(X, 1/f)$, respectively. Hence part (f) of the lemma can be improved slightly by noting that the underlying surfaces of $\partial_S D$ and $\partial_{E\setminus S} D$ coincide not just as topological, but as Riemann surfaces too.

6.1 Partial Duals, Delta-Matroids and the Galois Action

Given a dessin $D = (X, f)$, the absolute Galois group acts on it and its partial duals. It appears that the relationship between the Belyĭ function of D and $\partial_j D$ is very complicated. For if D is a tree, its Belyĭ function is a polynomial; however, the Belyĭ function of $\partial_j D$, for any edge j, clearly is no longer polynomial. More worryingly, Example 13 shows that the Riemann surface of $\partial_j D$ can be a point of a completely different moduli space than the one of D!

Nevertheless, some *nice* behaviour can be observed. For example, we shall prove that D always has a partial dual defined over its field of moduli by using a correspondence between delta-matroids and partial duals established in [12, Theorem 4.8].

We start with a simple proposition.

Proposition 3 *Let $D = (\sigma, \alpha, \varphi)$ be a map, E its set of edges and S some subset of E. Then* $\mathrm{Mon}(D)$ *is abelian if, and only if* $\mathrm{Mon}(\partial_S D)$ *is abelian.*

Proof By Lemma 1 it is enough to consider $S = \{1\}$. Let c_1 be the corresponding cycle in α. Then

$$\sigma\alpha = \alpha\sigma \iff \sigma\alpha c_1 = \alpha\sigma c_1 \iff (\sigma c_1)\alpha = \alpha(\sigma c_1),$$

since c_1 commutes with α.

It was shown in [22][3] than any dessin with abelian monodromy group is defined over $\mathbb{Q}$. Therefore the following corollary is obvious.

Corollary 2 *Let $D = (\sigma, \alpha, \varphi)$ be a map such that* $\mathrm{Mon}(D)$ *is abelian. Then D and its partial duals are all defined over $\mathbb{Q}$.*

Remark 7 Proposition 3 is no longer true if D is a hypermap. For if c is a non-trivial cycle in α which is not a transposition, then $c^{-1}\hat{\alpha} = c^{-1}\alpha c^{-1} = c^{-2}\alpha$. Furthermore, if $\mathrm{Mon}(D)$ is abelian we have

$$\begin{aligned}(\sigma c)(c^{-1}\hat{\alpha}) = (c^{-1}\hat{\alpha})(\sigma c) &\iff \\ \sigma\alpha c^{-1} = c^{-2}\alpha\sigma c &\iff \\ \sigma c^{-1}\alpha = c^{-2}\sigma c\alpha &\iff \\ c^2\sigma = \sigma c^2.\end{aligned}$$

The last equality does not hold always, of course. For example, if

$$D = ((1\ 2)(3\ 4)(5\ 6), (1\ 3\ 5)(2\ 4\ 6), (1\ 6\ 3\ 2\ 5\ 4))$$

is a dessin (see Fig. 18) then $\mathrm{Mon}(D) \cong \mathbb{Z}_6$, however for $c = (1\ 3\ 5)$ we have $\sigma c^2 \neq c^2\sigma$.

Given a delta-matroid $\Delta(E)$ on some set E with $\mathcal{F}$ as its collection of feasible sets, one can easily see that for some subset S of E the collection

$$\mathcal{F} \vartriangle S = \{F \vartriangle S \mid F \in \mathcal{F}\}$$

[3] For an alternative argument, see the discussion after Proposition 3 in [13] as well.

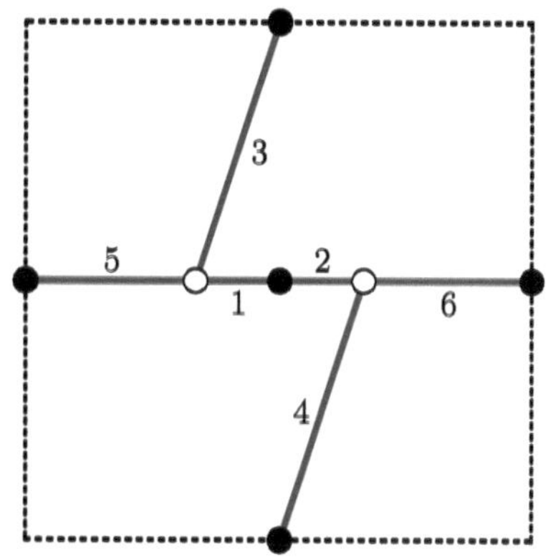

Fig. 18 The dessin D from Remark 7

satisfies the symmetric axiom of Definition 6. This motivates the following.

Definition 11 Let $\Delta(E)$ be a delta-matroid on E with $\mathcal{F}$ as its collection of feasible sets. Let S be a subset of E. The delta-matroid on E with $\mathcal{F} \triangle S$ as its collection of feasible sets is called *the twist of* $\Delta(E)$ *with respect to* S and is denoted by $\Delta(E) * S$.

Similarly as before, when D is a map, we shall use the notation $\Delta(D) * S$. The following lemma from [12] gives a correspondence between delta-matroids and partial duals.

Lemma 2 *Let D be a map, E its set of edges and S some subset of E. Then*

$$\Delta(\partial_S D) = \Delta(D) * S.$$

Proof It is sufficient to show the lemma for $S = \{j\}$ since the general result will then follow from Lemma 1(c).

If j is in no base, then it is a contractible loop in D and in $\partial_j D$ it is a pendant, i.e. an edge incident to a degree 1 vertex. In that case, the lemma follows easily.

So suppose that B is a base of D with $j \in B$. Moreover, suppose that j is not a loop. If j is a pendant, then the lemma is again obvious. Therefore, suppose that both vertices incident to j have degree at least 2.

By our construction, j is a loop in $\partial_j D$. Therefore, D/j is the same map as $(\partial_j D)\backslash j$. The underlying surface of D/j is the surface of D, hence $B\backslash j$ does not disconnect it. Therefore, $B\backslash j$ is a base of $(\partial_j D)\backslash j$ as well.

Let us now adjoin the loop j back to $(\partial_j D)\backslash j$. If we were forced to add a handle, then j^* will not disconnect the underlying surface since it will split the new handle into two sleeves and leave the rest of the surface unaffected. Therefore, $(B\backslash j) \cup j^* = B \triangle \{j, j^*\}$ will be a base of $\partial_j D$. Furthermore, if F is the feasible set of $\Delta(D)$ with $F = E \cap B$, then

$$F \triangle j = E \cap (B \triangle \{j, j^*\})$$

is a feasible set of $\Delta(\partial_j D)$.

If a new handle was not needed, then $\partial_j D$ and $(\partial_j D) \setminus j$ are on the same surface X. Since $(\partial_j D) \setminus j$ is a map on X with at least one face, adjoining j to it will clearly split some face into two new faces. Hence j^* must be a contractible segment on X since its endpoints are in the two faces with j as a common boundary. Therefore, $B \triangle \{j, j^*\}$ is a base of $\partial_j D$ and, by passing to feasible sets, we conclude that $F \triangle j$ is a feasible set in $\Delta(\partial_j D)$, if F is a feasible set in $\Delta(D)$.

Now suppose that j is a loop. Since $j \in B$, it cannot be contractible. If D and $\partial_j D$ are on the same surface, then, topologically, $j \in D$ and $j^* \in \partial_j D$ are the same loop. Therefore, $B \triangle \{j, j^*\}$ must be a base of $\partial_j D$. Otherwise, by removing a handle, Euler's formula implies that $\partial_j D$ gained an additional face. By construction, j must be on the boundary of the additional face, and at least one other face since other edges in D do not contribute to the partial dual. Therefore, j^* is contractible and $B \triangle \{j, j^*\}$ a base for $\partial_j D$.

So far we have shown that $\Delta(D) * j \subseteq \Delta(\partial_j D)$. The other inclusion is obtained by noting that if $F \in \Delta(\partial_j D)$, then

$$(F \triangle j) \in \Delta(\partial_j D) * j.$$

However, by using the just proven inclusion we have

$$(F \triangle j) \in \Delta(\partial_j \partial_j D) = \Delta(D).$$

Moreover, since $F = (F \triangle j) \triangle j$, we must have $F \in \Delta(D) * j$.

Remark 8 The proof of the preceding lemma is somewhat more natural in the language of ribbon graphs, as it can be seen in [12, Theorem 4.8]. However, in this paper, we prefer to work with maps instead.

We finish this section by demonstrating that partial duals with respect to feasible sets can be defined over their fields of moduli.

Theorem 8 *Let D be a clean dessin and E its set of edges. Then D has a partial dual which can be defined over its field of moduli.*

Proof Recall that by Theorem 3 a dessin can be defined over its field of moduli if it has a black vertex, or a white vertex, of a face center which is unique for its type and degree. If D has precisely one face, then that face is the unique face of some degree and therefore both $\partial_\emptyset D = D$ and $\partial_E D = D^*$ can be defined over their corresponding fields of moduli (which coincide).

Otherwise, let $F \neq E$ be a feasible set of $\Delta(D)$ and set $S = E \setminus F$. Then by Lemma 2 the map $\partial_S D$ has $S \triangle F = E$ as a feasible set. Therefore, E is a base of $\partial_S D$. Furthermore, if X_S is the underlying surface of $\partial_S D$, then $X_S \setminus \partial_S D$ is connected. This implies that $\partial_S D$ has precisely one face. As before, Theorem 3 implies that $\partial_S D$ can be defined over its field of moduli.

Corollary 3 *Let D be a clean dessin and $\Delta(D)$ its delta-matroid. If F is a feasible set of $\Delta(D)$, then both $\partial_F D$ and $\partial_{E\setminus F} D$ can be defined over their fields of moduli. Moreover, the two fields coincide.*

Proof The case for $\partial_{E\setminus F} D$ was discussed in the proof the previous theorem. The second case follows from Lemma 1 (d), that is

$$\partial_E(\partial_{E\setminus F} D) = \partial_F D.$$

Since the fields of definition of a map and its dual map coincide, and both maps can be defined over their field of moduli, then the fields of moduli coincide as well.

7 Maps, Their Partial Duals and Tropical Curves

In this section we informally comment on a simple relationship between the monodromy groups of dessins, partial duals and tropical curves. To the best knowledge of the author, this relationship has not been noted in the literature yet. We do not assume any knowledge of tropical geometry, however the reader is referred to [26] for an introduction.

Let $D = (\sigma, \alpha, \varphi)$ be a clean dessin with

$$\sigma = v_1 \cdots v_j,\ \alpha = c_1 \cdots c_n,\ \varphi = f_1 \cdots f_k,$$

and consider the planar graph G obtained from the triple $(\sigma, \alpha, \varphi)$ in the following way.

- Mark the integer points in the segment $[0, n+1]$.
- Place j vertices, one for each cycle in σ, vertically above 0.
- To a vertex i attach an open segment of length 1 and label it with the cycle v_i.
- Choose a cycle $(p\ q)$ in α.
  - If p and q are in the cycles v_p and v_q, respectively, above 1 join the edges with labels v_p and v_q into a single edge of length 1/2, so that a degree 3 vertex above 1 is formed. Label the edge with the cycle $\sigma_p\sigma_q(p\ q)$.
  - If p and q are in the same cycle, say v_r, above 1 split the edge with label v_r into two edges of length 1/2, so that a degree 3 vertex above 1 is formed. Label the two edges with the cycles in $\sigma_r(p\ q)$.
  - Extend all other edges so that their ends are above 3/2.
- Repeat the previous step until all the cycles of α are exhausted. Above $n+1$ there are k vertices, one for each cycle of φ. The edges incident with the final vertices have labels corresponding to the cycles in φ^{-1}.

Planar graphs obtained in this fashion are called *monodromy graphs* [9, 23]. Let us look at an example.

Example 14 Let $D = (\sigma, \alpha, \varphi)$ be the map B from Fig. 10. It can be represented by the triple

$$((1\,3\,5\,7\,8\,6\,2\,4), (1\,2)(3\,4)(5\,6)(7\,8), (1\,6\,3\,2\,4)(5\,8)(7)).$$

Therefore, above 0 we should have one vertex, and above 5 we should have three vertices. A monodromy graph obtained by multiplying σ in the order (1 2), (3 4), (5 6) and (7 8) is given in Fig. 19.

Multiplying σ with the cycles of α in a different order may produce a different monodromy graph. For example, if we multiply in the order (1 2), (5 6), (3 4), (7 8), the resulting monodromy graph shown in Fig. 20 will not be isomorphic to the previous one since it will have a cycle of length 3.

Irregardless of the order in which we multiply the cycles of σ with the cycles of α, monodromy graphs capture all of the information contained in the passport of a clean dessin D. Clearly the number and the degrees of black vertices and face centers correspond to the number of vertices and the lengths of the labels of edges above 0 and $n + 1$, and the genus of D corresponds to the genus of the graph, which is defined as the first Betti number of the graph (this fact is a simple consequence of the handshaking lemma). Moreover, the vertices of the graphs correspond precisely

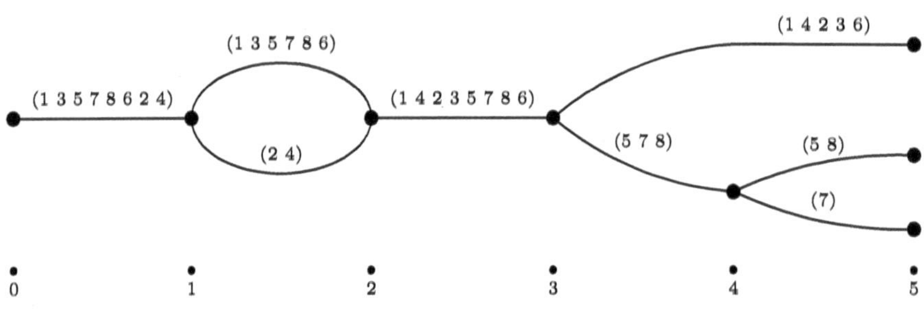

Fig. 19 A monodromy graph for the map $D = (\sigma, \alpha, \varphi)$ from Example 14

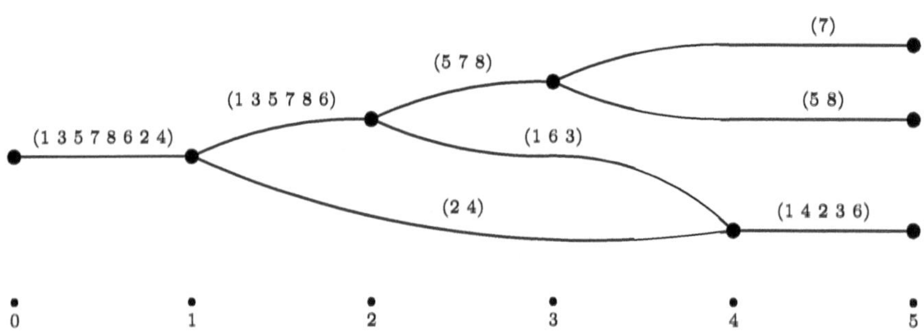

Fig. 20 A monodromy graph for the map $D = (\sigma, \alpha, \varphi)$ from Example 14 not isomorphic to the monodromy graph in Fig. 19

to the partial duals of D and two trivalent vertices v and w are adjacent if, and only if $\partial_j v = w$ or $\partial_j w = v$ for some edge j. Furthermore, monodromy graphs transfer dessins into the realm of tropical geometry.

Definition 12 An *abstract tropical curve* is a connected graph without vertices of degree 2 and with edges decorated by the elements of the set $\langle 0, \infty]$. The decorations on the edges are called *lengths*. Edges incident to degree 1 vertices have length ∞ and all other edges have finite length.

It is easy to see how to pass from a clean dessin D to an abstract tropical curve: first form a monodromy graph for D and decorate each edge with the length of its corresponding cycle. Finally, decorate the edges incident to degree 1 vertices with ∞. Tropical curves obtained in this way capture most information contained in the passport, and since they depend only on the monodromy group of the dessin, the following is clear.

Theorem 9 *Let D and D' be clean dessins and $\mathcal{T}$ and $\mathcal{T}'$ the sets of abstract tropical curves obtained from the monodromy graphs of D and D', respectively. If D and D' are conjugate, then any two curves $T \in \mathcal{T}$ and $T' \in \mathcal{T}'$ have*

- *The same number of finite edges and the same number of infinite edges.*
- *The same number of degree 3 vertices.*
- *The same* genus, *which is defined as the genus of the underlying monodromy graph. In particular, if $D \simeq D'$ is a tree, then T and T' are tropical trees.*

The invariants above most likely do not improve on the already known invariants. However, they may serve as a motivation for studying tropical curves in the context of the theory of dessins d'enfants.

Acknowledgments The author would like to thank the organisers and participants of the SIGMAP14 conference (5th Workshop SIGMAP—Symmetries In Graph, Maps And Polytopes. Sponsored by the Open University, London Mathematical Society and British Combinatorial Committee. Dates: 7th–11th July 2014. Location: ELIM Conference Centre, West Malvern, U.K. http://mcs.open.ac.uk/SIGMAP/) for the opportunity to present a talk on which this work is based, and for the lovely and informative presentations and discussions throughout. The author is also grateful for the invaluable guidance provided by his PhD advisor Prof. Alexandre Borovik.

References

1. Adrianov, N. M. et al.: Catalog of Dessins d'Enfants with no more than 4 edges. Journal of Mathematical Sciences **158**, 22–80 (2009).
2. Birch, B.: Noncongruence subgroups, covers and drawings. In: Schneps, L. (ed.) The Grothendieck Theory of Dessins dEnfants, pp. 25–46. Cambridge University Press, Cambridge (1994).
3. Belyĭ, G. V.: On Galois Extensions of a Maximal Cyclotomic Field. Math. USSR Izvestija, **14**, 247–256 (1980).
4. Belyĭ, G.V.: A New Proof of the Three Point Theorem. Sb. Math. **193(3–4)**, 329–332 (2002).

5. Bouchet, A.: Greedy algorithm and symmetric matroids. Mathematical Programming **38**, 147–159 (1987).
6. Bouchet, A.: Maps and Δ-matroids. Discrete Mathematics **78**, 59–71 (1989).
7. Borceux, F., Janelidze, G.: Galois Theories. Cambridge University Press, Cambridge (2001).
8. Borovik, A. V., Gelfand, I. M., White, N.: Coxeter Matroids. Birkhäuser, Boston, Mass (2003).
9. Cavalieri, R., Johnson, P., Markwig, H.: Tropical Hurwitz numbers. Journal of Algebraic Combinatorics **32**, 241–265 (2010).
10. Chmutov, S.: Generalized duality for graphs on surfaces and the signed Bollobás-Riordan polynomial. Journal of Combinatorial Theory, Ser. B. **99**, 617–638 (2009).
11. Chmutov, S., Vignes-Tourneret F.: Partial duality of hypermaps. http://arxiv.org/abs/1409.0632v1, version 1 (2014).
12. Chun, C., Moffatt, I., Noble, S. D., Rueckriemen, R.: Matroids, Delta-matroids and Embedded Graphs. http://arxiv.org/abs/1403.0920v1, version 1 (2014).
13. Conder, M.D.E., Jones, G. A., Streit, M., Wolfart, J.: Galois actions on regular dessins of small genera. Rev. Mat. Iberoam, **28**, 1–19 (2012).
14. Gelfand, I. M., Serganova, V. V.: On a general definition of a matroid and a greedoid. Soviet Math. Dokl., **35**, 6–10 (1987).
15. Gelfand, I. M., Serganova, V. V.: Combinatorial geometries and torus strata on homogeneous compact manifolds. Russian Math. Surveys, **42**, 133–168 (1987).
16. Girondo, E., González-Diez, G.: Introduction to Compact Riemann Surfaces and Dessins d'Enfants. Cambridge University Press, Cambridge (2012).
17. Goldring, W.: A new proof of Belyĭ's Theorem. J. Number Theory, **135**, 151–154 (2014).
18. González-Diez, G., Jaikin-Zapirain, A.: The absolute Galois group acts faithfully on regular dessins and Beauville surfaces. https://www.uam.es/personal_pdi/ciencias/gabino/Jule03.pdf (2013).
19. Gordon, G., McNulty, J.: Matroids: A Geometric Introduction. Cambridge University Press, Cambridge (2012).
20. Grothendieck, A.: Esquisse d'un Programme/Sketch of a Programme. In: Schneps L., Lochak P. (eds.) Geometric Galois Actions 1. Around Grothendieck's Esquisse d'un Programme, pp. 5–48 and 243–284. Cambridge University Press, Cambridge (1997).
21. Guillot, P.: An elementary approach to dessins d'enfants and the Grothendieck-Teichmüller group. Enseign. Math., **60**, 293–375 (2014).
22. Hidalgo, R.: Homology closed Riemann surfaces. Quarterly Journal of Math. (2011) doi:10.1093/qmath/har026.
23. Johnson, P.: Hurwitz numbers, ribbon graphs, and tropicalization. In: Athorne, C., Maclagan, D., Strachan, I. (eds.) Tropical Geometry and Integrable Systems, pp. 55–72. American Mathematical Society (2012).
24. Jones, G., Streit, M.: Galois groups, monodromy groups and cartographic groups. In: Schneps L., Lochak P. (eds.) Geometric Galois Actions 2. The Inverse Galois Problem, Moduli Spaces and Mapping Class Groups, pp. 25–65. Cambridge University Press, Cambridge (1997).
25. Lando, S. K., Zvonkin, A.: Graphs on Surfaces and their Applications. Springer-Verlag, Berlin (2004).
26. Maclagan, D.: Introduction to tropical algebraic geometry. In: Athorne, C., Maclagan, D., Strachan, I. (eds.) Tropical Geometry and Integrable Systems, pp. 1–20. American Mathematical Society (2012).
27. Matchett-Wood, M.: Belyĭ-Extending Maps and the Galois Action on Dessins d'Enfants. In: Publications of the Research Institute for Mathematical Sciences, **42**, 721–738 (2006).
28. Mohar, B., Thomassen C.: Graphs on Surfaces. John Hopkins University Press, Baltimore and London (2001).
29. Oxley, J. G.: Matroid Theory. Oxford University Press, Oxford (1992).
30. Schneps, L. (ed.): The Grothendieck Theory of Dessins d'Enfants. Cambridge University Press, Cambridge (1994).
31. Schneps, L., Lochak, P. (eds.): Geometric Galois Actions 1. Around Grothendieck's Esquisse d'un Programme. Cambridge University Press, Cambridge (1997).

32. Sijsling, J., Voight, J.: On explicit descent of marked curves and maps. http://arxiv.org/abs/1504.02814, version 2 (2015).
33. Szamuely, T.: Galois Groups and Fundamental Groups. Cambridge University Press, Cambridge (2009).
34. Whitney, H.: Planar Graphs. Fund. Math., **21**, 73–84 (1933).
35. Whitney, H.: On The Abstract Properties Of Linear Dependence. Amer. J. Math., **57**, 509–533 (1935).
36. Wolfart, J.: The 'obvious' part of Belyĭ's theorem and Riemann surfaces with many automorphisms. In: Schneps, L., Lochak, P. (eds.) Geometric Galois Actions 1. Around Grothendieck's Esquisse d'un Programme, pp. 97–112. Cambridge University Press, Cambridge (1997).
37. Wolfart, J.: ABC for polynomials, dessins denfants, and uniformization - A survey. In: Schwarz W., Steuding, J. (eds.) Elementare und Analytische Zahlentheorie (Tagungsband), Proceedings ELAZ-Conference, pp. 313–345. Springer (2006).

[illegible]

Faithful Embeddings of Planar Graphs on Orientable Closed Surfaces

Seiya Negami

Abstract A graph G is said to be *faithfully embeddable* on a closed surface F^2 if G can be embedded on F^2 in such a way that any automorphism of G extends to an auto-homeomorphism of F^2. It has been known that every 3-connected planar graph is faithfully embeddable on the sphere. We shall show that every 3-connected planar graph is faithfully embeddable on a suitable orientable closed surface other than the sphere unless it is one of seven exceptions.

1 Introduction

Let G be a graph regarded as a 1-dimensional topological space and let F^2 be a closed surface. Intuitively, an embedding of G on F^2 is a drawing of the graph G on the surface without edge crossings. Technically, we regard an embedding to be an injective continuous map $f : G \to F^2$.

Two embeddings f_1 and $f_2 : G \to F^2$ are said to be *congruent* if there exist an automorphism $\tau : G \to G$ and a homeomorphism $h : F^2 \to F^2$ with $hf_1 = f_2\tau$. In this case, the images $f_1(G)$ and $f_2(G)$ look the same up to homeomorphism if we neglect the labels of vertices of G. A graph G is said to be *uniquely embeddable* on F^2 (up to congruence) if all embeddings of G on F^2 are pairwise congruent.

On the other hand, an embedding $f : G \to F^2$ is said to be *faithful* if, for any automorphism $\tau : G \to G$, there exists a homeomorphism $h_\tau : F^2 \to F^2$ with $h_\tau f = f\tau$. Roughly speaking, in a faithful embedding, any symmetry of the graph can be realized as an action over the surface. A graph G is said to be *faithfully embeddable* on a closed surface F^2 if G has a faithful embedding on F^2.

It is well-known that every 3-connected planar graph has a unique embedding on the sphere, which follows from the uniqueness of its combinatorial dual, proved by Whitney [5]. The author [2] has introduced two notions of the uniqueness and

S. Negami (✉)
Faculty of Environment and Information Sciences, Yokohama National University, 79-2 Tokiwadai, Hodogaya-ku, Yokohama 240-8501, Japan
e-mail: negami@ynu.ac.jp

J. Širáň and R. Jajcay (eds.), *Symmetries in Graphs, Maps, and Polytopes*, Springer Proceedings in Mathematics & Statistics 159,
DOI 10.1007/978-3-319-30451-9_12

faithfulness of embeddings as above and pointed out that Whitney's uniqueness is equivalent with the composition of these two. That is, we can restate Whitney's result by saying that any 3-connected planar graph is uniquely and faithfully embeddable on the sphere.

A graph G embedded on a closed surface F^2, other than the sphere, is said to be *r-representative* if any non-contractible simple closed curve on F^2 meets G in at least r points and the *representativity* of G is defined as the minimum number r such that G is r-representative. It is known that a graph is uniquely and faithfully embeddable on a closed surface if it can be embedded there with sufficiently large representativity. On the other hand, no planar graph is 3-representative on any closed surface other than the sphere. (See [4] for these facts.)

In this situation, a natural question will arise; can we embed a 3-connected planar graph faithfully on a closed surface other than the sphere? For example, adding a handle to each face of a 3-connected planar graph embedded on the sphere yields such an embedding. To avoid such a trivial answer, we assume that any embedding of a graph is *cellular*, that is, each face must be homeomorphic to an open 2-cell. With this assumption, we shall give the complete answer to the question in the orientable case:

Theorem 1 *Every 3-connected planar graph can be embedded faithfully on an orientable closed surface other than the sphere unless it is isomorphic to the 1-skeleton of one of the following polyhedra:*

(i) the tetrahedron, the octahedron, the dodecahedron,
(ii) the truncated tetrahedron, the truncated dodecahedron, the truncated icosahedron,
(iii) the icosidodecahedron.

The first three are three of the five Platonic solids and the other are Archimedean solids. As a graph, the tetrahedron is K_4, while the octahedron is $K_{2,2,2}$. In general, the *truncation* of a convex polyhedron is another polyhedron obtained from the polyhedron by cutting off a small part around each vertex with the plane. The truncated dodecahedron consists of 12 decagons and 20 triangles, while the truncated icosahedron has 20 hexagons and 12 pentagons resembling a soccer ball. These two can be obtained from the dodecahedron and the icosahedron by truncation. The icosidodecahedron consists of 12 pentagons and 20 triangles and can be obtained from the truncated dodecahedron by shrinking each edge joining two triangles. We often call a planar graph embedded on the sphere by the name of a convex polyhedron whose 1-skeleton is isomorphic to the graph.

We shall prove that almost all 3-connected planar graphs have faithful embeddings on suitable orientable closed surfaces different from the sphere, introducing a useful method in Sect. 2. Sections 3 and 4 are devoted to the 3-regular and 4-regular cases. Section 5 presents the proof of our main theorem and some comments related to this topic.

2 Faithful Rotation Schemes

To present an embedding of a graph G on an orientable closed surface combinatorially, we used a *rotation scheme* of G, which define a cyclic order over the neighborhood of each vertex v of G, called a *rotation* around v. It may be useful to use a function $\rho_v : N(v) \to N(v)$ over the neighborhood $N(v)$ of a vertex v to present the rotation around v. That is, $\rho_v(u)$ presents the successor of u in the cyclic order given as a rotation around v. The rotation scheme of G can be regarded as the collection of these functions, $\rho = \{\rho_v : v \in V(G)\}$.

Trace edges to form a closed walk, according to the following rule; if one comes from a neighbor u of a vertex v to v along an edge uv, then he goes toward the successor $w = \rho_v(u)$ of u in the rotation around v along the edge vw. Each of the closed walks so constructed corresponds to the boundary walk of a face in the embedding of G derived by the given rotation scheme. (A general description on the rotation scheme can be found in [1].)

Put $\rho^{-1} = \{\rho_v^{-1} : v \in V(G)\}$ and call it the *inverse* of a rotation scheme $\rho = \{\rho_v : v \in V(G)\}$ of G. Then ρ^{-1} determines another rotation scheme of G. It is clear that any closed walk derived from ρ^{-1} with the above rule coincide with one derived from ρ if we reverse its direction. This means that ρ^{-1} presents the same embedding of G on the same surface as ρ does.

Here, we shall consider a rotation scheme which presents a faithful embedding. Let G be a graph and assume that a rotation scheme ρ presents an embedding $f : G \to F^2$ of G on an orientable closed surface F^2. For convenience, we denote the image $f(G)$ of G by G itself. The rotation scheme induces not only the local orientation around v, but also a global orientation over F^2 consistently. Thus, we assume that F^2 is oriented with this orientation. In general, a cyclic order arbitrarily given around a vertex v is said to be *coherent* with the orientation of F^2, or with the rotation scheme ρ, if it induces the same cyclic permutation over the neighborhood of v as ρ_v. It is said to be *reverse* to ρ if its inverse order is coherent with ρ.

Take any automorphism $\tau \in \mathrm{Aut}(G)$. Let v be any vertex of G and let $u_0, u_1, \ldots, u_{k-1}$ be its neighbors such that $\rho_v(u_i) = u_{i+1}$ with indices taken modulo k. That is, the cyclic order derived from ρ_v reads $u_0 u_1 \cdots u_{k-1}$. Then τ naturally induces a cyclic order $\tau(u_0)\tau(u_1)\cdots\tau(u_{k-1})$ over the neighborhood of $\tau(v)$ and this corresponds to a function $\tau\rho_{\tau^{-1}(v)}\tau^{-1} : N(\tau(v)) \to N(\tau(v))$. We call it the *rotation around* $\tau(v)$ *induced by* τ. Put $\rho^\tau = \{\tau\rho_v\tau^{-1} : v \in V(G)\}$ and call it the *rotation scheme of* G *induced by* τ.

If the rotations belonging to ρ^τ are all coherent with (or reverse to) the original rotation ρ, then the boundary walk of each face of G derived from ρ can be translated into the boundary walk of a face derived from ρ^τ (or one traced in the reverse direction) via τ. Thus, we can define a one-to-one correspondence between the sets of faces in the embeddings of G derived from ρ and ρ^τ. This implies that τ extends to a homeomorphism $h_\tau : F^2 \to F^2$ over the surface, which is orientation-preserving or -reversing, depending on whether the rotations in ρ^τ are all coherent or all reverse. Therefore, we have the following criterion for a rotation scheme to present a faithful embedding:

Lemma 1 *The embedding of G derived from a rotation scheme ρ is faithful if and only if the rotations belonging to the rotation scheme induced by τ are all coherent with ρ or all reverse to ρ for any automorphism $\tau \in \mathrm{Aut}(G)$.*

Proof Let G be a graph embedded on the oriented closed surface F^2. We may assume that this embedding is derived from a given rotation scheme ρ and that the rotations in ρ are all coherent to the orientation of F^2. The inclusion map $i_G : G \to F^2$ can be regarded as its embedding map.

Let $\tau : G \to G$ be any automorphism of G. Suppose that the rotation in ρ^τ are all coherent with or all reverse to ρ, then τ extends to a homeomorphism $h_\tau : F^2 \to F^2$ with $h_\tau|_G = \tau$, as shown in the previous. This implies that $h_\tau i_G = i_G \tau$ and hence the embedding i_G is faithful if this assumption holds for all automorphism τ. Thus, the sufficiency follows.

Conversely, if the inclusion map $i_G : G \to F^2$ is faithful, then any automorphism $\tau : G \to G$ extends to a homeomorphism $h_\tau : F^2 \to F^2$ with $h_\tau|_G = \tau$. Since any homeomorphism over the surface preserves a cyclic rotation around any point, the rotation around $\tau(v)$ induced by τ for any vertex of G must be coherent with (or reverse to) $\rho_{\tau(v)}$ if h_τ is orientation-preserving (or -reversing), Thus, the necessity follows. □

Call a rotation scheme of a graph G *faithful* if it satisfies the necessary condition in the lemma. Thus, it suffices to construct a faithful rotation scheme for each 3-connected planar graph, if any, so that the derived surface is not homeomorphic to the sphere to answer our question mentioned in introduction.

First, we should know the fact that any unnatural rotation scheme for a 3-connected planar graph exhibits its embedding on an orientable closed surfaces other than the sphere.

Lemma 2 *A 3-connected planar graph admits only two rotation schemes which present embeddings on the sphere, namely one induced by a planar embedding and its inverse.*

Proof Let G be a 3-connected planar graph and assume that G is embedded on the oriented sphere S^2 in one way. This embedding is presented by the inclusion map $i_G : G \to S^2$. Embed G on the sphere in another way via an embedding map $f : G \to S^2$. Since G is uniquely and faithfully embedded on the sphere, f extends to a homeomorphism $h : S^2 \to S^2$. This implies that the same set of closed walks exhibit the boundary cycles of faces in both embeddings and each of such closed walks determine the corners of faces, which induces a unique rotation around each vertex lying at the corners. Such a rotation is coherent with the orientation of S^2 or is reverse, depending on whether f preserves the orientation of the sphere or not. Therefore, the two embeddings i_G and f of G can be presented by the same rotation scheme or by two rotation schemes each of which is the inverse of the other. □

Let S be a subset of $V(G)$ and $\overline{S} = V(G) - S$ its complement. The pair $(S, \overline{S})$ is called an *equivariant partition* of G (under $\mathrm{Aut}(G)$) if both S and $\overline{S}$ are not empty and if either $\tau(S) = S$ or $\tau(S) = \overline{S}$ for any automorphism $\tau \in \mathrm{Aut}(G)$.

Lemma 3 *If a 3-connected planar graph G has an equivariant partition, then it admits a faithful embedding on an orientable closed surface other than the sphere.*

Proof Let G be a 3-connected planar graph embedded on the oriented sphere having a rotation scheme ρ and let $(S, \overline{S})$ be an equivariant partition of G. Then the rotation around each vertex of G in ρ is coherent with the orientation of the sphere. Replace the rotation around a vertex v with its inverse if v belongs to $\overline{S}$. Denote the resulting rotation scheme by ρ_S. Since ρ_S is different from ρ and ρ^{-1}, it exhibits an embedding of G on another orientable closed surface different from the sphere by Lemma 2.

Take any automorphism $\tau \in \mathrm{Aut}(G)$. Since G is faithfully embedded on the sphere, τ extends to an auto-homeomorphism h_τ over the sphere. We should consider how τ acts on the rotation scheme ρ_S. There are four cases depending on whether h_τ is orientation-preserving or not and on whether $\tau(S) = S$ or not.

We shall prove only the case where h_τ is orientation-preserving and that $\tau(S) = S$; the other case are similar. Then τ induces the same rotation around a vertex $\tau(v)$ as in the rotation scheme ρ for each vertex v. If $v \in S$, then $\tau(v) \in S$ and both rotations around these two vertices v and $\tau(v)$ in ρ_S are coherent with ρ. Otherwise, $v, \tau(v) \in \overline{S}$ and their rotations in ρ_S are reverse to ρ. These observations imply that each rotation in the rotation scheme $(\rho_S)^\tau$ induced by τ is coherent with ρ_S. Therefore, ρ_S is faithful by Lemma 1. □

By the above lemma, we can easily construct a faithful embedding of a 3-connected planar graph if we can find an equivariant partition, as in the following two corollaries:

Corollary 1 *Every 3-connected bipartite planar graph admits a faithful embedding on an orientable closed surface other than the sphere.*

Proof A 3-connected bipartite graph G has a proper coloring with black and white. Let S be the set of black vertices. Then its complement $\overline{S}$ consists of the white vertices. It is clear that $(S, \overline{S})$ forms an equivariant partition and the corollary follows from Lemma 3. □

In general, a graph is said to be *vertex-transitive* if for any pair (u, v) of vertices, there exists an automorphism $\tau \in \mathrm{Aut}(G)$ with $\tau(u) = v$. If a graph is vertex-transitive, then it must be *regular*, that is, all vertices have the same degree. We say that two vertices u and v are *equivalent* to each other (under $\mathrm{Aut}(G)$) if there is an automorphism $\tau \in \mathrm{Aut}(G)$ with $\tau(u) = v$.

Corollary 2 *If a 3-connected planar graph is not vertex-transitive, then it admits a faithful embedding on an orientable closed surface other than the sphere.*

Proof Assume that a 3-connected planar graph G is not vertex-transitive. Then we can choose two vertices u and v so that they are not equivalent. Let S be the set of vertices equivalent to u. Then its complement $\overline{S}$ contains v and is not empty. It is clear that $(S, \overline{S})$ is equivariant since any automorphism of G leaves each of S and $\overline{S}$ invariant. □

Combining this corollary with arguments on degrees of vertices in planar graphs, we can establish the following theorem, which will restrict the exceptions in our main theorem:

Theorem 2 *A* 3*-connected planar graph admits a faithful embedding on an orientable closed surface other than the sphere unless it is either* 3*- or* 4*-regular.*

Proof Let G be a 3-connected planar graph. Suppose that G is not regular. Then G is not vertex-transitive and hence its admits a faithful embedding on an orientable closed surface other than the sphere by Corollary 2. So we may assume that G is regular. Since any planar graph has a vertex of degree at most 5, then G must be either 3-, 4- or 5-regular. However, the first two cases are excluded as the exceptional cases in the theorem.

Now suppose that G is 5-regular and let ρ be a planar rotation scheme of G. Let v be any vertex of G and let $u_0, \dots, u_4$ be its five neighbors lying around v according to ρ_v. Since ρ_v is a cyclic permutation over five vertices, also its square $(\rho_v)^2$ is a cyclic permutation over $N(v)$. Therefore, $\rho^2 = \{(\rho_v)^2 : v \in V(G)\}$ determines another rotation scheme and exhibits an embedding of G on an orientable closed surface other than the sphere by Lemma 2 since $(\rho_v)^2$ coincides with neither ρ_v nor ρ_v^{-1}.

Since $\tau(\rho_v)^2\tau^{-1} = (\tau\rho_v\tau^{-1})^2$, any automorphism τ induces a rotation around $\tau(v)$ coherent with or reverse to $(\rho_v)^2$, which depends on whether the extension of τ preserves the orientation of the sphere or not. This implies that ρ^2 is a faithful rotation scheme. □

3 The 3-Regular Case

In this section, we shall discuss the 3-regular planar graphs to recognize some of the exceptional cases in Theorem 1. Fortunately, we can characterize those 3-connected 3-regular planar graphs that have faithful embeddings on orientable closed surfaces other than the sphere, using the notion of equivariant partitions, as follows.

Let G be a 3-regular connected planar graph embedded on the oriented sphere or the plane. To make a rotation scheme ρ exhibiting another surface, we assign "black" or "white" to each vertex of G and set ρ_v to be clockwise (or anticlockwise) for black (or white) vertices v. Since there are only two ways to define a rotation around each vertex of degree 3, any rotation scheme of G can be obtained in this way.

Lemma 4 *A* 3*-connected* 3*-regular planar graph admits a faithful embedding on an orientable closed surface other than the sphere if and only if it has an equivariant partition.*

Proof Let G be a 3-connected 3-regular planar graph embedded on the oriented sphere. Since the sufficiency follows from Lemma 3, it suffices to show the necessity.

Suppose that G has a rotation scheme ρ which exhibits a faithful embedding on an oriented closed surface F^2, other than the sphere. Then the vertices of G are colored by black and white, according to the rotations in ρ. Let S and $\overline{S}$ be the sets of black vertices and of white vertices, respectively. Since F^2 is not the sphere, both of S and $\overline{S}$ are not empty by Lemma 2 and $V(G) = S \cup \overline{S}$.

Take any automorphism $\tau \in \mathrm{Aut}(G)$. Then τ acts on F^2, preserving or reversing its orientations. If τ carries a black vertex v to another black vertex u, then τ induces a rotation around u coherent with ρ and hence τ is orientation-preserving over F^2 and carries any vertex to a vertex of the same color. We have $\tau(S) = S$ in this case. On the other hand, if τ carries a black vertex to a white vertex, then τ is orientation-reversing over F^2 and we conclude that $\tau(S) = \overline{S}$. Therefore, $(S, \overline{S})$ is an equivariant partition. □

Zelinka [7] has already classified the vertex-transitive 3-regular planar graphs which may have multiple edges. To establish the following theorem, it suffices to choose only non-bipartite simple ones from his classification. However, we can recognize those easily as in our proof below.

Lemma 5 *Let G be a 3-connected 3-regular planar graph embedded on the sphere and suppose that G is not bipartite and is vertex-transitive. Then G is isomorphic to one of the following polyhedra:*

(i) the tetrahedron, the dodecahedron,
(ii) the truncations of the tetrahedron, the cube, the dodecahedron and the icosahedron.

Proof Let v be any vertex with three neighbors u_0, u_1 and u_2, and denote each face having the corner $u_i v u_{i+1}$ by A_i for $i \equiv 0, 1, 2 \pmod 3$. For a face A, let $|A|$ denote its size, that is, the length of its boundary cycle.

CASE 1: *The three faces A_0, A_1 and A_2 have the same size.* Since G is vertex-transitive, all faces of G must have the same size. This implies that G is isomorphic to one of the five Platonic solids. Since G is 3-regular, G is isomorphic to either the tetrahedron, the cube or the dodecahedron. However, the cube is excluded since it is bipartite, which yields (i).

CASE 2: *Only two of the three faces A_0, A_1 and A_2 have the same size, say $|A_0| = |A_1| \neq |A_2|$.* Put $r = |A_2|$. Since G is vertex-transitive, there is one r-gonal face incident to each vertex and the boundary cycles of such r-gonal faces cover all vertices of G. Each vertex on one of the r-gonal cycles is joined to a vertex on another r-gonal cycle by an edge. In this situation, we find that G is isomorphic to the truncation of a Platonic solid, which is r-regular, and that $|A_0| = |A_1|$ is an even number. If r also is even, then all faces are bounded by even cycles and G would be bipartite, contrary to our assumption in the lemma. Therefore, r must be odd and the octahedron is excluded, which yields (ii).

CASE 3: *The three faces A_0, A_1 and A_2 have all different sizes.* In this case, faces of two different sizes $|A_i|$ and $|A_j|$ lie alternately around each face of size $|A_k|$ for $\{i, j, k\} = \{0, 1, 2\}$. This implies that $|A_k|$ is an even number and hence all faces are bounded by even cycles, which contradicts that G is not bipartite. Therefore, this is not the case. □

Discussing the existence of equivariant partitions, we can conclude the following theorem from the above lemma.

Theorem 3 *A 3-connected 3-regular planar graph admits a faithful embedding on an orientable closed surface other than the sphere if and only if it is not isomorphic to any of the following polyhedra:*

(i) the tetrahedron, the dodecahedron,
(ii) the truncation of the tetrahedron, the dodecahedron and the icosahedron.

Proof First, notice that the truncation of the cube is missing from the lists in Lemma 5. The cube has a proper coloring by black and white. Color the vertices on a triangle added by truncation by the same color as its corresponding vertex of the cube has, and let S and $\overline{S}$ be the sets of black vertices and of white vertices in the truncated cube. It is easy to see that $(S, \overline{S})$ forms an equivariant partition since any automorphism of the truncated cube carries each triangle to a triangle. Thus, the truncation of the cube has a faithful embedding on an orientable closed surface other than the sphere by Lemma 4.

To complete the proof, it suffices to show that each of the graphs listed in the theorem has no equivariant partition, and use Lemma 4. Let G be any of them. Then there is a set of odd cycles bounding faces which covers all vertices of G. Such a set of odd cycles covers G doubly for each in (i) while it consists of all triangles or pentagons created by truncation for each in (ii). It is clear that there is an automorphism τ_C which carries u_i to u_{i+1}, along each odd cycle $C = u_0 u_1 \cdots u_{k-1}$ in the set.

Let $(S, \overline{S})$ be any partition of $V(G)$ with $S \neq \emptyset$ and $\overline{S} \neq \emptyset$. Choose any cycle C from the covering set constructed above. If both $V(C) \cap S$ and $V(C) \cap \overline{S}$ are not empty, then we can find a vertex on C, say u_j, with $u_j \in S$ and $u_{j+1} \in \overline{S}$. Since $\tau(u_j) = u_{j+1}$, if $(S, \overline{S})$ is equivariant, then we have $\tau(S) = \overline{S}$ and $\tau(V(C) \cap S) = V(C) \cap \overline{S}$. This implies that $|V(C) \cap S| = |V(C) \cap \overline{S}|$. However, it is impossible since C consists of an odd number of vertices. Therefore, one of $V(C) \cap S$ and $V(C) \cap \overline{S}$ must be empty.

For each graph in (i), using the above argument and the fact that the odd cycles form a connected spanning subgraph of G, we conclude that $V(G) = S$ or $= \overline{S}$ and hence S or $\overline{S}$ would be empty, a contradiction. For each graph in (ii), we conclude that odd cycles in the covering set are classified into two nonempty groups, one corresponding to S and the other to $\overline{S}$. However, such a partition $(S, \overline{S})$ does not fit to the symmetry of G since its original graph is not bipartite and has an automorphism which rotates a face of odd size. Therefore, $(S, \overline{S})$ is not equivariant and hence G has no equivariant partition. □

4 The 4-Regular Case

In this section, we shall discuss the 4-regular planar graphs. We can use the result on equivariant partitions in Sect. 2, but need slightly complicated arguments since there are may ways to define a rotation around a vertex of degree 4. First, we shall list up candidates for the exceptional cases in Theorem 1. Note that the classification of the vertex-transitive 4- and 5-regular planar graphs which may have multiple edges can be found in Zelinka [6].

Lemma 6 *Let G be a 3-connected 4-regular planar graph embedded on the sphere which is vertex-transitive. Then G is isomorphic to one of the following polyhedra:*

(i) the octahedron,
(ii) the antiprisms,
(iii) the cuboctahedron, the icosidodecahedron,
(iv) the rhombicuboctahedron, the rhombicosidodecahedron.

Proof Let V, E and F denote the number of vertices, edges and faces of such a graph G embedded on the sphere. Since G is 4-regular, we have $4V = 2E$. Substituting this to Euler's formula $V - E + F = 2$, we obtain $4V = 4F - 8$. If all faces would have size at least 4, then we have $4F \leq 2E$ and thus $2E = 4V = 4F - 8 \leq 2E - 8$, which is a contradiction. Therefore, G has a triangular face. Since G is vertex-transitive, at least one triangular face is incident to each vertex of G. Let v be any vertex of G with four neighbors u_0, u_1, u_2, u_3 lying around it in this cyclic order.

CASE 1: *Four triangular faces are incident to v.* It is easy to see that G is isomorphic to the octahedron in this case. This appears in (i).

CASE 2: *Only three triangular faces are incident to v.* We may assume that these triangles are vu_0u_1, vu_1u_2, vu_2u_3. Look at u_2 and consider the three triangular faces incident to u_2. Two of them are vu_1u_2 and vu_2u_3. If the third triangle were incident to the edge u_1u_2, then Case 2 would not hold at the third vertex of this triangle. Thus, the third triangle around u_2 must be incident to the edge u_2u_3. Carrying out the same argument, we find a sequence of triangles and finally conclude that the whole of G consists of triangles $x_ix_{i+1}x_{i+2}$ for $i = 0, 1, \ldots$ after relabeling vertices. Then G has two disjoint cycles $x_0x_2 \cdots x_{2n-2}$ and $x_1x_3 \cdots x_{2n-1}$ and is isomorphic to the antiprism with n-gonal base, which appears in (ii).

CASE 3: *Only two triangular faces are incident to v.* If the two triangles were vu_0u_1 and vu_1u_2, then Case 3 would not hold at u_2. Thus, the two triangles incident to v do not share any edge. We may assume that they are vu_0u_1 and vu_2u_3. Let A_1 and A_2 be the other two faces incident to v having corners u_0vu_3 and u_1vu_2 and let A_3 be the non-triangular face incident to the edge u_0u_1, which meets A_1 and A_2 at u_0 and u_1, respectively. If $|A_1| \neq |A_2|$, the we have $|A_1| \neq |A_3|$ and $|A_2| = |A_3|$ since v and u_0 are equivalent. However, since v and u_1 are equivalent, the last equality would imply that $|A_1| = |A_2|$, contrary to the assumption. Thus, we have $|A_1| = |A_2| = |A_3|$ and hence all non-triangular faces have the same size, say $r \geq 4$. In this situation, we can construct a 3-regular Platonic solid with faces of size r, by placing a vertex in each triangular face and by joining each pair of vertices lying in two triangular faces meeting at a vertex. Such a Platonic solid is either the cube or the dodecahedron. Conversely, G can be obtained from it by truncating it and by shrinking each edge joining two triangles created by truncation. The cuboctahedron comes from the cube and the icosidodecahedron comes from the dodecahedron. They appear in (iii).

CASE 4: *Only one triangular face is incident to v.* Let $v_0v_1v_2$ denote the cycle bounding such a triangular face A. There are six faces A_0, B_0, A_1, B_1, A_2, B_2 surrounding A such that A_i meets A at v_i and B_i shares the edge v_iv_{i+1} with A. None of them is triangular. Since v_0, v_1 and v_2 are all equivalent, we conclude that $|A_0| = |A_1| = |A_2|$ $(= a \geq 4)$ and $|B_0| = |B_1| = |B_2|$ $(= b \geq 4)$.

Let F_3 denote the number of triangular faces of G and let F_A and F_B be the number of faces corresponding to A_i's and B_i's. Then we have $V = 3F_3 = aF_A$ and $2V = bF_B$. On the other hand, we have $F = F_3 + F_A + F_B = V + 2$ by Euler's formula. Combining these two equalities, we obtain:

$$\left(\frac{1}{3} + \frac{1}{a} + \frac{2}{b} - 1\right)V = 2$$

Since the coefficient of V in the above must be positive, we conclude that $1/a + 2/b > 2/3$. It is easy to see that this inequality has only two solutions $(a, b) = (4, 4)$ and $(5, 4)$. The first corresponds to the rhombicuboctahedron while the second corresponds to the rhombicosidodecahedron. They can be obtained from the cube and the dodecahedron, respectively, by replacing their vertices with triangles and edges with squares. They appear in (iv). □

Let G be a 3-connected 4-regular planar graph embedded on the sphere (or on the plane) and suppose that G has an embedding on an orientable closed surface F^2 other than the sphere. Let ρ be the rotation scheme to exhibit the embedding of G on F^2, not on the sphere. To present ρ, we add the following marks to the picture of G on the plane.

The rotation scheme ρ induces a cyclic order around each vertex v of G, which may not be coherent with or not be reverse to the orientation over the plane. Let u_0, u_1, u_2, u_3 be the four neighbors of v lying clockwise around v in this cyclic order according to the orientation over the place with indices taken modulo 4. Then each sequence u_ivu_{i+1} represents a corner of a face incident to v in the planar embedding of G. If u_{i+1} is the immediate successor (or predecessor) of u_i in the rotation ρ_v, that is, if $\rho_v(u_i) = u_{i+1}$ (or $\rho_v(u_{i+1}) = u_i$), then we draw a black dot (or a white dot) at the corner u_ivu_{i+1}. We call the picture of G with such black and white dots a *dot scheme* here.

It is clear that there are three cases around each vertex v in any dot scheme:

(i) There are four black dots.
(ii) There are four white dots.
(iii) There are one black dot and one white dot placed in opposite angles.

The rotation ρ_v induces the clockwise (or anticlockwise) rotation around v of type (i) (or (ii)) while the cyclic order induced by ρ_v is $(u_0u_1u_3u_2)$ for example if v is of type (iii).

Now assume that G is vertex-transitive and that ρ exhibits a faithful embedding in addition. Take any automorphism τ of G, which extends to an auto-homeomorphism over the sphere. It is clear that if a vertex v is of type (i), then $\tau(v)$ is of type (i) or (ii). Thus, if S consists of all vertices of type (i), then $(S, \overline{S})$ becomes an equivariant partition. Its complement $\overline{S}$ contains all vertices of type (ii) and is not empty; otherwise, F^2 would be the sphere. Therefore, we can conclude the following lemma:

Lemma 7 *Let G be a 3-connected 4-regular planar graph embedded on the sphere. Then G has a faithful embedding on an orientable closed surface other than the sphere if and only if either G has an equivariant partition, or there is a dot scheme of G such that all vertices of G are of type (iii) and any automorphism of G sends dots to dots, either preserving all colors or exchanging all colors.*

Proof First suppose that G has a faithful embedding on an orientable closed surface F^2. This embedding is derived from a faithful rotation scheme ρ. Draw the dot scheme which presents ρ. If this dot scheme contains at least one vertex of type (i) or (ii), then G has an equivariant partition, as shown in the previous. Thus, we may assume that all vertices are of type (iii) in the dot scheme.

Let v be a vertex in G and let u_0, u_1, u_2 and u_3 be its neighbors lying clockwise around v in the dot scheme, that is, the cyclic permutation $(u_0u_1u_2u_3)$ is coherent with the orientation of the sphere S^2. Take any automorphism τ of G. Since the inclusion map $i_G : G \to S^2$ is faithful on the sphere, the cyclic permutation $(\tau(u_0)\tau(u_1)\tau(u_2)\tau(u_3))$ induced around $\tau(v)$ by τ is coherent with or reverse to the orientation of the sphere.

Since v is of type (iii), we may assume that the faithful rotation ρ induces a cyclic permutation $\rho_v = (u_0u_1u_3u_2)$ around v. That is, there is a black dot at the corner u_0vu_1 and a white dot at the corner u_2vu_3, and the other two corners contain no dot. The automorphism τ translates $(u_0u_1u_3u_2)$ into a cyclic permutation $(\tau(u_0)\tau(u_1)\tau(u_3)\tau(u_2))$ around $\tau(v)$. Since the embedding derived from ρ is faithful, the latter must be coherent with or reverse to $\rho_{\tau(v)}$ and hence the corners $\tau(u_0)\tau(v)\tau(u_1)$ and $\tau(u_2)\tau(v)\tau(u_3)$ have dots in the dot scheme. Since the layout of black and white dots determines a unique rotation ρ_v around each vertex v and since all ρ_v's are coherent to an orientation of F^2, if τ carries a black dot at one corner to a black dot at another, then it carries all black dots in the dot scheme to black dots. Thus, τ either preserves all colors or exchanges all colors. The necessity follows.

If G has an equivariant partition, then G has a faithful embedding on an orientable closed surface other than the sphere by Lemma 3. On the other hand, if there is a dot scheme of G with rotation ρ which satisfies the condition in the lemma, then the cyclic permutation around $\tau(v)$ induced by any automorphism τ are coherent with or reverse to $\rho_{\tau(v)}$ for each vertex v and one of these two options, coherent or reverse, holds for all vertices. Therefore, ρ is a faithful rotation. The sufficiency follows. □

We call a dot scheme satisfying the condition in the above lemma *a faithful dot scheme.*

Theorem 4 *A* 3*-connected* 4*-regular planar graph admits a faithful embedding on an orientable closed surface other than the sphere if and only if it is isomorphic to neither the octahedron nor the icosidodecahedron.*

Proof It suffices to decide which graphs listed in Lemma 6 have faithful embeddings on orientable closed surfaces other than the sphere. We start by constructing equivariant partitions of the antiprism and the rhombicuboctahedron, as follows.

The antiprism consists of two disjoint cycles $x_0x_2 \cdots x_{2n-2}$ and $x_1x_3 \cdots x_{2n-1}$ with edges x_ix_{i+1} for $i = 0, 1, \ldots$ between them. Let S be the set of vertices lying along one of these cycles and $\overline{S}$ its complement. It is clear that $(S, \overline{S})$ forms an equivariant partition.

The rhombicuboctahedron comes from the cube and its triangular faces correspond to the vertices of the cube. Consider a proper coloring of vertices in the cube by black and while. Let S be the set of vertices on triangles corresponding to black vertices in the cube and let $\overline{S}$ be the set of vertices corresponding to white vertices. Since any automorphism of the rhombicuboctahedron sends those triangles to those, $(S, \overline{S})$ is an equivariant partition.

It is easy to construct faithful dot schemes for the cuboctahedron and the rhombicosidodecahedron. The former is covered by eight triangles and they can be separated into two groups each of which contains four disjoint triangles. Put black dots at the three corners of each triangle in one group and white dots similarly in the other group. This gives us a faithful dot scheme.

The rhombicosidodecahedron is covered by 30 disjoint triangles corresponding to the vertices of the dodecahedron. Also it is covered by 12 pentagons corresponding to the faces. Put black dots at the three corners of each triangle and white dots at the five corners of each pentagon. This is a faithful dot scheme.

By Lemma 7, the four graphs discussed above have faithful embeddings on orientable closed surfaces other than the sphere. To complete the proof, we shall show that the octahedron and the icosidodecahedron have neither equivariant partitions nor faithful dot schemes. Since each of them has a set of odd cycles which covers all vertices, we can carry out the same argument as in the proof of Theorem 3 to conclude that they have no equivariant partition, namely the argument in Case (i) of Theorem 3 works for the octahedron while the argument in Case (ii) works for the icosidodecahedron.

Finally, we shall show that the octahedron and the icosidodecahedron have no faithful dot scheme. The proof is easy for the former since any two faces are equivalent; if one corner of a face of the octahedron has a dot, then the automorphisms carry the dot to all corners of all faces. This implies that there would be no vertex of type (iii).

Now let G be the icosidodecahedron. This is covered by the set of 30 triangles and by the set of 12 pentagons. Suppose that there is a faithful dot scheme of G giving a faithful embedding of G on an orientable closed surface F^2. Choose a face A which contains a dot in the scheme, black or white, at one of its corners. Then there is an automorphism τ of period $|A| = 3$ or 5 which rotates the boundary cycle of A. Since its period is odd, τ should extend to an orientation-preserving auto-homeomorphism over F^2 and hence τ preserves the colors of any dot in the scheme.

In particular, the compositions $\tau, \tau^2, \tau^3, \ldots$ carry one dot to the dots placed at all corners of A and they have the same color, say black. This implies that any face sharing a vertex with A has white dots at all of its corners. Such a situation should hold for any pair of faces sharing a vertex, but it is impossible since such faces form a ring of odd length, 3 or 5. Therefore, G does not have any faithful dot scheme. □

5 Conclusion

Now we have prepared all we need to prove our main theorem. Combining Theorems 2, 3 and 4, we can conclude that the exceptions listed in our main theorem are exceptions indeed. We conclude our paper with a proof of Theorem 1.

Proof of Theorem 1 Let G be a 3-connected planar graph and suppose that G is not faithfully embeddable on any orientable close surface other than the sphere. By Theorem 2, G is either 3-regular or 4-regular. If G is 3-regular, then G is isomorphic to the tetrahedron, the dodecahedron, the truncated tetrahedron, the truncated dodecahedron or the truncated icosahedron by Theorem 3. If G is 4-regular, then G is isomorphic to the octahedron or the icosidodecahedron. These seven graphs coincide with the exceptions in the statement of the theorem. □

The results of this paper lead us to the following two questions:

- Define the *faithfully embeddable genus* of a planar graph as the minimum positive genus of orientable closed surfaces where it is faithfully embeddable. Is there a method to determine the faithfully embeddable genus of a given planar graph?
- Can we characterize those planar graphs that are faithfully embeddable on a fixed orientable closed surface, say the torus?

One might wonder if a 3-connected planar graph can be faithfully embedded on a nonorientable closed surface. For example, the tetrahedron K_4 can be embedded on the projective plane with three quadrilateral faces. It is easy to see that this embedding is faithful. This embedding can be regarded as the Petrie dual of the planar embedding of K_4.

In fact, in [3], the author gives a sufficient condition for a 3-connected planar graph to have a faithful embedding on a nonorientable closed surface and exhibits infinitely many 3-connected planar graphs that have no faithful embedding on nonorientable closed surfaces.

Acknowledgments The author would like to express his thanks to all participants of SIGMAP 2014 who gave him many good advices around his arguments on maps on surfaces. In particular, the notion of "Petrie duals" led him to a similar work on this topic with nonorientable closed surfaces. Also he appreciates Gašper Fijavž's helpful discussion on Lemma 2 and an anonymous referee who taught him about Zelinka's works.

References

1. J.L. Gross and T.W. Tucker, *"Topological Graph Theory"*, John Wiley & Sons, 1987.
2. S. Negami, Uniqueness and faithfulness of embedding of toroidal graphs, *Discrete Math.* **44** (1983), 161–180.
3. S. Negami, Faithful embeddings of planar graphs on nonorientable closed surfaces, preprint.
4. N. Robertson and R. Vitray, Representativity of surface embeddings, In: *Paths, flows, and VLSI layout*, B. Korte, L. Lovász, H.J. Prömel, and A. Schrijver, eds., Springer-Verlag, Berlin, Heidelberg, 1990, 293–328.
5. H. Whitney, Congruent graphs and the connectivity of graphs, *Amer. J. Math.* **54** (1932), 150–168.
6. B. Zelinka, Finite vertex-transitive planar graphs of the regularity degree four or five, *Matematický časopis* **25** (3) (1975), 271–280.
7. B. Zelinka, Finite planar vertex-transitive graphs of the regularity degree three, Časopis pro pěstování **102** (1) (1977), 1–9.

The Higher Dimensional Hemicuboctahedron

Daniel Pellicer

Abstract The paper describes the first known infinite sequence of 2-orbit d-polytopes in $\mathbb{R}^d$ with $d \geq 3$. The sequence has the remarkable property that its d-dimensional member has vertex-figures isomorphic to the $(d-1)$-dimensional member.

1 Introduction

Highly symmetric polytopes have been studied since antiquity, starting with the regular polygons and the Platonic solids. A lot of attention has been given to regular polytopes, which are those showing the highest degree of symmetry in terms of the action of their symmetry groups on the flags. At first, only convex regular polytopes were considered, in particular the Platonic solids were regarded as the only regular polyhedra. Kepler and Poinsot in the seventeenth and nineteenth centuries respectively, added four star polyhedra to the list of regular polyhedra. By doing this they dropped the requirement of convexity and admitted self-intersections. In the twentieth century Grünbaum no longer required the faces to lie on a plane and found the remaining nine finite regular polyhedra in $\mathbb{R}^3$. In the nineteenth century Schläfli gave a list of convex regular polytopes and Hess studied the star polytopes. Van Oss proved in the early twentieth century that these lists are complete. Finally, McMullen describes in this century all regular polytopes of rank d in $\mathbb{R}^d$ according to a definition in the spirit of Grünbaum's polyhedra ([19, Sect. 1A] and [16] for a more complete history on the topic).

Although the cuboctahedron and some other 2-orbit polyhedra have been known for a long time, their systematic study started only recently. The abstract theory was developed by Hubard and Schulte in [10, 13], while the classification of convex

D. Pellicer (✉)
Centro de Ciencias Matemáticas, UNAM, Antigua Carretera a Pátzcuaro 8701, 58089 Morelia, Mich, Mexico
e-mail: pellicer@matmor.unam.mx

J. Širáň and R. Jajcay (eds.), *Symmetries in Graphs, Maps, and Polytopes*, Springer Proceedings in Mathematics & Statistics 159,
DOI 10.1007/978-3-319-30451-9_13

2-orbit polytopes can be found in [15]. However, there is no final classification of finite 2-orbit polyhedra in $\mathbb{R}^3$, and even less of finite 2-orbit d-polytopes in $\mathbb{R}^d$.

The search for 2-orbit d-polytopes in $\mathbb{R}^d$ took a new turn with the discovery in [15] that there is no 2-orbit convex d-polytope in $\mathbb{R}^d$ for $d \geq 4$. In the same work it is also proved that there is no 2-orbit tessellation by convex tiles of $\mathbb{R}^d$ for $d \geq 4$. Similar behaviour was observed in [8], where 2-orbit abstract polytopes (combinatorial structures generalising the notion of polytope in this paper) arise as products of partially ordered sets only for rank 3. Moreover, in [12] it was proven that no 2-orbit tessellation by cubes of the 3-torus exists even though there are several types of 2-orbit tessellations by squares of the 2-torus. In the same work it is stated as a possibility that there are no 2-orbit tessellations by d-cubes of the d-torus for $d \geq 3$. These results highlight the relevance of the question of existence of 2-orbit d-polytopes in $\mathbb{R}^d$ for every $d \geq 4$.

In this paper we exhibit the first known family of 2-orbit d-polytopes in $\mathbb{R}^d$ for every $d \geq 4$. In Sects. 2 and 3 we recall basic definitions of regular and 2-orbit polytopes. The main construction is explained in Sect. 4.

2 Regular Polytopes

Polygons, polyhedra and polytopes admit a number of definitions, not all of them equivalent. Here we are interested in symmetry of polytopes on Euclidean spaces, and hence we shall use appropriate definitions to obtain a rich theory in this direction. In our definition of polytope we follow the ideas in [9].

A *polygon* (or 2-polytope) in $\mathbb{R}^d$ consists of a set of points called *vertices* and a set of line segments between pairs of vertices called *edges*, with the property that the induced graph is connected, every vertex belongs to precisely two edges, and every compact subset of $\mathbb{R}^d$ intersects only finitely many vertices. If the polygon is finite then the graph induced by the vertex and edge sets is a cycle; on the other hand, if the polygon is infinite then the induced graph is isomorphic to a two-sided infinite path. By convention, the vertices have rank 0 and the edges have rank 1.

Remark 1 In [9] the requirement of discreteness of polygons in $\mathbb{R}^d$ states that every compact subset of $\mathbb{R}^d$ intersects finitely many edges. This is harder to generalise appropriately to higher rank polytopes. In any case, for finite polytopes, like the ones considered in this paper, the discreteness requirement is superfluous.

There is no restriction regarding intersections of the edges of a polygon in interior points. Also there is no disk spanned into the polygon. In fact, in the definition above we can instead define the edges to consist of pairs of vertices as opposed to line segments, thereby avoiding the association of an i-dimensional geometric object to a rank i element of the polygon.

For $n \geq 3$ we define *n-polytopes*, or polytopes of rank n, in a Euclidean space recursively. An n-polytope $\mathcal{P}$ in $\mathbb{R}^d$ consists of a collection $\mathcal{P}_{n-1}$ of $(n-1)$-polytopes in $\mathbb{R}^d$ called *facets* satisfying the following properties.

(I) **Diamond condition.** For every facet F of $\mathcal{P}_{n-1}$ and every facet G of F there exists a unique $F' \in \mathcal{P}_{n-1} \setminus \{F\}$ such that G is a facet of F'.
(II) **Connectivity.** For every pair of elements G and G' of rank $n-2$ there exists a sequence $(G \equiv G_0, F_1, G_1, F_2, \ldots, F_k, G_k = G')$ where $F_i \in \mathcal{P}_{n-1}$, G_i has rank $n-2$, and F_i contains G_{i-1} and G_i for $i \in \{1, \ldots, k\}$.
(III) **Discreteness.** Every compact subset of $\mathbb{R}^d$ intersects finitely many vertices.

Remark 2 The definitions of n-polytope in a Euclidean space above and of polyhedron in [9] are more general than faithful realizations of abstract polytopes and of abstract polyhedra, respectively (see [19, Chaps. 1, 5]). Faithful realizations of abstract polytopes require strong flag-connectivity, whereas our current definition only implies flag-connectivity. In any case, the choice of kind of connectivity has no effect on the contents of this paper since the structures described satisfy both definitions.

We shall call the building blocks of $\mathcal{P}$ the elements of $\mathcal{P}$, and provide them with the partial order given by inclusion. The elements of rank i are called i-faces, the 0-faces are the vertices, the 1-faces are the edges, and the $(n-1)$-faces are the facets. The *vertex-figure* at a vertex v is the $(n-1)$-polytope constructed recursively as follows. Its vertices are the neighbours of v, each corresponding to an edge incident with v. For $i \geq 1$, each i-face F of the vertex figure corresponds to an $(i+1)$-face G_F of $\mathcal{P}$ containing v, and contains precisely the $(i-1)$-faces that correspond to the i-faces of $\mathcal{P}$ contained in G_F and containing v. The *k-skeleton* of $\mathcal{P}$ consists of all faces of $\mathcal{P}$ of rank at most k.

A *flag* of $\mathcal{P}$ is a maximal totally ordered set and contains precisely n elements, one of each rank in $\{0, \ldots, n-1\}$. By construction, for every $i \in \{0, \ldots, n-1\}$ and every flag Φ of $\mathcal{P}$ there exists a unique *i-adjacent flag* Φ^i differing with Φ only on the i-face.

A *symmetry* of $\mathcal{P}$ is an isometry of the ambient space preserving $\mathcal{P}$. The group of symmetries of $\mathcal{P}$ is denoted by $Sym(\mathcal{P})$.

We say that $\mathcal{P}$ is *regular* whenever $Sym(\mathcal{P})$ acts transitively on the flags of $\mathcal{P}$. The classical examples given by the Platonic solids and the regular convex d-polytopes satisfy this definition.

A lot is known about regular polytopes. There are 48 regular polyhedra in $\mathbb{R}^3$ including 18 finite ones, 6 infinite polyhedra whose ambient space can be understood as any plane $\mathbb{R}^2$ contained in $\mathbb{R}^3$, and 24 infinite polyhedra that do not fit on a plane (see [6, 7, 18]). Finite regular d-polytopes and infinite $(d+1)$-polytopes in $\mathbb{R}^d$ are described by McMullen in [16]. In various other papers McMullen studies other families of regular polytopes (see for example [17]).

To conclude this section we describe in detail the cross-polytope of dimension d and two groups of isometries related to it.

Let $\mathcal{C} = \{e_1, \ldots, e_d\}$ be the canonical basis of $\mathbb{R}^d$. We shall abuse notation and denote by e_i the points of $\mathbb{R}^3$ instead of the corresponding vectors. The d-dimensional *cross-polytope* $\mathcal{O}_d$ is the convex hull of $\{\pm e_1, \ldots, \pm e_d\}$. For $k \leq n-1$, its k-faces are simplices determined by a set of $k+1$ vertices not containing *antipodal pairs* (that is, vertices e_i and $-e_i$ for some i). A more detailed description of cross-polytopes can be found in [2, Sect. 7.2].

The symmetry group of $\mathcal{O}_d$ is a Coxeter group of type B_d. In other words, it is the semidirect product of the group $G_1 \cong \mathbb{Z}_2^d$ generated by the reflections with respect to all canonical hyperplanes, and the group $G_2 \cong S_d$ permuting the coordinate axes. Clearly the group G_2 acts on G_1 by conjugation. The group $Sym(\mathcal{O}_d)$ has then $2^d d!$ elements, and $\mathcal{O}_d$ has $2^d d!$ flags (hence $\mathcal{O}_d$ is a regular d-polytope).

The group $Sym(\mathcal{O}_d)$ has an index 2 subgroup of Coxeter type D_n, that is, it is isomorphic to $G_1^+ \rtimes G_2$ where G_1^+ is the subgroup of G_1 consisting of those elements that change the sign to an even number of coordinate axes. Note that conjugation by G_2 preserves the group G_1^+. For further information the reader is referred to [14].

3 2-orbit Polytopes

We say that $\mathcal{P}$ is a 2-*orbit* polytope whenever $Sym(\mathcal{P})$ induces two orbits on the flags. Polytopes with this property are those with highest degree of symmetry among the ones that are not regular.

If some pair of i-adjacent flags of a 2-orbit polytope $\mathcal{P}$ of rank n belong to the same flag-orbit then Φ and Φ^i are in the same flag-orbit for every flag Φ of $\mathcal{P}$ (see [10, Lemma 2] for the version for abstract polytopes, and note that the proof holds also in this geometric setting). This allows us to define the class 2_I for $I \subset \{0, \dots, n-1\}$ consisting of all 2-orbit n-polytopes $\mathcal{P}$ for which any two i-adjacent flags are in the same flag-orbit if and only if $i \in I$. Here we do not consider $I = \{0, \dots, n-1\}$ since this would imply that $\mathcal{P}$ is regular. The combinatorial analogue of this definition can be found in [10].

There are not many examples known of 2-orbit polyhedra so far. In [15] Matteo shows that there are only four 2-orbit convex polytopes of rank $d \geq 3$, namely the cuboctahedron, the icosidodecahedron, the rhombic dodecahedron and the rhombic triacontahedron, all in rank 3. In particular, there is no convex 2-orbit d-polytope for $d \geq 4$. The cuboctahedron and the icosidodecahedron are in class $2_{\{0,1\}}$, whereas the rhombic dodecahedron and the rhombic triacontahedron are in class $2_{\{1,2\}}$.

Polyhedra in class $2_\emptyset$ (also called class 2) are called *chiral* and were classified in $\mathbb{R}^3$ by Schulte in [21, 22]. All chiral polyhedra in $\mathbb{R}^3$ are infinite and they are organised in six families depending on their kind of faces and vertex-figures. Recently a finite 2-orbit 4-polytope in class $2_\emptyset$ was discovered (see [1]).

Finite 2-orbit polyhedra in $\mathbb{R}^3$ that are combinatorially regular but cannot be realized in ordinary space in a flag-transitive manner were classified in [4, 5].

Some of the uniform polyhedra described in [3] are examples of finite 2-orbit polyhedra in $\mathbb{R}^3$. One of them is of particular interest for this paper and can be described as follows. The triangular faces of a regular octahedron $\mathcal{O}_3$ admit a bipartition in which two faces share an edge if and only if they are in distinct parts. The faces of the 2-orbit polyhedron are then the triangles of $\mathcal{O}_3$ in one of the parts, together with the three (planar) equatorial squares of the octahedron. The centres of all equatorial

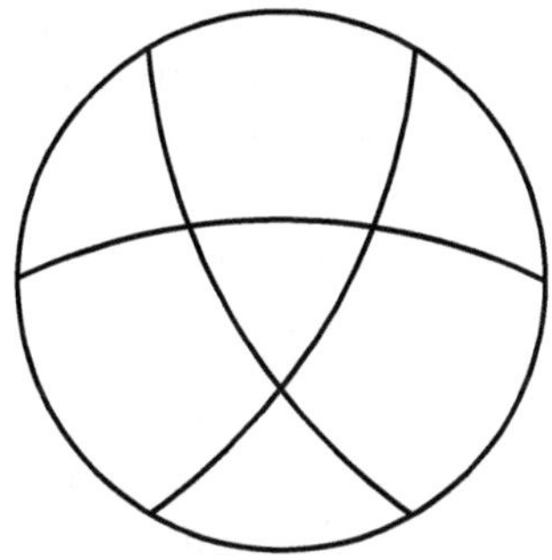

Fig. 1 The hemicuboctahedron on the projective plane

squares coincide with the centre of $\mathcal{O}_3$, and each of them is determined by two pairs of antipodal vertices of $\mathcal{O}_3$.

The 2-orbit polyhedron previously described is combinatorially equivalent to the *hemicuboctahedron*, a map on the projective plane obtained by antipodal identification of the symmetric drawing of the cuboctahedron on the sphere (see Fig. 1, where the points in the outer circle are identified). This can be seen easily by noting that in the polyhedron and in the map every triangle shares an edge with every square and every vertex belongs to two triangles and two squares in an alternating manner.

4 The Higher Dimensional Hemicuboctahedron

To the author's knowledge, no 2-orbit d-polytope for $d \geq 4$ appears in a published work besides the chiral 4-polytope in [1]. In any case, there is no known family of 2-orbit polytopes containing a polytope of each dimension $d \geq 4$ in $\mathbb{R}^d$. In what follows we describe a 2-orbit d-polytope $\mathcal{H}_d$ in $\mathbb{R}^d$ for every $d \geq 3$, generalising the idea of the hemicuboctahedron in $\mathbb{R}^3$.

Recall that $\mathcal{C} = \{e_1, \ldots, e_d\}$ is the canonical basis of $\mathbb{R}^d$ and let $\mathcal{O}_d$ be the d-dimensional cross polytope with vertex set $\{\pm e_i \mid i \in \{1, \ldots, d\}\}$. Let $\mathcal{X}_1$ be the set of simplicial facets of $\mathcal{O}_d$ whose vertex sets contain an even number of points with a coordinate -1, and let $\mathcal{X}_2$ be the set of "equatorial" $(d-1)$-dimensional cross polytopes obtained by intersecting $\mathcal{O}_d$ with the canonical hyperplanes.

Before constructing the 2-orbit polytope we state the following easy observations.

Remark 3 The sets $\mathcal{X}_1$ and $\mathcal{X}_2$ have 2^{d-1} and d elements, respectively.

Remark 4 Every element of $\mathcal{X}_1$ shares one and only one $(d-2)$-simplex with each element of $\mathcal{X}_2$, and vice versa.

We construct $\mathcal{H}_d$ by adjoining $\mathcal{X}_1 \cup \mathcal{X}_2$ to the $(d-2)$-skeleton of $\mathcal{O}_d$ as the set of $(d-1)$-faces.

Theorem 5 *The combinatorial structure $\mathcal{H}_d$ just defined is a finite 2-orbit d-polytope in class $2_{\{0,1,\ldots,d-2\}}$ with symmetry group of Coxeter type D_d.*

Proof Since the $(d-2)$-skeletons of $\mathcal{O}_d$ and $\mathcal{H}_d$ coincide we only need to verify properties (I) and (II) in order to show that $\mathcal{H}_d$ is a polytope. (Note here that the vertex sets of $\mathcal{O}_d$ and $\mathcal{H}_d$ are the same, implying that $\mathcal{H}_d$ is discrete.)

A $(d-2)$-face of $\mathcal{H}_d$ is also a $(d-2)$-face of $\mathcal{O}_d$, that is, a $(d-2)$-simplex determined by a set of $d-1$ vertices containing no antipodal pair. Such a $(d-2)$-face G belongs to precisely two $(d-1)$-faces of $\mathcal{O}_d$, namely the simplices determined by the vertices of G and each of the vertices of the antipodal pair with no representative in G. Clearly the number of vertices with an entry -1 is even in one of these $(d-1)$-simplices and is odd in the other one. Consequently, G belongs to a unique facet of $\mathcal{H}_d$ in $\mathcal{X}_1$. Furthermore, G can be extended to a unique face in $\mathcal{X}_2$, namely to the $(d-1)$-dimensional cross polytope determined by the vertices of G and their antipodes. Hence G belongs to precisely two facets of $\mathcal{H}_d$ and the diamond condition holds.

Let G_1 and G_2 be two $(d-2)$-faces of $\mathcal{O}_d$ and $F_0 \in \mathcal{X}_1$. The discussion above shows that there exist F_1 and F_2 in $\mathcal{X}_2$ incident to G_1 and G_2, respectively. By Remark 4, F_1 (resp. F_2) shares a $(d-2)$-face H_1 (resp. H_2) with F_0. Then the connectivity follows from the sequence $(G_1, F_1, H_1, F_0, H_2, F_2, G_2)$.

When considering $Sym(\mathcal{O}_d) \cong G_1 \rtimes G_2$ as in Sect. 2, the group G_2 permutes the elements of $\mathcal{X}_1$, whereas all generating reflections of G_1 map them to the facets of $\mathcal{O}$ not in $\mathcal{X}_1$. It follows that $Sym(\mathcal{H}_d)$ contains S_d and the subgroup of $\mathbb{Z}_2^d$ consisting of products of an even number of the generating reflections. Hence $Sym(\mathcal{H}_d)$ is of Coxeter type D_d.

The number of flags on each facet of $\mathcal{H}_d$ in $\mathcal{X}_1$ is $d!$, since they are $(d-1)$-simplices. The facets of $\mathcal{H}_d$ in $\mathcal{X}_2$ are $(d-1)$-dimensional cross-polytopes and therefore each of them has $2^{d-1}(d-1)!$ flags. It follows from Remark 3 that the number of flags of $\mathcal{H}$ is

$$2^{d-1}d! + d \cdot 2^{d-1}(d-1)! = 2^d d!.$$

Since the Coxeter groups of type D_d have half as may elements as the symmetry group of the d-dimensional cross-polytope, we conclude that $\mathcal{H}_d$ is a 2-orbit polytope for every $d \geq 3$. Furthermore, flags in different kinds of facets must be in different orbits, implying that flags in the same kind of facets are in the same orbit. Since every $(d-2)$-simplex belongs precisely to one simplicial facet and to one a $(d-1)$-dimensional cross-polytope, $\mathcal{H}_d$ is in class $2_{\{0,1,\ldots,d-2\}}$. ■

Part of the combinatorial structure of the polytope $\mathcal{H}_d$ is given by Remarks 3 and 4. Next we describe the vertex-figure of $\mathcal{H}_d$

Proposition 6 *For $d \geq 4$ the vertex figure of $\mathcal{H}_d$ is $\mathcal{H}_{(d-1)}$.*

Proof The neighbours of any given vertex v of $\mathcal{H}_d$ are all vertices of $\mathcal{H}_d$ except for v and $-v$. Moreover, for $k \leq d-1$ any set of k vertices containing v but no antipodal pairs induces a $(k-1)$-face of $\mathcal{H}_d$, and therefore a $(k-2)$-face of the vertex-figure at v. Finally, there are two kinds of facets ($(d-2)$-faces) of the vertex-figure at v. One of them consists of simplicial facets of $\mathcal{H}_d$ containing v, and therefore they are simplices. The facets of the other kind can be obtained by removing v and $-v$ to all $(d-1)$-dimensional cross-polytopes that contain v. It is easy to see that such a vertex-figure can be alternatively obtained from the construction above when considering only the $2(d-1)$ vertices of $\mathcal{H}_d$ different from v and $-v$. Hence the vertex-figure is precisely $\mathcal{H}_{d-1}$. ■

Remark 7 Proposition 6 provides an easy recursive proof that $\mathcal{H}_d$ is strongly flag-connected as defined in [19, Sect. 2A]. Hence $\mathcal{H}_d$ is a realization of an abstract polytope.

The polytope $\mathcal{H}_d$ can be visualised as a combinatorial structure in $\mathbb{S}^{d-1}$ as follows. Project the simplicial facets of $\mathcal{H}_d$ from the origin to the $(d-1)$-sphere containing the vertices of $\mathcal{H}_d$, and understand each $(d-1)$-cross polytope $\mathcal{O}$ as tessellations by $(d-2)$-simplices of the $(d-2)$-sphere spanned by the vertices of $\mathcal{O}$. For example, $\mathcal{H}_3$ can be viewed in $\mathbb{S}^2$ by dividing it in the 8 triangles arising from the projection of the octahedron and considering as 2-faces half of the triangles (one part of the bipartition) as well as the three equatorial squares. In this setting, every edge belongs to a triangle and to an equatorial square. Note that this visualization does not induce a tessellation of $\mathbb{S}^3$.

Whenever d is even the polytope $\mathcal{H}_d$ is invariant under the isometry $-Id$ and therefore it admits antipodal identification into de projective space $P^{d-1}(\mathbb{R})$. The facets of these objects are simplices and halves of cross-polytopes (that is, their images under antipodal identification). It is worth mentioning that the case $d = 4$ corresponds to the Tomotope (see [20]), whose facets are four tetrahedra and four hemioctahedra, and whose vertex-figures are hemicuboctahedra.

5 Conclusions

The polytopes $\mathcal{H}_d$ generalising the hemicuboctahedron are the first 2-orbit polytopes known to exist in ranks $d \geq 5$, and $\mathcal{H}_4$ is among the first ones in rank 4. Given a family of polytopes with certain properties, one can often find more polytopes satisfying the same set of properties by applying operations like duality and the Petrie operation (see [16, 19, Chap. 7]). However, it seems that no other 2-orbit d-polytope is related to $\mathcal{H}_d$ in these standard ways.

For example, there is a construction of a dual for the Platonic solids and other convex polyhedra where the vertices of the dual are the centres of the facets of the original polytope. This fails for $\mathcal{H}_d$ since all faces in $\mathcal{X}_2$ are centred at the origin. This originates a collapse in the 2-faces, which are no longer polygons.

The Petrie operation was defined in [16, p. 4] for regular polytopes and in [11, p. 8] for arbitrary polytopes. When applied to a d-polytope $\mathcal{P}$, the Petrie operation yields a d-polytope $\mathcal{P}^\pi$ where two flags Φ and Ψ are $(d-3)$-adjacent in $\mathcal{P}^\pi$ if and only if $\Phi^{d-3,d-1} = \Psi$ when viewed as flags of $\mathcal{P}$. The polytopes $\mathcal{P}$ and $\mathcal{P}^\pi$ have the same $(d-2)$-skeleton, but their facets in general are not isomorphic. It is not hard to see that the facets of $(\mathcal{H}_4)^\pi$ fail the diamond condition, since the edges belong to four rank 2 faces in the same facet. By Proposition 6, this situation carries over to higher ranks and the diamond condition in $(\mathcal{H}_d)^\pi$ fails for $d \geq 4$.

Thus, the full classification of 2-orbit polytopes of rank d in $\mathbb{R}^d$ is far from being complete. The polytopes $\mathcal{H}_d$ provide examples satisfying the additional property that all k-faces lie on a k-dimensional affine subspace of $\mathbb{R}^d$. This is the analogue of requiring flat 2-faces for polyhedra. We therefore propose the following open problem as an intermediate step for the full classification of 2-orbit polytopes.

Open problem Determine all 2-orbit polytopes in $\mathbb{R}^d$ such that all k-faces lie on a k-dimensional affine subspace.

Acknowledgments The author wants to thank Barry Monson for helpful discussion, and the anonymous referees for their suggestions of improvements. This work was supported by PAPIIT-UNAM under project IN101615 and CONACyT under project 166951.

References

1. Javier Bracho, Isabel Hubard, and Daniel Pellicer. A finite chiral 4-polytope in $\mathbb{R}^4$. *Discrete Comput. Geom.*, 52(4):799–805, 2014.
2. H. S. M. Coxeter. *Regular polytopes*. Dover Publications, Inc., New York, third edition, 1973.
3. H. S. M. Coxeter, M. S. Longuet-Higgins, and J. C. P. Miller. Uniform polyhedra. *Philos. Trans. Roy. Soc. London. Ser. A.*, 246:401–450 (6 plates), 1954.
4. Anthony M. Cutler. Regular polyhedra of index two, II. *Beitr. Algebra Geom.*, 52(2):357–387, 2011.
5. Anthony M. Cutler and Egon Schulte. Regular polyhedra of index two, I. *Beitr. Algebra Geom.*, 52(1):133–161, 2011.
6. Andreas W. M. Dress. A combinatorial theory of Grünbaum's new regular polyhedra. I. Grünbaum's new regular polyhedra and their automorphism group. *Aequationes Math.*, 23(2-3):252–265, 1981.
7. Andreas W. M. Dress. A combinatorial theory of Grünbaum's new regular polyhedra. II. Complete enumeration. *Aequationes Math.*, 29(2-3):222–243, 1985.
8. Ian Gleason and Isabel Hubard. Products of abstract polytopes. In preparation.
9. Branko Grünbaum. Regular polyhedra—old and new. *Aequationes Math.*, 16(1-2):1–20, 1977.
10. Isabel Hubard. Two-orbit polyhedra from groups. *European J. Combin.*, 31(3):943–960, 2010.
11. Isabel Hubard, Alen Orbanić, and Asia Ivić Weiss. Monodromy groups and self-invariance. *Canad. J. Math.*, 61(6):1300–1324, 2009.
12. Isabel Hubard, Alen Orbanić, Daniel Pellicer, and Asia Ivić Weiss. Symmetries of equivelar 4-toroids. *Discrete Comput. Geom.*, 48(4):1110–1136, 2012.
13. Isabel Hubard and Egon Schulte. Two-orbit polytopes. *In preparation.*
14. James E. Humphreys. *Reflection groups and Coxeter groups*, volume 29 of *Cambridge Studies in Advanced Mathematics*. Cambridge University Press, Cambridge, 1990.
15. Nicholas Matteo. Two-orbit convex polytopes and tilings. *Discrete Comput. Geom.*, In press.
16. Peter McMullen. Regular polytopes of full rank. *Discrete Comput. Geom.*, 32(1):1–35, 2004.
17. Peter McMullen. Regular polytopes of nearly full rank. *Discrete Comput. Geom.*, 46(4):660–703, 2011.
18. Peter McMullen and Egon Schulte. Regular polytopes in ordinary space. *Discrete Comput. Geom.*, 17(4):449–478, 1997. Dedicated to Jörg M. Wills.
19. Peter McMullen and Egon Schulte. *Abstract Regular Polytopes*. Cambridge University Press, 2002.

20. Barry Monson, Daniel Pellicer, and Gordon Williams. The tomotope. *Ars Math. Contemp.*, 5(2):355–370, 2012.
21. Egon Schulte. Chiral polyhedra in ordinary space. I. *Discrete Comput. Geom.*, 32(1):55–99, 2004.
22. Egon Schulte. Chiral polyhedra in ordinary space. II. *Discrete Comput. Geom.*, 34(2):181–229, 2005.

Groups of Order at Most 6,000 Generated by Two Elements, One of Which Is an Involution, and Related Structures

Primož Potočnik, Pablo Spiga and Gabriel Verret

Abstract A (2,*)-*group* is a group that can be generated by two elements, one of which is an involution. We describe the method we have used to produce a census of all (2,*)-groups of order at most 6,000. Various well-known combinatorial structures are closely related to (2,*)-groups and we also obtain censuses of these as a corollary.

1 Introduction

The objects that play a central role in our paper are $(2,*)$-*groups*, that is, groups that can be generated by two (not necessarily distinct) elements, one of which is an involution. We will also need the notion of a $(2,*)$-triple, which we now define.

Definition 1 A $(2,*)$-*triple* is a triple (G, x, g) such that G is a $(2,*)$-group, $\{x, g\}$ is a generating set for G and x is an involution. Two $(2,*)$-triples (G_1, x_1, g_1) and (G_2, x_2, g_2) are *isomorphic* if there exists a group isomorphism from G_1 to G_2 mapping x_1 to x_2 and g_1 to g_2.

P. Potočnik (✉)
Faculty of Mathematics and Physics, University of Ljubljana,
Jadranska 21, 1000 Ljubljana, Slovenia
e-mail: primoz.potocnik@fmf.uni-lj.si

P. Potočnik
IAM, University of Primorska, Muzejski trg 2, 6000 Koper, Slovenia

P. Spiga
Dipartimento di Matematica Pura e Applicata, University of Milano-Bicocca,
Via Cozzi 55, 20126 Milano, Italy
e-mail: pablo.spiga@unimib.it

G. Verret
Centre for the Mathematics of Symmetry and Computation,
The University of Western Australia, 35 Stirling Hwy, Crawley 6009, Australia
e-mail: gabriel.verret@uwa.edu.au

G. Verret
FAMNIT, University of Primorska, Glagoljaška 8, 6000 Koper, Slovenia

J. Širáň and R. Jajcay (eds.), *Symmetries in Graphs, Maps, and Polytopes*,
Springer Proceedings in Mathematics & Statistics 159,
DOI 10.1007/978-3-319-30451-9_14

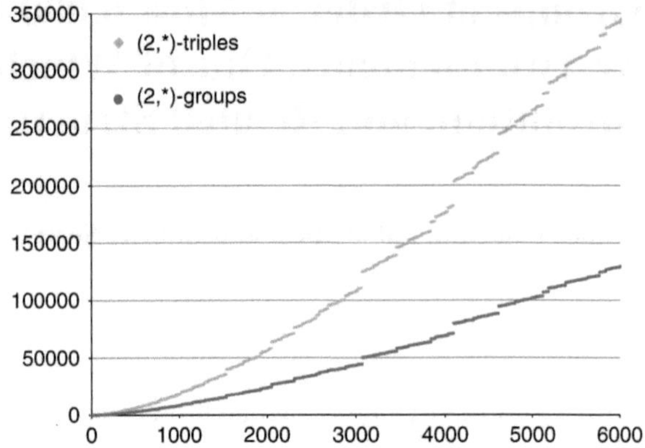

Fig. 1 Number of (2, $*$)-groups and triples up to a given order

The first aim of this paper is to announce a complete determination of all (2, $*$)-groups of order at most 6,000. The methods we used and how they improve on the ones used by previous authors are discussed in Sect. 2. Here, we just state the overall enumeration result.

Theorem 1 *Up to isomorphism, there are precisely* 129,340 (2, $*$)*-groups and* 345,070 (2, $*$)*-triples of order at most* 6,000.

The database of all (2, $*$)-groups and triples in a form readable by MAGMA [2] is available at [16].

The second aim of this paper is to prove an asymptotic enumeration result for (2, $*$)-groups and (2, $*$)-triples. Let $f(n)$ and $f_t(n)$ denote the number (up to isomorphism) of (2, $*$)-groups and the number of (2, $*$)-triples, respectively, of order at most n. The graphs of $f(n)$ and $f_t(n)$ are depicted in Fig. 1. A quick look at this picture might suggest that both $f(n)$ and $f_t(n)$ grow polynomially in n. This is not the case. In fact, in Sect. 3, we show the following:

Theorem 2 *There exist positive constants a and b such that, for $n \geq 2$, we have*

$$n^{a \log n} \leq f(n) \leq f_t(n) \leq n^{b \log n}.$$

The problem of optimising the constants a and b in Theorem 2 is beyond the scope of this article and is related to the problem of enumerating the normal subgroups of finite index in certain finitely presented groups (see, for example, [11, Chap. 2]).

The third aim of the paper is a discussion of a relationship between (2, $*$)-groups and several highly symmetrical geometric and combinatorial objects; for example, cubic Cayley graphs, arc-transitive digraphs of out-valence 2, and rotary maps (both chiral and reflexible, on orientable and non-orientable surfaces). These relationships are explained in Sects. 4 and 5. Together with our census of (2, $*$)-triples they have allowed us to generate complete lists of:

- cubic Cayley graphs generated by an involution and a non-involution, with at most 6,000 vertices;
- digraphs of out-valence 2 admitting an arc-regular group of automorphisms, with at most 3,000 vertices;
- rotary maps (both chiral and reflexive) on orientable surfaces, with at most 3,000 edges;
- regular maps on non-orientable surfaces, with at most 1,500 edges.

Databases of these objects are also available at [16].

2 Constructing the Census of Small (2, $*$)-Groups

In this section, we are concerned with the problem of generating a complete list of (2, $*$)-triples (G, x, g) with $|G| \leq m$ for some prescribed constant m. Let us discuss a few possible approaches to this problem.

2.1 Using a Database of Small Groups

If m is sufficiently small, then a database of all the groups of order at most m might be available. For example, at the time of writing of this article, all groups of order 2,000 have been known, and all except those of order 1,024 have been available in standard distributions of GAP [20] and MAGMA [2]. One might thus try to search through such a database and, for each group G in the database, determine all possible generating pairs (x, g) with x being an involution, up to conjugacy in $\mathrm{Aut}(G)$.

While this approach is rather straight-forward, it has an obvious downside in that it requires iterating over all the groups of order at most m. Namely, getting access to the groups of order 1,024 is difficult at the moment and the groups of order 2,048 will probably remain out of reach in the near future. Even if one had access to these groups, their number would make it inconvenient to iterate over them. (There are more than 10^{15} groups of order 2,048 [5].)

These considerations should make it clear that, to make any significant progress, one should find a way to avoid having to consider all groups of order at most m.

2.2 Using the Magma LowIndexNormalSubgroups Algorithm

Observe that every (2, $*$)-group is an epimorphic image of the free product $U := C_2 * C_\infty = \langle x, g \mid x^2 \rangle$ and can thus be obtained as a quotient of U by a normal subgroup N not containing x. Note that this yields not only the (2, $*$)-group U/N, but also the $(2, *)$-triple $(U/N, Nx, Ng)$. In order to find all $(2, *)$-triples of order at most m it thus suffices to find all normal subgroups of U of index at most m.

Firth and Holt [6] have developed a very efficient algorithm for determining normal subgroups of bounded index in a finitely presented group. The current implementation of this algorithm in MAGMA can, in principle, compute all normal subgroups of index at most 500,000. However, for certain finitely presented groups the practical limitations of the algorithm (or at least its current implementation in MAGMA) make the computation unfeasible, even for much smaller indices.

An approach along these lines (in the language of rotary maps; see Sect. 5.2) has been successfully used by Conder [3] to determine all normal subgroups of U of index at most 2,000, but computations took several months.

2.3 Using Group Extensions

Finally, we describe the approach that we used to compile a complete list of $(2, *)$-groups and triples of order at most 6,000. The method is inductive and constructs $(2, *)$-groups as extensions of smaller ones. The general idea is not new (see, for example, [7]), but our recent implementation proved to be more efficient than recent efforts using the `LowIndexNormalSubgroups` algorithm.

Let us first set some terminology. If N is a normal subgroup of a group G and Q is a group isomorphic to the quotient G/N, then we say that G is an *extension of* Q *by* N. (Some authors call G an extension of N by Q.) If N is a minimal normal subgroup of G, then we shall say that the extension is *direct*, and if N is elementary abelian, then we say that the extension is *elementary abelian*. The *soluble radical* of a group is its (unique) largest normal soluble subgroup.

Lemma 1 *If G is a $(2, *)$-group, then either G has a trivial soluble radical, or G is a direct elementary abelian extension of a smaller $(2, *)$-group or of a cyclic group of odd order.*

Proof As G is a $(2, *)$-group, we have $G = \langle x, g\rangle$, for some involution $x \in G$ and some $g \in G$. Suppose that the soluble radical S of G is non-trivial. Let N be a minimal normal subgroup of G contained in S. Since S is soluble, N is elementary abelian and hence G is a direct elementary abelian extension of N by G/N. If $G/N = \langle xN, gN\rangle$ is not a $(2, *)$-group, then $xN = N$ and gN has odd order, that is, $G/N = \langle gN\rangle$ is cyclic of odd order.

Lemma 1 suggests an inductive procedure to construct $(2, *)$-groups from smaller ones. The base case of this inductive process are $(2, *)$-groups with trivial soluble radical and cyclic groups of odd order. If G is a finite group with trivial soluble radical, then soc(G) (that is, the group generated by the minimal normal subgroups of G) is isomorphic to a direct product of non-abelian simple groups and, moreover, G acts faithfully on soc(G) by conjugation and thus G embeds into Aut(soc(G)).

This allows one to use a database of small simple groups (available, say, in MAGMA or GAP) to construct all groups of order at most m with trivial soluble radical.

For example, it is an easy computation to determine that there are precisely 23 groups with trivial soluble radical of order at most 6,000. For a given group G with trivial soluble radical, one can find all $(2, *)$-triples (G, x, g) by determining all epimorphisms from the group $C_2 * C_\infty$ to G (where two epimorphisms are considered equivalent if they differ by some automorphism of G).

Let us now discuss the inductive step. Suppose we are given a group Q of order n (which, for our purposes, can be taken to be either a $(2, *)$-group or cyclic of odd order) and would like to find all direct elementary abelian extensions of Q of order at most m. In view of the general theory of group extensions, it suffices to find all irreducible $\mathbf{Z}_p Q$-modules N, with N isomorphic to an elementary abelian group $\mathbf{Z}_p^d$, such that $p^d n \leq m$ and then, for each such module N, compute the cohomology group $H^2(Q, N)$. Each element of $H^2(Q, N)$ then gives rise to a direct extension of Q by N, and conversely, each direct elementary abelian extension of Q of order at most m can be obtained in this manner. Efficient algorithms for computing the irreducible modules of a given group and the corresponding second cohomology group are known (see for example [7]) and are implemented in MAGMA.

It is not surprising that computationally the hardest case is the extension of 2-groups by 2-groups. Fortunately, in this case some parts of the inductive step can be simplified. Namely, when Q is a 2-group and $p = 2$, the only irreducible $\mathbf{Z}_2 Q$-module is the (trivial) 1-dimensional $\mathbf{Z}_2 Q$-module $\mathbf{Z}_2$ and hence only the cohomology group $H^2(Q, \mathbf{Z}_2)$ needs to be considered. This shortcut speeds up the determination of $(2, *)$- 2-groups considerably.

Once the direct elementary abelian extensions G of Q are determined, one needs to check which of them are $(2, *)$-groups and, for those which are, find all pairs (x, g) such that (G, x, g) is a $(2, *)$-triple. This can be done by first computing the automorphism group $\mathrm{Aut}(G)$, then choosing a representative of each orbit of $\mathrm{Aut}(G)$ on the set of involutions of G then, for each representative x, computing the stabiliser $\mathrm{Aut}(G)_x$ of x in $\mathrm{Aut}(G)$, choosing a representative g from each orbit of $\mathrm{Aut}(G)_x$ on G and, finally, discarding the pairs (x, g) that do not generate G.

As mentioned at the beginning of the section, this method is the one that we used in order to obtain the complete list of $(2, *)$-groups and triples of order at most 6,000. The computation took a few weeks on a computer with a 2.93 GHz Intel Xeon processor and 56 GB of memory.

3 Proof of Theorem 2

Since every $(2, *)$-group gives rise to a $(2, *)$-triple, we have $f(n) \leq f_t(n)$. On the other hand, if G is a $(2, *)$-group of order n, then there are at most n^2 choices for $(x, g) \in G \times G$ hence at most n^2 $(2, *)$-triples with first coordinate G. This shows that $f_t(n) \leq n^2 f(n)$. In particular, it suffices to prove that there exist positive constants a and b such that, for $n \geq 2$, we have

$$n^{a \log n} \leq f(n) \leq n^{b \log n}.$$

Clearly, $f(n)$ is at most the number of groups (up to isomorphism) of order at most n generated by 2 elements. By a celebrated theorem of Lubotzky [10, Theorem 1], the latter is at most $n^{b \log n}$. (Some information on the constant b can be found in [10, Sect. 3, Remark 1].)

The lower bound follows easily from a theorem of Müller and Schlage-Puchta: let A be a cyclic group of order 2, let B be a cyclic group of order 3 and let G be the free product of A and B, that is, $G = A * B$. Let $C = A \times B$, let $\pi : G \to C$ be the natural projection and let N be the kernel of π.

Observe that, since π is surjective, $N \cap A = N \cap B = 1$. By Bass-Serre theory, N is a free group (see [17, Theorem 4, p. 27]). Observe also that G has a natural action as a transitive group of automorphisms of the infinite 3-valent tree $\mathscr{T}$. As $N \trianglelefteq G$ and $|G : N| = |C| = 6$, we see that N has at most 6 orbits on the vertices of $\mathscr{T}$. Assume that N is cyclic and let α be a generator of N. From [21, Proposition 3.2(iii)], the element α acts as a translation on some infinite path of $\mathscr{T}$. As $\mathscr{T}$ has valency 3, from this it follows immediately that N has infinitely many orbits on the vertices $\mathscr{T}$, a contradiction. Therefore N is non-cyclic and hence is a free group of rank at least 2.

For each $n \in \mathbb{N}$, define

$$\mathscr{N}_n = \{M \mid M \trianglelefteq G, M \leq N, |G : M| \leq n\}.$$

As N is a free group of rank at least 2, [12, Theorem 1] yields that there exists a positive constant a' with $|\mathscr{N}_n| \geq n^{a' \log n}$ for $n \geq 2$. Observe that, for every group $M \in \mathscr{N}_n$, the quotient G/M is a $(2, *)$-group of order at most n.

Fix $M \in \mathscr{N}_n$, the number of $M' \in \mathscr{N}_n$ with $G/M' \cong G/M$ is exactly the number of surjective homomorphisms from G to G/M. Since G is 2-generated and $|G/M| \leq n$, the number of such homomorphisms is at most $|G/M|^2 \leq n^2$. We conclude that $f(n) \geq |\mathscr{N}_n|/n^2 \geq (n^{a' \log n})/n^2$ and the result follows.

4 (2, ∗)-Groups and Graphs

4.1 Cubic Cayley Graphs

Let G be a group and let S be a generating set for G which is inverse-closed and does not contain the identity. The *Cayley graph* Cay(G, S) *on G with connection set* S is the graph with vertex-set G and two vertices u and v adjacent if $uv^{-1} \in S$. It is easy to see that Cay(G, S) is a connected vertex-transitive graph of valency $|S|$.

Cayley graphs form one of the most important families of vertex-transitive graphs. In fact, at least for graphs of small order, the overwhelming majority of vertex-transitive graphs are Cayley graphs. This makes them crucial in any project of enumeration of vertex-transitive graphs.

With respect to valency, the first non-trivial case is the case of cubic graphs. Let $\Gamma = \mathrm{Cay}(G, S)$ be a cubic Cayley graph. Note that S is an inverse-closed set of size three and thus must consist either of three involutions or have the form $\{x, g, g^{-1}\}$ where x is an involution and g is not. In the latter case, we say that Γ has *type I*. In this case, (G, x, g) is a $(2, *)$-triple and thus type I graphs arise from $(2, *)$-triples.

While constructing type I Cayley graphs from the catalogue of $(2, *)$-triples is computationally easy, reduction modulo graph isomorphism requires a careful choice of computational tools. For example, MAGMA failed to finish the computation in reasonable time but the SAGE package [19] performed considerably better and yielded the result in a few hours. We would like to thank Jernej Azarija for his help in this matter, which allowed us to conclude that:

Theorem 3 *There are precisely 274,171 connected cubic Cayley graphs of type I with at most 6,000 vertices.*

Moreover, by Theorem 2, there are at most $n^{b \log n}$ type I Cayley graphs of order at most n. On the other hand, non-isomorphic $(2, *)$-triples may give rise to isomorphic Cayley graphs. In general, it is very hard to control when two non-isomorphic $(2, *)$-triples give rise to isomorphic Cayley graphs and thus the lower bound in Theorem 2 does not immediately give a lower bound on the number of Cayley graphs of type I. (See for example [15] for more details on such lower bounds.)

Recently, we published a census of all cubic vertex-transitive graphs of order at most 1,280 [13]. The method we used to construct the Cayley graphs of type I was a mix of the ones described in Sects. 2.1 and 2.3 (see [13, Sect. 3]) and would have been difficult to extend to orders greater than 2,000. The methods described in the current paper thus constitute an improvement, as they allowed us to reach order 6,000.

4.2 Arc-Transitive Digraphs of Out-Valency Two

A *digraph* is an ordered pair (V, A) where V is a finite non-empty set and $A \subseteq V \times V$ is a binary relation on V. If $\Gamma = (V, A)$ is a digraph, then we shall refer to the set V and the relation A as the *vertex-set* and the *arc-set* of Γ, and denote them by $\mathrm{V}(\Gamma)$ and $\mathrm{A}(\Gamma)$, respectively. Members of V and A are called *vertices* and *arcs*, respectively. For a vertex v of Γ, the number $|\{w \in \mathrm{V}(\Gamma) \mid (v, w) \in A(\Gamma)\}|$ is called the *out-valency* of v.

An *automorphism* of a digraph Γ is a permutation of $\mathrm{V}(\Gamma)$ which preserves the arc-set $\mathrm{A}(\Gamma)$. Let G be a subgroup of the automorphism group $\mathrm{Aut}(\Gamma)$ of Γ. We say that Γ is *G-arc-transitive* provided that G acts transitively on $\mathrm{A}(\Gamma)$. In this case, if Γ is connected, then each of its vertices has the same out-valency, say d, and we say that Γ has *out-valency d*.

If Γ is an arc-transitive digraph, then its arc-set $\mathrm{A}(\Gamma)$ is either *symmetric* (that is, for every arc $(u, v) \in \mathrm{A}(\Gamma)$, also $(v, u) \in \mathrm{A}(\Gamma)$), or *asymmetric* (that is, for every $(u, v) \in \mathrm{A}(\Gamma)$, we have $(v, u) \notin \mathrm{A}(\Gamma)$). We will think of a digraph with a symmetric arc-set as a *graph*.

Let Γ be a connected G-arc-transitive digraph of out-valency two. It is easily seen that, for a vertex v of Γ, the vertex-stabiliser G_v has order 2^s for some $s \geq 1$. Moreover, $s = 1$ if and only if G acts regularly on $\mathrm{A}(\Gamma)$. In this case, let x be the involution generating G_v and let g be an element of G mapping (u, v) to (v, w), where (u, v) and (v, w) are arcs of Γ. It is not hard to show that $\{x, g\}$ generates G and thus (G, x, g) is a $(2, *)$-triple. Note also that $\langle x \rangle$ is not central in G (as it is the point-stabiliser of a transitive permutation group). Every digraph of out-valency 2 with an arc-regular group of automorphisms thus arises from a $(2, *)$-triple with x not central.

Conversely, given a $(2, *)$-triple (G, x, g) such that $\langle x \rangle$ is not central in G, one can recover a G-arc-regular digraph of out-valency 2 by the well-known coset graph construction: the vertices are the right cosets of $H = \langle x \rangle$ in G with (Ha, Hb) being an arc whenever $ba^{-1} \in HgH$.

As in the previous section, checking for digraph isomorphism requires some computational work which was performed by Katja Berčič as a part of her doctoral thesis [1]. This allowed us to obtain:

Theorem 4 *There are precisely* 165,952 *asymmetric connected digraphs of out-valency* 2 *on at most* 3,000 *vertices, with an arc-regular group of automorphisms.*

This census of digraphs was used in our recent census of all arc-transitive digraphs of out-valency two with at most 1,000 vertices [14].

As in Sect. 4.1, Theorem 2 implies that, up to isomorphism, there are at most $n^{b \log n}$ digraphs of out-valency 2 and order at most n with an arc-regular group of automorphisms but, again, non-isomorphic $(2, *)$-triples may give rise to isomorphic digraphs and thus lower bounds are harder to obtain.

Finally, we note that the underlying graph of an asymmetric G-arc-transitive digraph Γ of out-valency d is a $2d$-valent graph on which G acts *half-arc-transitively* (that is, vertex- and edge- but not arc-transitively). Moreover, this process can be reversed (see for example [14, Sect. 2.2]) and we thus obtain the following:

Theorem 5 *There are precisely* 76,200 *connected* 4*-valent graphs on at most* 3,000 *vertices that admit a half-arc-transitive group of automorphisms with vertex-stabiliser of order* 2.

5 (2, *)-groups and Maps

Intuitively, a map is a drawing of a graph onto a surface or, slightly more formally, it is an embedding of a graph onto a closed surface (either orientable or non-orientable) which decomposes the surface into open, simply connected regions, called *faces*.

Each face can be decomposed further into *flags*, that is, triangles with one vertex in the centre of the face, one vertex in the centre of an edge and one in a vertex of the embedded graph. An automorphism of a map is then defined as a permutation of the flags induced by a homeomorphism of the surface that preserves the embedded graph.

It is well known that this geometric notion can also be viewed algebraically. In this paper, we adopt this algebraic point of view and use the geometric interpretation only as a source of motivation. For a more thorough discussion on different aspects of maps, and the relationship between their geometric and algebraic description, we refer the reader to [8, 9], or to the excellent survey [18].

Enumeration of maps, especially those exhibiting many symmetries, has a long history, going back to the Ancient Greeks and the classification of the Platonic solids. In this section, we shall be interested in the enumeration and construction of all rotary maps (both reflexible and chiral, orientable and non-orientable) with a small number of edges. Such an enumeration was first attempted by Wilson in [22] for the case of oriented rotary maps on at most 100 edges. More recently, a complete list of all rotary maps on at most 1,000 edges was obtained by Conder [3].

This section has no ambition to be a survey on maps and their symmetries; its main purpose is to show how the database of $(2, *)$-groups was used to extend Conder's database [3] up to 3,000 edges in the orientable case and up to 1,500 edges in the non-orientable case.

5.1 Monodromy Groups of Maps

A faithful transitive action of a $(2, *)$-group on a set $\mathscr{D}$ can be interpreted as the *monodromy group* of a map on an orientable surface. More precisely, if (G, x, g) is a $(2, *)$-triple acting faithfully and transitively on a finite set $\mathscr{D}$ in such a way that x has no fixed points, then one can construct a map with faces, edges and vertices corresponding to the orbits of the groups $\langle g\rangle$, $\langle x\rangle$ and $\langle xg\rangle$, respectively, and with incidence between these objects given in terms of non-empty intersection. Conversely, every map on a closed orientable surface can be obtained in this way from a $(2, *)$-triple. By considering transitive faithful actions of $(2, *)$-groups, one can thus obtain all graph embeddings into orientable surfaces.

5.2 Oriented Rotary Maps

An *automorphism* of the map $\mathscr{M}$ associated with a $(2, *)$-triple (G, x, g) acting on $\mathscr{D}$ is any permutation of $\mathscr{D}$ that commutes with x and g, and thus the automorphism group $\mathrm{Aut}(\mathscr{M})$ equals the centraliser of G in $\mathrm{Sym}(\mathscr{D})$.

A very special case occurs when $\mathrm{Aut}(\mathscr{M})$ is transitive on the dart-set $\mathscr{D}$, which occurs if and only if G (and thus also $\mathrm{Aut}(\mathscr{M})$) acts regularly on $\mathscr{D}$. In that case one can identify $\mathscr{D}$ with the elements of G in such a way that x and g act upon $\mathscr{D} = G$ as permutations $a \mapsto xa$ and $a \mapsto ga$ for all $a \in G$, respectively. The centraliser $\mathrm{Aut}(\mathscr{M})$ of G in $\mathrm{Sym}(\mathscr{D})$ is then generated by the permutation $a \mapsto ax$ and $a \mapsto ag$. In this sense we may view the group G as the automorphism group $\mathrm{Aut}(\mathscr{M})$ (rather than the monodromy group) acting regularly with right multiplication on the set of darts $\mathscr{D} = G$. In this setting, the elements $R = g$ and $S = g^{-1}x$ act as one step-rotations around the centre of a face and around a vertex incident to that face, respectively. We shall always assume that the underlying surface of the map is oriented in such a way that R and S rotate one step in the clock-wise sense; note that the same map but with the opposite orientation is obtained from the triple (G, g^{-1}, gxg^{-1}), giving rise to the rotations R^{-1} and S^{-1}. This justifies the following terminology.

Definition 2 An *oriented rotary map* is a triple (G, R, S) such that G is a group, $\{R, S\}$ is a generating set for G and RS is an involution. Two oriented rotary maps (G_1, R_1, S_1) and (G_2, R_2, S_2) are *isomorphic* if there exists a group isomorphism from G_1 to G_2 mapping R_1 to R_2 and S_1 to S_2.

Given an oriented rotary map (G, R, S) one can reverse the process and construct the associated $(2, *)$-triple (G, RS, R). Moreover, two oriented rotary maps are isomorphic if and only if the associated $(2, *)$-triples are isomorphic. Thus, there is a bijective correspondence between the isomorphism classes of $(2, *)$-triples and the isomorphism classes of oriented rotary maps.

Let us now define a few invariants and operations on oriented rotary maps that are motivated by their geometric interpretations as embeddings of graphs on surfaces. Let (G, R, S) be an oriented rotary map. A right coset of $\langle R\rangle$ in G is called a *face*, a coset of $\langle S\rangle$ a *vertex*, and a coset of $\langle RS\rangle$ an *edge* of the map. The orders of $|R|$ and $|S|$ of R and S are thus called the *face-length* and the *valence* of the map, respectively, while the symbol $\{|R|, |S|\}$ is called the *type* of the map. Furthermore, since $|\langle RS\rangle| = 2$, it follows that a the oriented rotary map (G, R, S) has $|G|/2$ edges. The *mirror image* of (G, R, S) is the oriented rotary map (G, R^{-1}, S^{-1}). If an oriented rotary map is isomorphic to its mirror image, it is called *reflexible* and is *chiral* otherwise. Our enumeration of $(2, *)$-triples (see Theorem 1) yields the following result.

Theorem 6 *There are precisely 345,070 oriented rotary maps with at most 3,000 edges, of which 122,092 are chiral and 222,978 are reflexible.*

The number of reflexible and chiral oriented rotary maps with up to a given number of edges is depicted in Fig. 2.

5.3 Regular Maps

Let $\mathscr{M} = (G, R, S)$ be a reflexible oriented rotary map. By definition, there exists an automorphism τ of G, called the *reflector* of $\mathscr{M}$, with $\tau(R) = R^{-1}$ and $\tau(S) = S^{-1}$.

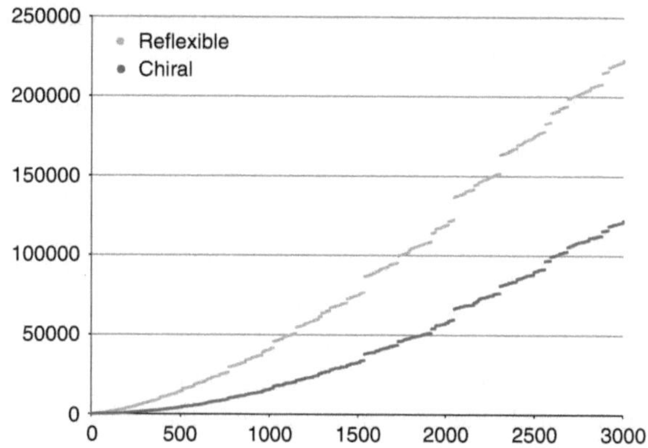

Fig. 2 Number of chiral and reflexible oriented rotary maps with up to a given number of edges

(Since R and S generate G, the reflector is unique and of order at most 2). Let C_2 be a group of order 2, let b be its generator, let $\vartheta : C_2 \to \mathrm{Aut}(G)$ be the homomorphism mapping b to τ, and let

$$A = G \rtimes_\vartheta C_2.$$

Further, let $a = Rb$ and $c = bS$, and observe that a and c are involutions such that $a \neq c$. Moreover, $ac = Rb \cdot bS = RS$ and $ca = bS \cdot Rb = S^{-1}R^{-1} = RS$ because RS is an involution; in particular, $\langle a, c \rangle$ is the Klein 4-group. Note also that $R = ab$ and $S = bc$, and therefore $\langle ab, bc \rangle$ has index 2 in A.

Geometrically, the group A can be viewed as the automorphism group of the orientable (but unoriented) map arising from (G, R, S), with $\langle R, S \rangle$ corresponding to the group of orientation preserving automorphisms and b acting as the orientation reversing automorphism which reflects about the axis through the vertex corresponding to $\langle S \rangle$ and the centre of the face corresponding to $\langle R \rangle$. In this setting, the automorphism c can be viewed as the reflection over the edge $\{v, v^{R^{-1}}\}$, where v is the vertex corresponding to $\langle S \rangle$, while a reflects over the line perpendicular to that edge. The group A then acts regularly on the set of flags of the oriented rotary map. This motivates the following definition:

Definition 3 A *regular map* is a quadruple (A, a, b, c) such that A is a group, a, b, c are involutions generating A and $|\langle a, c \rangle| = 4$. Two regular maps (A_1, a_1, b_1, c_1) and (A_2, a_2, b_2, c_2) are *isomorphic* if there exists a group isomorphism from A_1 to A_2 mapping (a_1, b_1, c_1) to (a_2, b_2, c_2).

If a regular map $\mathcal{M}' = (A, a, b, c)$ is obtained from a reflexible oriented rotary map $\mathcal{M} = (G, R, S)$ by the procedure described above, then we shall say that $\mathcal{M}'$ is an *orientable regularisation* of $\mathcal{M}$. It should be observed at this point that the geometric interpretation of the reflexible oriented rotary map $\mathcal{M}'$ and its orientable regularisation $\mathcal{M}$ are the same, and that the oriented rotary map $\mathcal{M}' = (G, R, S)$ can be reconstructed from $\mathcal{M} = (A, a, b, c)$ by letting $R = ab$, $S = bc$, and $G = \langle ab, bc \rangle$.

Geometrically, the group G corresponds to the orientation-preserving automorphisms of $\mathcal{M}$ and has index 2 in G.

Definition 4 A regular map (A, a, b, c) is called *orientable* if $\langle ab, bc\rangle$ has index 2 in A and *non-orientable* otherwise.

The above discussion shows that a regular map is orientable if and only if it arises as the orientable regularisation of some reflexible oriented rotary map. Since the orientable regularisations $\mathcal{M}_1'$ and $\mathcal{M}_2'$ of $\mathcal{M}_1$ and $\mathcal{M}_2$ are isomorphic if and only if $\mathcal{M}_1$ and $\mathcal{M}_2$ are isomorphic (as oriented rotary maps), there is a bijective correspondence between the isomorphism classes of reflexible oriented rotary maps and the isomorphism classes of orientable regular maps. In particular, our enumeration immediately yields a census of orientable regular maps with at most 3,000 edges (see Theorem 6).

Besides orientable regularisation, there is also a different procedure that can be applied to certain oriented rotary maps, which yields all non-orientable regular maps.

Let $\mathcal{M} = (G, R, S)$ be an oriented rotary map. If b is an involution of G such that $R^b = R^{-1}$ and $S^b = S^{-1}$, then we say that b is an *antipodal reflector of* $\mathcal{M}$; we shall follow the terminology of [4] and call $\mathcal{M}$ *antipodal* in this case.

If b is an antipodal reflector of $\mathcal{M}$, then one can form a non-orientable regular map (G, Rb, b, bS), which we shall call the *non-orientable regularisation of* $\mathcal{M}$ *with respect to* b. Conversely, if $\mathcal{M}' = (G, a, b, c)$ is a non-orientable regular map, then $\mathcal{M} = (G, ab, bc)$ is an oriented rotary map admitting an antipodal reflector b, and $\mathcal{M}'$ is the non-orientable regularisation of $\mathcal{M}$ with respect to b.

Note that non-orientable regularisations of $\mathcal{M}$ that correspond to distinct antipodal reflectors are never isomorphic. Indeed, if b_1 and b_2 are two antipodal reflectors of (G, R, S) and if the corresponding non-orientable regularisations (G, Rb_1, b_1, b_1S) and (G, Rb_2, b_2, b_2S) are isomorphic via an automorphism φ of G, then $b_2 = \varphi(b_1)$, and thus $\varphi(R) = \varphi(R){b_2}^2 = \varphi(Rb_1)b_2 = R{b_2}^2 = R$; similarly $\varphi(S) = S$, and since $G = \langle R, S\rangle$, this shows that φ is trivial and $b_2 = b_1$. Moreover, two antipodal reflectors always differ by a central involution, implying that the number of non-isomorphic non-orientable regularisations arising from an antipodal oriented rotary map (G, R, S) is one more than the number of involutions in the centre of G. This phenomenon was first observed in [23].

Let us point out here that a non-orientable regular map (G, a, b, c) also has a geometric interpretation, in which vertices, edges and faces correspond to the cosets of the subgroups $\langle b, c\rangle$, $\langle a, c\rangle$, and $\langle b, c\rangle$ in G, respectively, and with the incidence between these objects given with non-empty intersection. The underlying surface of the map is in this case non-orientable.

With this geometric interpretation in mind, the non-orientable regularisation $\mathcal{M}'$ of an antipodal oriented rotary map $\mathcal{M}$ is obtained as the quotient by a central involution in $\mathrm{Aut}(\mathcal{M})$ that acts as an orientation reversing homeomorphism of the underlying surface (see [4, Proof of Theorem]), and conversely, $\mathcal{M}$ is the unique orientable smooth 2-cover of $\mathcal{M}'$.

The discussion above suggests an obvious strategy to construct all non-orientable regular maps: construct all oriented rotary maps then, for each oriented rotary map,

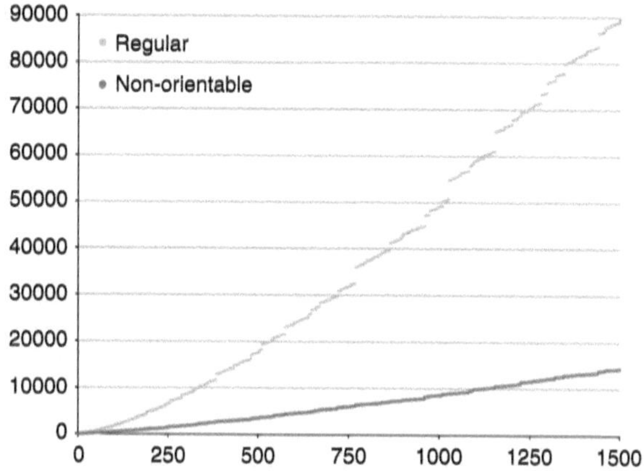

Fig. 3 Number of all and of non-orientable regular maps with up to a given number of edges

find all of its antipodal reflectors and then, for each such reflector, construct the corresponding non-orientable regularisation.

In this correspondence, an antipodal oriented rotary map with m edges yields a non-orientable regular map with $m/2$ edges. Hence our database of oriented rotary maps with at most $3{,}000$ edges yields a complete list of non-orientable regular maps with at most $1{,}500$ edges. The following theorem summarises the results of our computations.

Theorem 7 *There are precisely 14,375 non-orientable regular maps with at most 1,500 edges.*

The number of regular maps with up to a given number of edges is shown in Fig. 3.

Acknowledgments The first author is supported by the Slovenian Research Agency (research projects P1-0294, J1-5433 and J1-6720). The third author is supported by The University of Western Australia as part of the Australian Research Council grant DE130101001. We would like to thank the anonymous referees for their helpful comments.

References

1. K. Berčič, Konstrukcije in katalogizacije simetričnih grafov, Ph.D. thesis, University of Ljubljana, 2015.
2. W. Bosma, J. Cannon and C. Playoust, The Magma algebra system. I: The user language, *J. Symbolic Comput.* **24** (1997), 235–265.
3. M. Conder, Rotary maps on closed surfaces with up to 1000 edges, http://www.math.auckland.ac.nz/~conder/, accessed April 2015.
4. M. Conder, S. Wilson, Inner reflectors and non-orientable regular maps, *Discrete Math.* **307** (2007), 367–372.
5. J. H. Conway, H. Dietrich, E. A. O'Brien, Counting Groups: Gnus, Moas, and other Exotica, *Math. Intelligencer* **30** (2008), 6–15.

6. D. Firth, An algorithm to find normal subgroups of a finitely presented group, up to a given finite index, Ph.D. thesis, University of Warwick, 2005.
7. D. F. Holt, B. Eick, E. O'Brien, Handbook of computational Group Theory, *Discrete Mathematics and its applications*, CRC Press (2005).
8. G. A. Jones, D. Singerman, Theory of maps on orientable surfaces, *Proc. London Math. Soc.* **37** (1978), 273–307.
9. G. A. Jones, D. Singerman, Maps, hypermaps and triangle groups. The Grothendieck theory of dessins d'enfants (Luminy, 1993), *London Math. Soc. Lecture Note Ser.* **200**, Cambridge Univ. Press, Cambridge (1994), 115–145.
10. A. Lubotzky, Enumerating Boundedly Generated Finite Groups, *J. Algebra* **238** (2001), 194–199.
11. A. Lubotzky, D. Segal, *Subgroup growth*, Progress in Mathematics 212, Birkhäuser Verlag, 2003.
12. T. W. Müller, J. -C. Schlage-Puchta, Normal growth of large groups, II, *Arch. Math.* **84** (2005), 289–291.
13. P. Potočnik, P. Spiga, G. Verret, Cubic vertex-transitive graphs on up to 1280 vertices, *J. Symbolic Comput.* **50** (2013), 465–477.
14. P. Potočnik, P. Spiga, G. Verret, A census of 4-valent half-arc-transitive graphs and arc-transitive digraphs of valence two, *Ars Math. Contemporanea* **8** (2015), 133–148.
15. P. Potočnik, P. Spiga, G. Verret, Asymptotic enumeration of vertex-transitive graphs of fixed valency, http://arxiv.org/abs/1210.5736arXiv:1210.5736 [math.CO].
16. P. Potočnik, P. Spiga, G. Verret, *Primož Potočnik's home page*, http://www.fmf.uni-lj.si/~otocnik/work.htm, accessed April 2015.
17. J.-P. Serre, Trees, Springer-Verlag Berlin Heidelberg 1980.
18. J. Širáň, Regular Maps on a Given Surface: A Survey, *Topics in Discrete Mathematics Algorithms and Combinatorics* **26** (2006), 591–609.
19. W. A. Stein et al., Sage Mathematics Software (Version 6.4.1), *The Sage Development Team* (2015) http://www.sagemath.org.
20. The GAP Group, GAP—Groups, Algorithms, and Programming, Lehrstuhl D für Mathematik, RWTH Aachen and School of Mathematical and Computational Sciences, University of St Andrews (2000), http://www.gap-system.orgwww.gap-system.org.
21. J. Tits, Sur le groupe des automorphismes d'un arbre, *Essays on topology and related topics*, Springer New York, 1970, 188–211.
22. S. E. Wilson, *New Techniques For the Construction of Regular Maps*, PhD Dissertation, University of Washington (1976).
23. S. E. Wilson, Non-orientable regular maps, *Ars Combin.* **5** (1978), 213–218.

Even-Integer Continued Fractions and the Farey Tree

Ian Short and Mairi Walker

Abstract Singerman introduced to the theory of maps on surfaces an object that is a universal cover for any map. This object is a tessellation of the hyperbolic plane together with a certain subset of the ideal boundary. The 1-skeleton of this tessellation comprises the edges of an infinite tree whose vertices belong to the ideal boundary. Here we show how this tree can be used to give a beautiful geometric representation of even-integer continued fractions. We use this representation to prove some of the fundamental theorems on even-integer continued fractions that are already known, and we also prove some new theorems with this technique, which have familiar counterparts in the theory of regular continued fractions.

1 Introduction

In [13], Singerman introduced a tessellation of the hyperbolic plane that can be used as a universal cover for any map on a surface (see also [5]). To describe this universal tessellation, we first define the well known *Farey graph*, written as $\mathscr{G}$. We use the upper half-plane model of the hyperbolic plane, denoted by $\mathbb{H}$, along with the ideal boundary of $\mathbb{H}$, which is the extended real line $\mathbb{R}_\infty$ (that is, the real line $\mathbb{R}$ with the point ∞ attached). The Farey graph is a subset of $\mathbb{H} \cup \mathbb{R}_\infty$, which can be viewed as a planar graph. The vertices of $\mathscr{G}$ all belong to $\mathbb{R}_\infty$: they are the rationals together with the point ∞. From now on, we assume that every rational a/b is in reduced form, meaning that a and b are coprime, and b is positive. The edges of $\mathscr{G}$ are hyperbolic geodesics in $\mathbb{H}$: two rationals a/b and c/d are joined by an edge of $\mathscr{G}$ if and only if $|ad - bc| = 1$ (with the convention that ∞ is identified with $1/0$). The Farey

I. Short (✉)
Department of Mathematics and Statistics, The Open University,
Milton Keynes, MK7 6AA, UK
e-mail: ian.short@open.ac.uk

M. Walker
School of Mathematics, The University of Edinburgh,
Edinburgh, EH9 3FD, UK
e-mail: mairi.walker@ed.ac.uk

J. Širáň and R. Jajcay (eds.), *Symmetries in Graphs, Maps, and Polytopes*,
Springer Proceedings in Mathematics & Statistics 159,
DOI 10.1007/978-3-319-30451-9_15

graph induces a tessellation of the hyperbolic plane (different from the tessellation mentioned earlier) that also appears in [13], as a universal cover for any triangular map on a surface. Part of the Farey graph is shown in Fig. 1 (both grey and black lines).

The *Farey tree*, which we denote by $\mathscr{F}$, is obtained by removing from $\mathscr{G}$ all vertices that as rationals in reduced form have odd numerator and denominator. It is a tree with a countably infinite number of vertices, and a countably infinite number of edges incident to each vertex. The vertices adjacent to ∞ are the even integers. Part of the Farey tree is shown in black in Fig. 1, and there is another illustration of $\mathscr{F}$ in Fig. 2 without the distraction of the Farey graph. The Farey tree induces a tessellation of the hyperbolic plane, which is Singerman's universal tessellation—although the definition in [13] is slightly different to this one. (We remark that in some other works 'Farey tree' refers to a different subgraph of $\mathscr{G}$ than $\mathscr{F}$.)

There are other ways to define $\mathscr{G}$ and $\mathscr{F}$. Here is one such way. Let ℓ denote the hyperbolic geodesic in $\mathbb{H}$ between 0 and ∞. Then the edges of $\mathscr{G}$ are the images of ℓ under the modular group Γ (and the vertices of $\mathscr{G}$ are the images of ∞ under Γ). We can describe $\mathscr{F}$ in a similar manner. Let Θ denote the group generated by the

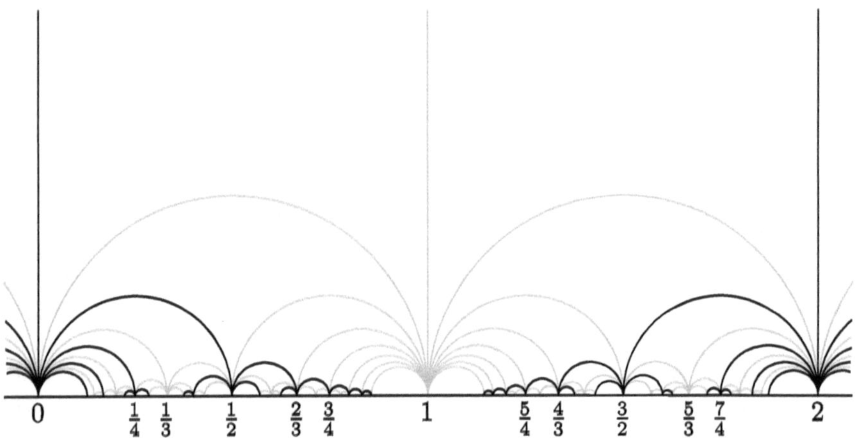

Fig. 1 The Farey tree superimposed with the Farey graph

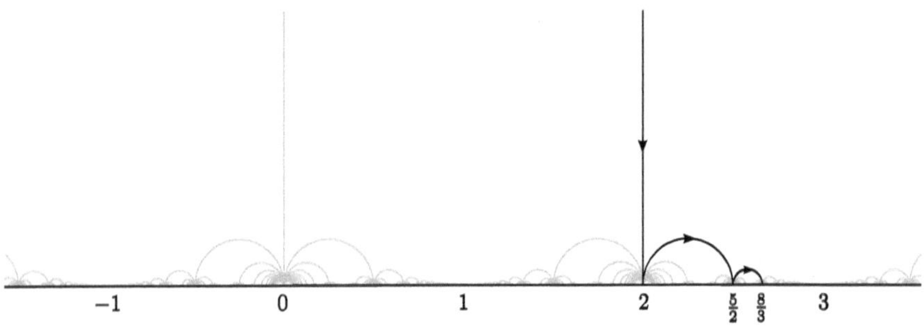

Fig. 2 A path in the Farey tree

transformations $s(z) = -1/z$ and $h(z) = z+2$. This Fuchsian group, called the *theta group*, is a subgroup of the modular group of index 3. It consists of those Möbius transformations $z \mapsto (az+b)/(cz+d)$, where $a, b, c, d \in \mathbb{Z}$ and $ad - bc = 1$, such that

$$\begin{pmatrix} a & b \\ c & d \end{pmatrix} \equiv \begin{pmatrix} 1 & 0 \\ 0 & 1 \end{pmatrix} \text{ or } \begin{pmatrix} 0 & 1 \\ 1 & 0 \end{pmatrix} \pmod 2$$

(see [7, Corollary 4]). The edges of $\mathscr{F}$ are the images of ℓ under Θ (and the vertices of $\mathscr{F}$ are the images of ∞ under Θ).

We call the vertices of the Farey tree *∞-rationals*. They are reduced rationals whose numerator and denominator differ in parity, together with the point ∞. The ∞-rationals are the fixed points of one of the two conjugacy classes of parabolic elements in Θ. The vertices of $\mathscr{G}$ that are not vertices of $\mathscr{F}$ are called *1-rationals* because they consist of the images of 1 under Θ. They are the reduced rationals with odd numerator and denominator (called *face-centre points* in [13]), and they are the fixed points of the other of the two conjugacy classes of parabolic elements in Θ. It can easily be shown that Θ acts on $\mathscr{F}$, and in fact each element of Θ is a graph automorphism of $\mathscr{F}$.

This paper is about an attractive connection between the Farey tree and *even-integer continued fractions*. An even-integer continued fraction (or, more briefly, an EICF) is a sequence of even integers $b_1, b_2, \dots$, which may be finite or infinite (or empty), such that all terms except possibly b_1 are nonzero. We denote this continued fraction by $[b_1, b_2, \dots]$ (and sometimes by $[b_1, \dots, b_n]$ if it is finite). The number

$$b_1 + \cfrac{1}{b_2 + \cfrac{1}{b_3 + \dots + \cfrac{1}{b_n}}},$$

is called the *value* of the finite EICF $[b_1, \dots, b_n]$. The *convergents* of a finite or infinite EICF $[b_1, b_2, \dots]$ are the values of $[b_1, \dots, b_n]$ for $n = 1, 2, \dots$. If the sequence of convergents of an infinite EICF converges in $\mathbb{R}_\infty$ to a point x, then we say that the EICF *converges* and has *value* x. Sometimes we abuse notation and use $[b_1, b_2, \dots]$ to represent its value; this is quite natural—in fact, the distinction between continued fractions and their values is blurred in most works on continued fractions. An EICF *expansion* of a real number x is an EICF with value x.

In either of the graphs $\mathscr{F}$ or $\mathscr{G}$, we say that two vertices are *adjacent* or *neighbours* if they are incident to the same edge. A *path* in one of these graphs is a sequence of *distinct* vertices $v_1, v_2, \dots$ such that v_i and v_{i+1} are adjacent for $i = 1, 2, \dots$. The path is said to be *finite* if the sequence has finite length, and otherwise it is *infinite*. We say that an infinite path $v_1, v_2, \dots$ *converges* to a real number x if the sequence converges to x in $\mathbb{R}_\infty$. In these circumstances, we describe $v_1, v_2, \dots$ as a path *from* v_1 to x.

Let

$$t_n(z) = b_n + \frac{1}{z} \quad \text{and} \quad T_n = t_1 \circ t_2 \circ \cdots \circ t_n, \quad n = 1, 2, \ldots,$$

where $b_1, b_2, \ldots$ are even integers and all except possibly b_1 are nonzero. Notice that the convergents of the EICF $[b_1, b_2, \ldots]$ are $T_1(\infty), T_2(\infty), \ldots$. Now, 0 and ∞ are adjacent in $\mathcal{F}$, and it is easy to check that adjacency is preserved by the maps t_n, so $T_n(0)$ and $T_n(\infty)$ are also adjacent in $\mathcal{F}$. But

$$T_n(0) = T_{n-1}t_n(0) = T_{n-1}(\infty),$$

so any two consecutive vertices in the sequence $\infty, T_1(\infty), T_2(\infty), \ldots$ are adjacent. Furthermore, the condition $b_n \neq 0$ for $n \geqslant 2$ implies that this walk in $\mathcal{F}$ never 'backtracks': it is a path. Conversely, a short argument shows that the vertices of a path with initial vertex ∞ are the convergents of a unique EICF. Thus we see that there is a *correspondence between even-integer continued fractions and paths in $\mathcal{F}$ with initial vertex ∞*. Finite continued fractions correspond to finite paths, and infinite continued fractions correspond to infinite paths (and the empty continued fraction corresponds to the path consisting of the vertex ∞ alone).

For example, the EICF expansion of the rational $8/3$ is $[2, 2, -2]$, and this continued fraction corresponds to the path in $\mathcal{F}$ represented by the black directed edges in Fig. 2. The vertices of this path are, in order, $\infty, 2, 5/2, 8/3$ and the final three of these are the convergents of the continued fraction.

There is a similar correspondence between *integer* continued fractions and paths in the Farey *graph* that is well known (and the proofs of the validity of the correspondence are similar); see, for example, [1, 9]. However, there are two reasons why the tree $\mathcal{F}$ is better to work with than the graph $\mathcal{G}$: (i) all infinite paths in the tree converge, and (ii) there is an (almost) unique path from ∞ to each real number (in particular, as $\mathcal{F}$ is a tree there is a unique finite path from ∞ to each ∞-rational). In terms of even-integer continued fractions, these statements are (i) all infinite EICFs converge, and (ii) each real number has an (almost) unique EICF expansion. We explain the meaning of the qualification 'almost' later on. Both (i) and (ii) fail for integer continued fractions, but they do hold for *regular* continued fractions (the most familiar type of continued fractions, with positive integer coefficients). Here we will show that in fact much of the theory of regular continued fractions (from, for example, [6, Chaps. I and II] or [4, Chap. X]) can be reformulated using even-integer continued fractions. To an extent, this is already known, and has been demonstrated in works such as [8, 10]. The novelty of our approach is that we develop the theory of even-integer continued fractions geometrically using elementary properties of the Farey tree.

In Sects. 2–4 we prove some of the more fundamental theorems on even-integer continued fractions using the Farey tree, covering material that is similar (although not identical) to part of [8]. Sections 5–6 contain results that appear to be new. To keep this account concise, we omit certain relevant topics such as the EICF expansions of quadratic irrationals and the Hurwitz constant for the theta group (see [11] for a treatment of the latter topic in the spirit of this paper). Furthermore, for the sake of brevity, we sometimes skip the details of elementary geometric arguments, so that the reader gets a feel for the geometric approach without getting bogged down in details.

2 Infinite Continued Fractions

In this section we prove that every infinite EICF converges. There are several ways to do this; for example, we could invoke a more general theorem on the convergence of continued fractions, or we could use algebraic relationships between the convergents to estimate the distance between consecutive convergents. Our approach is to use the Farey tree to establish the following theorem.

Theorem 1 *Every infinite EICF converges to an irrational or a* 1*-rational.*

To prove the theorem, consider any infinite EICF, and let γ be the corresponding infinite path in $\mathscr{F}$ with initial vertex ∞. First we will show that γ cannot accumulate at an ∞-rational. Suppose, on the contrary, that γ does accumulate at a vertex x of $\mathscr{F}$. By applying an element of Θ to γ if necessary (which will change the initial vertex of γ) we can assume that $x \neq \infty$. Furthermore, by removing the first so many terms from γ we can assume that it does not pass through x (remember that a path passes through a vertex at most once). Let a be the initial vertex of γ.

Like all vertices of $\mathscr{F}$, the vertex x has infinitely many neighbours, which accumulate on the left and right of x. Choose any two neighbours u and v such that $u < x < v$, and such that the vertex a lies outside the real interval (u, v), as shown in Fig. 3. Edges of $\mathscr{F}$ do not intersect in $\mathbb{H}$, so we see that because γ accumulates at x, it must pass through one of u, x and v. However, because $\mathscr{F}$ is a tree, any path from a to a neighbour of x must pass through x itself, unless that neighbour happens to lie on the unique path between a and x. Providing we choose u and v sufficiently close to x that they do not lie on this path, we can be sure that γ passes through x. This contradicts an earlier assumption, so γ cannot accumulate at a vertex of $\mathscr{F}$ after all.

We have just seen that the path γ cannot accumulate at an ∞-rational. Suppose, in order to reach a contradiction, that γ accumulates at two numbers x and y, each of which is either irrational or a 1-rational, and $x < y$. Now, the vertices of $\mathscr{F}$ that lie inside the real interval (x, y) are connected in $\mathscr{F}$ to the vertices that lie outside this interval, so there must be an edge of $\mathscr{F}$ with one end vertex u inside the interval and the other v outside. Edges of $\mathscr{F}$ do not intersect in $\mathbb{H}$, so we see that because γ

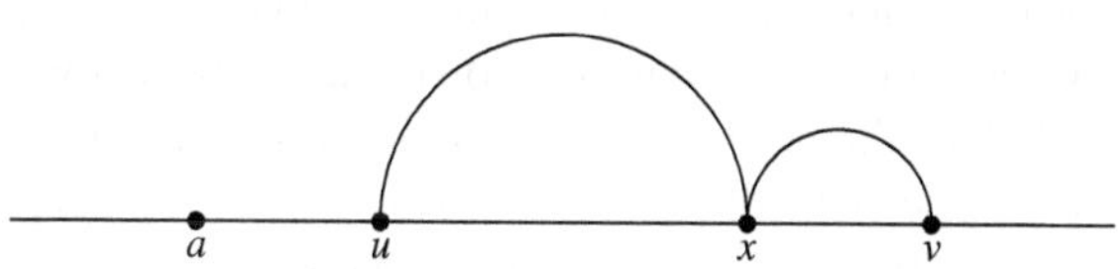

Fig. 3 Two neighbours u and v of the vertex x, and another vertex a

accumulates at both x and y, it must pass through at least one of u or v infinitely many times, which is impossible. Thus, contrary to our assumption, γ cannot accumulate at two numbers, so it converges. The proof of Theorem 1 is now complete.

3 Representing Real Numbers by Even-Integer Continued Fractions

The next fundamental result is about the existence and uniqueness of EICF expansions of real numbers. It is unoriginal (see, for example, [8], where there are a number of results similar to parts of this one); however, our method of proof using the Farey tree is original, and it is simple and elegant.

Theorem 2

(i) The value of any finite EICF is an ∞-rational, and each ∞-rational has a unique finite EICF expansion.
(ii) The value of an infinite EICF is either irrational or a 1-rational, and

(a) each irrational has a unique infinite EICF expansion,
(b) each 1-rational has exactly two infinite EICF expansions, each of which eventually alternates between 2 and -2.

As $\mathscr{F}$ is a tree, and the vertices are the ∞-rationals, we can immediately deduce statement (i) of the theorem using the correspondence between even-integer continued fractions and paths in $\mathscr{F}$. We now turn to statement (ii). The first part of statement (ii) follows from Theorem 1. It remains only to discuss statements (a) and (b).

We begin this discussion by looking at EICF expansions of the number 1; here are two of them:

$$1 = [0, 2, -2, 2, -2, \ldots] = [2, -2, 2, -2, \ldots].$$

We can check that the value x of the second continued fraction is 1 by observing that x must satisfy

$$x = 2 + \cfrac{1}{-2 + \cfrac{1}{x}},$$

and the only solution of this equation is $x = 1$. (The value of the first continued fraction can be obtained in a similar manner.) The paths in $\mathscr{F}$ corresponding to these two continued fractions are shown marked by arrows in Fig. 4.

In fact, the two EICF expansions that we have found are the *only* EICF expansions of 1. To see why this is so, let α denote the left-hand path (that passes through 0) and let β denote the right-hand path (that passes through 2). Observe that in the Farey

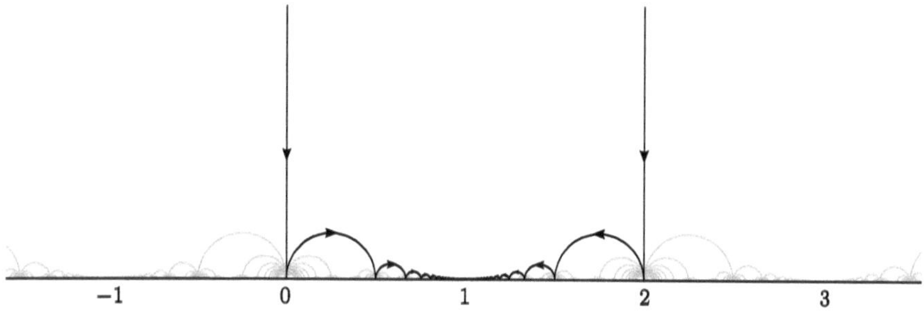

Fig. 4 Two paths that converge to 1

Fig. 5 Two neighbours of 1

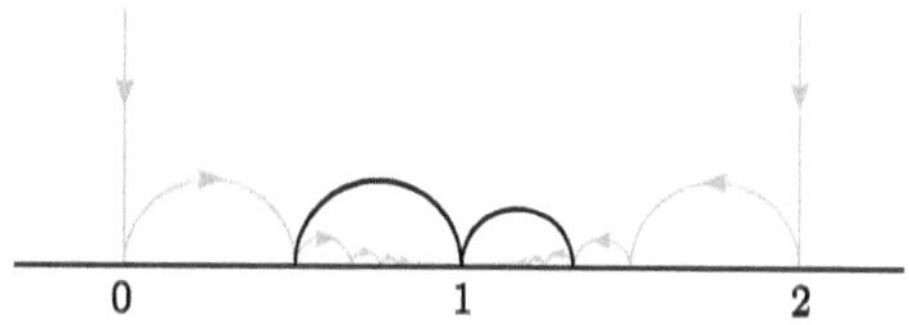

graph $\mathcal{G}$, every single one of the vertices in these two paths is connected to 1 by an edge (in fact, they are the full collection of neighbours of 1 in $\mathcal{G}$—see Fig. 1). Two such edges are shown in Fig. 5, on either side of 1.

Suppose now that γ is an infinite path in $\mathcal{F}$ from ∞ to 1. Aside from the initial vertex ∞, this path must lie entirely to the left or entirely to the right of 1 (because any path in $\mathcal{F}$ that passes from one side to the other of 1 must pass through ∞). Suppose that it lies to the left—the other case can be handled in a similar way. Then because edges in the Farey graph do not intersect, γ must pass through all of the vertices of α. There is only one such path that does this, namely α itself, so $\gamma = \alpha$.

We summarise this discussion in a lemma.

Lemma 1 *The number* 1 *has precisely two EICF expansions, namely*

$$[0, 2, -2, 2, -2, \ldots] \quad \textit{and} \quad [2, -2, 2, -2, \ldots].$$

If x is any 1-rational, then there is an element g of Θ such that $g(1) = x$. It follows that $g(\alpha)$ and $g(\beta)$ are infinite paths from $g(\infty)$ to 1. By connecting ∞ to $g(\infty)$ we obtain two walks from ∞ to 1 (each may have repeated vertices), which we can modify by adjusting a finite number of terms to give two paths from ∞ to 1. Thus we obtain two EICF expansions of x. We can reverse this argument to see that these are the only EICF expansions of x. This gives us the following corollary of Lemma 1.

Corollary 1 *Every* 1*-rational has precisely two EICF expansions.*

In the next section we will see that if x and y have infinite EICF expansions, and $g(x) = y$ for some transformation g in Θ, then it is possible to remove a finite number of consecutive terms from the start of the EICF expansions of x and y to give two expansions that agree. It follows that an EICF expansion of a 1-rational eventually alternates between 2 and -2. (Conversely, it is straightforward to show that any real number with an infinite EICF expansion that eventually alternates between 2 and -2 is a 1-rational.) Furthermore, one can check that the two continued fractions

$$[b_1, \dots, b_n, 2, -2, 2, \dots] \quad \text{and} \quad [b_1, \dots, b_{n-1}, b_n + 2, -2, 2, -2, \dots]$$

have the same value, so the two EICF expansions referred to in Corollary 1 are of these forms.

We have now proved statement (b) of Theorem 2, which leaves only statement (a). Let us prove the uniqueness assertion of (a). Suppose then that α and β are two infinite paths from ∞ to a real number x. The two paths may coincide for a certain number of vertices: let w be the final vertex for which they do so. Choose an element g of Θ such that $g(w) = \infty$. Let α' and β' be the paths obtained from $g(\alpha)$ and $g(\beta)$, respectively, after removing all vertices that occur before ∞. Then α' and β' are infinite paths from ∞ to $g(x)$, such that the second vertex u of α' is distinct from the second vertex v of β'. The vertices u and v are even integers, so there is an odd integer q (a 1-rational) that lies between them on the real line. Neither α' nor β' can pass from one side of q to the other, and since they converge to the same value, that value must be q. Therefore $g(x)$ is a 1-rational, so x is also a 1-rational.

This argument shows that each irrational has at most one EICF expansion. Let us now show that each irrational has at least one such expansion. One way to do this is to use an algorithm of a similar type to Euclid's algorithm: in this case the 'nearest even-integer algorithm' does the trick. However, we prefer to justify the existence of an expansion using the Farey graph and tree.

We define a *Farey interval* to be a real interval whose endpoints are neighbouring vertices in the Farey graph $\mathcal{G}$. If $[a/b, c/d]$ is a Farey interval (where, as usual, the fractions are given in reduced form), then it is easily seen that

$$[a/b, (a + c)/(b + d)] \quad \text{and} \quad [(a + c)/(b + d), b/d]$$

are both Farey intervals—let us call them the *Farey subintervals* of $[a/b, c/d]$. Now, any irrational x belongs to a Farey interval $[n, n + 1]$, where n is the integer part of x, and by repeatedly choosing Farey subintervals, we can construct a nested sequence of Farey intervals that contains x in its intersection. The width of one of these intervals $[a/b, c/d]$ is

$$\left| \frac{a}{b} - \frac{c}{d} \right| = \left| \frac{ad - bc}{bd} \right| = \frac{1}{bd},$$

so we see that the sequence of widths of this nested sequence of Farey intervals converges to 0.

Let us now restrict attention to those infinitely many Farey intervals $I_1 \supset I_2 \supset \cdots$ from the sequence for which one of the endpoints of I_n is a 1-rational v_n (and the other endpoint u_n must then be an ∞-rational). Let γ_n be the unique path from ∞

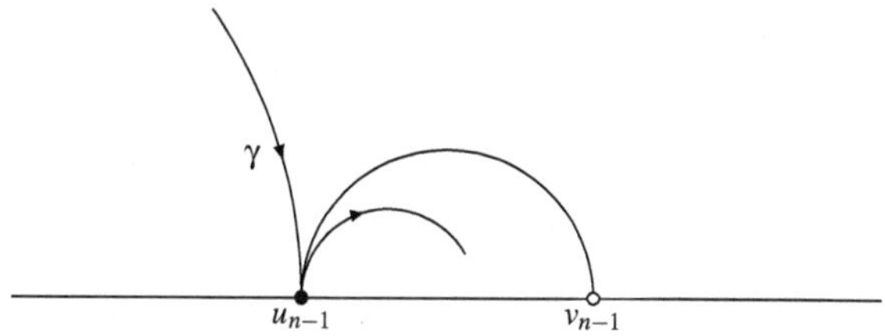

Fig. 6 The path γ passes through u_{n-1}

to u_n in $\mathscr{F}$. Any path in $\mathscr{F}$ from ∞ to a vertex inside I_{n-1} must pass through u_{n-1} (because u_{n-1} and v_{n-1} are neighbours in $\mathscr{G}$, as illustrated in Fig. 6, and edges of $\mathscr{G}$ do not intersect). Therefore γ_{n-1} is a subpath of γ_n. It follows that there is a unique infinite path γ that contains every path γ_n as a subpath. The path γ passes through all the vertices u_n, which accumulate at x, so γ must converge to x. This completes the proof of Theorem 2.

4 Serret's Theorem on Continued Fractions

This section is about a counterpart for even-integer continued fractions of a well-known theorem of Serret on regular continued fractions. Before we state our theorem, we must introduce the *extended theta group*, which is the group $\widetilde{\Theta}$ generated by the theta group and the transformation $r(z) = -z$. This group acts on $\mathbb{R}_\infty$, and it also acts on the set of ∞-rationals. In fact, elements of $\widetilde{\Theta}$ preserve adjacency in $\mathscr{F}$, so $\widetilde{\Theta}$ acts on the abstract graph underlying $\mathscr{F}$. We say that two real numbers are *equivalent* under the action of $\widetilde{\Theta}$ if they lie in the same orbit under this action.

Our version of Serret's theorem for even-integer continued fractions follows. It is similar to [8, Theorem 1], but not quite the same because even-integer continued fractions are defined differently in that paper.

Theorem 3 *Two real numbers x and y that are not ∞-rationals are equivalent under $\widetilde{\Theta}$ if and only if there are positive integers m and n such that the EICF expansions of x and y,*

$$x = [a_1, a_2, \ldots] \quad \textit{and} \quad y = [b_1, b_2, \ldots],$$

either satisfy $a_{m+i} = b_{n+i}$ for $i = 1, 2, \ldots$ or $a_{m+i} = -b_{n+i}$ for $i = 1, 2, \ldots$.

Serret's theorem for regular continued fractions is similar, but uses an extension of the modular group rather than the theta group, and the possibility $a_{m+i} = -b_{n+i}$ for $i = 1, 2, \ldots$ is absent.

Crucial to the proof of this theorem is the following lemma.

Lemma 2 *If a real number x has an EICF expansion $[a_1, a_2, \ldots]$, then an EICF expansion of $-x$ is $[-a_1, -a_2, \ldots]$.*

There is no obvious analogue of this lemma for regular continued fractions because the coefficients of regular continued fractions are (almost) all positive.

The lemma can be proven with the Farey tree by observing that the paths from ∞ to x and from ∞ to $-x$ are reflections of each other in the imaginary axis. However, in this case, we will prove the lemma using Möbius transformations. Let $t_a(z) = a + 1/z$, where a is even; this transformation belongs to $\widetilde{\Theta}$. Observe that $rt_ar = t_{-a}$. We are given that an EICF expansion of x is $[a_1, a_2, \ldots]$, which implies that $t_{a_1}t_{a_2}\cdots t_{a_n}(\infty) \to x$ as $n \to \infty$. Now

$$t_{-a_1}t_{-a_2}\cdots t_{-a_n}(\infty) = rt_{a_1}t_{a_2}\cdots t_{a_n}r(\infty) = rt_{a_1}t_{a_2}\cdots t_{a_n}(\infty).$$

So $t_{-a_1}t_{-a_2}\cdots t_{-a_n}(\infty) \to r(x) = -x$ as $n \to \infty$. Therefore an EICF expansion of $-x$ is $[-a_1, -a_2, \ldots]$.

Let us now prove Theorem 3. Suppose first that $y = g(x)$, where $g \in \widetilde{\Theta}$. We wish to prove that there are positive integers m and n such that $a_{m+i} = b_{n+i}$ for $i = 1, 2, \ldots$ or $a_{m+i} = -b_{n+i}$ for $i = 1, 2, \ldots$. Since $\widetilde{\Theta}$ is generated by the transformations $r(z) = -z$, $t(z) = 1/z$ and $h(z) = z + 2$, it suffices to prove the assertion when g is each of r, t, h and h^{-1}. It is straightforward to do so when g is one of the final three transformations, and the remaining case when g equals r is an immediate consequence of Lemma 2.

For the converse, suppose that $x = [a_1, a_2, \ldots]$, $y = [b_1, b_2, \ldots]$ and either (i) $a_{m+i} = b_{n+i}$ for $i = 1, 2, \ldots$, or (ii) $a_{m+i} = -b_{n+i}$ for $i = 1, 2, \ldots$. By replacing x by $-x$ if necessary, and invoking Lemma 2, we can assume that (i) holds. Observe that

$$x = t_{a_1}\cdots t_{a_m}([a_{m+1}, a_{m+2}, \ldots]) \quad \text{and} \quad y = t_{b_1}\cdots t_{b_n}([b_{n+1}, b_{n+2}, \ldots]).$$

Hence $y = t_{b_1}\cdots t_{b_n}t_{a_m}^{-1}\cdots t_{a_1}^{-1}(x)$, so x and y are equivalent under $\widetilde{\Theta}$. This completes the proof of Theorem 3.

5 An Alternative Characterisation of Convergents of Even-Integer Continued Fractions

In this section we describe an alternative way to characterise the convergents of the EICF expansion of any irrational x. The characterisation can easily be adapted to allow x to be rational.

Theorem 4 *A finite ∞-rational u is a convergent of the EICF expansion of an irrational x if and only if there is a 1-rational v adjacent to u in the Farey graph such that x lies between u and v on the real line.*

The second part of the theorem is illustrated in Fig. 7.

If there is a 1-rational v of this type, then the edge in the Farey graph $\mathscr{G}$ between u and v separates ∞ from any vertex of $\mathscr{G}$ that is sufficiently close to x on the real

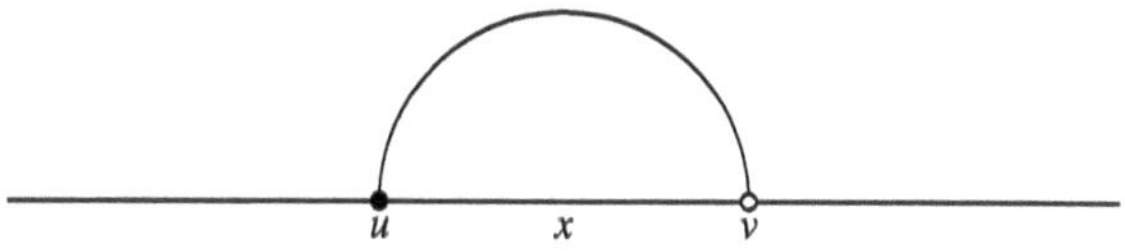

Fig. 7 The irrational x lies between the ∞-rational u and the 1-rational v

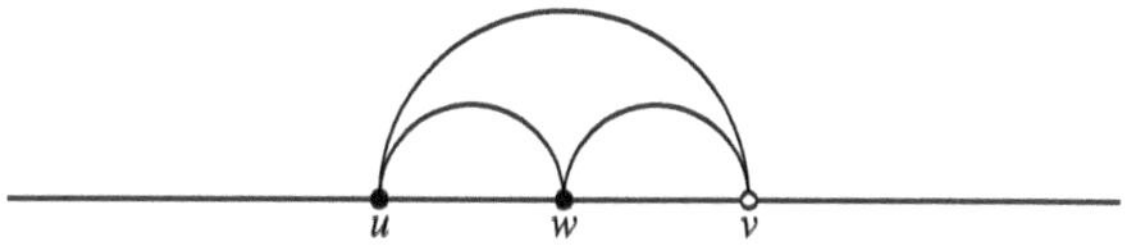

Fig. 8 A *triangle* in the Farey graph

line. So any path from ∞ to w must pass through one of u or v—and if the path lies in $\mathscr{F}$, then it must pass through u. In particular, this demonstrates that u must be a convergent of the EICF expansion of x.

The converse implication of Theorem 4 is a direct consequence of the following lemma (which is a slightly stronger statement).

Lemma 3 *Let u and w be two consecutive convergents in the EICF expansion of an irrational x, in that order. Then there is a* 1*-rational v adjacent to each of u and w in the Farey graph such that both w and x lie between u and v on the real line.*

Since u and w are adjacent in $\mathscr{F}$, they are also adjacent in $\mathscr{G}$. There are two other vertices in $\mathscr{G}$ that are adjacent to both u and w, precisely one of which (call it v) does *not* lie between u and w on the real line. Let γ be the path of convergents of the EICF expansion of x. If γ enters the interval between u and v, then it must pass through u to get there, and it cannot leave the interval. Similar comments apply to the interval between w and v. Now, u cannot lie in the interval between w and v because if it did, then, as we have just seen, the path γ would pass through w before it passed through u. So w lies in the interval between u and v (as illustrated in Fig. 8), and x lies in that interval too. This completes the proofs of Lemma 3 and Theorem 4.

6 Approximating Irrationals by Rationals

One of the principal uses of continued fractions is in the field of Diophantine approximation, which is concerned with approximating real numbers by rationals. In this section we prove an analogue for even-integer continued fractions of a classic result of Lagrange on regular continued fractions.

We call an ∞-rational a/b a *strong ∞-approximant* of a real number x if for each ∞-rational c/d such that $d \leqslant b$, we have

$$|bx - a| \leqslant |dx - c|,$$

with equality if and only if $c/d = a/b$.

Theorem 5 *An ∞-rational is a strong ∞-approximant of an irrational x if and only if it is a convergent of the EICF expansion of x.*

Lagrange's theorem for regular continued fractions is similar (see [6, Theorems 16 and 17]), but uses rationals rather than ∞-rationals.

There is no need to assume that x is irrational in the theorem—subject to minor modifications of the theorem we can allow x to be any real number—but it is the irrational case that interests us most, and the proof is marginally simpler with the assumption that x is irrational.

Our proof uses *Ford circles*, and is similar to the proof of Lagrange's theorem from [12]. Ford circles are a collection of horocycles in $\mathbb{H}$ used by Ford to study continued fractions in papers such as [2, 3]. We say that a horocycle is *based* at an element x of $\mathbb{R}_\infty$ if the horocycle is tangent to $\mathbb{R}_\infty$ at x. Given a reduced rational $u = a/b$, the Ford circle C_u is the horocycle based at u with Euclidean radius $\mathrm{rad}[C_u] = 1/(2b^2)$. There is one other Ford circle C_∞, which is the line $y = 1$ together with the point ∞. Two Ford circles intersect in at most a single point, and the interiors of the two circles are disjoint. In fact, one can check that the Ford circles $C_{a/b}$ and $C_{c/d}$ are tangent if and only if $|ad - bc| = 1$. Therefore the full collection of Ford circles is a model of the abstract graph underlying the Farey graph: the vertices of this graph are represented by Ford circles, and two vertices are adjacent if and only if the Ford circles are tangent. Similarly, the collection of Ford circles based at ∞-rationals is a model of the abstract graph underlying the Farey tree; this model is illustrated in Fig. 9. When studying even-integer continued fractions, it is helpful to consider both the Farey tree and this alternative model of the tree using Ford circles.

We now relate Ford circles to strong ∞-approximants. Let $u = a/b$. Notice that if $v = c/d$, then $d \leqslant b$ if and only if $\mathrm{rad}[C_u] \leqslant \mathrm{rad}[C_v]$. For any real number x, let

$$R_u(x) = \frac{1}{2}|bx - a|^2.$$

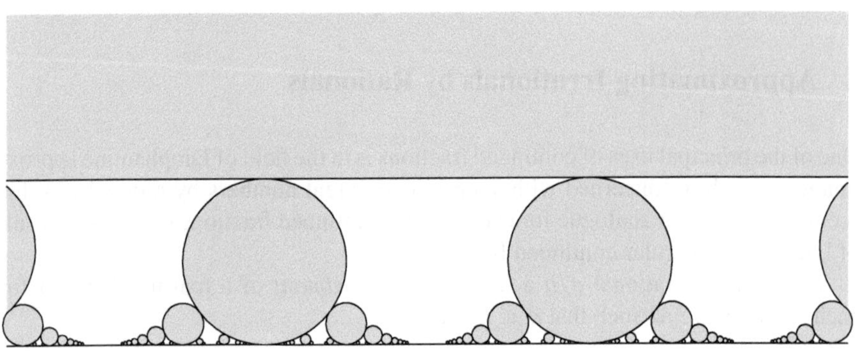

Fig. 9 Ford circles based at the ∞-rationals

Using elementary geometry, it can be shown that $R_u(x)$ is the Euclidean radius of the horocycle based at x that is externally tangent to C_u. With this terminology, we can describe a strong ∞-approximant of a real number x as an ∞-rational u such that for each ∞-rational w with $\mathrm{rad}[C_u] \leqslant \mathrm{rad}[C_w]$, we have $R_u(x) \leqslant R_w(x)$, with equality if and only if $w = u$. We will use this definition of strong ∞-approximants together with Theorem 4 to prove Theorem 5. Our proof omits several elementary geometric details.

Suppose first that u is a convergent of the EICF expansion of x. Theorem 4 tells us that there is a 1-rational v adjacent to u in the Farey graph such that x lies between u and v on the real line. If w is an ∞-rational distinct from u with $\mathrm{rad}[C_u] \leqslant \mathrm{rad}[C_w]$, then w must lie outside the real interval between u and v, so $R_u(x) < R_w(x)$, as illustrated in Fig. 10. Therefore u is a strong ∞-approximant of x.

Conversely, suppose that u is an ∞-rational that is not one of the convergents $w_1, w_2, \ldots$ of the EICF expansion of x. Choose a convergent w_n such that $\mathrm{rad}[C_{w_{n+1}}] < \mathrm{rad}[C_u] \leqslant \mathrm{rad}[C_{w_n}]$. By Lemma 3, there is a 1-rational v adjacent to each of w_n and w_{n+1} in the Farey graph such that both w_{n+1} and x lie between w_n and v on the real line. On the other hand, the radius of C_u is larger than that of $C_{w_{n+1}}$, so u does not lie between w_n and v, as illustrated in Fig. 11. Therefore $R_{w_n}(x) < R_u(x)$, so u is not a strong ∞-approximant of x. This completes the proof of Theorem 5.

Fig. 10 Ford circles based at u, v and w and horocycles based at x

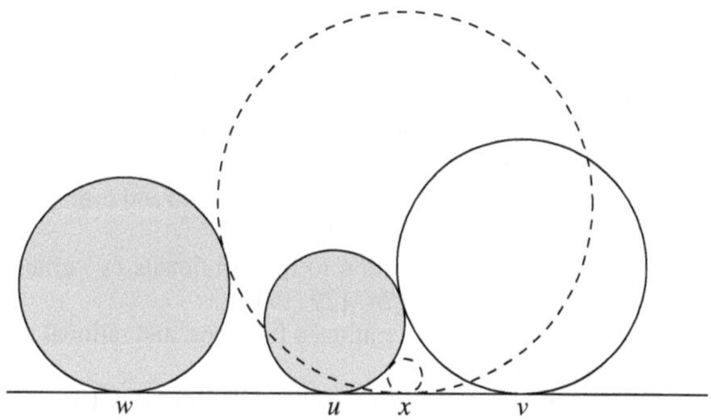

Fig. 11 Ford circles based at u, v, w_n and w_{n+1}

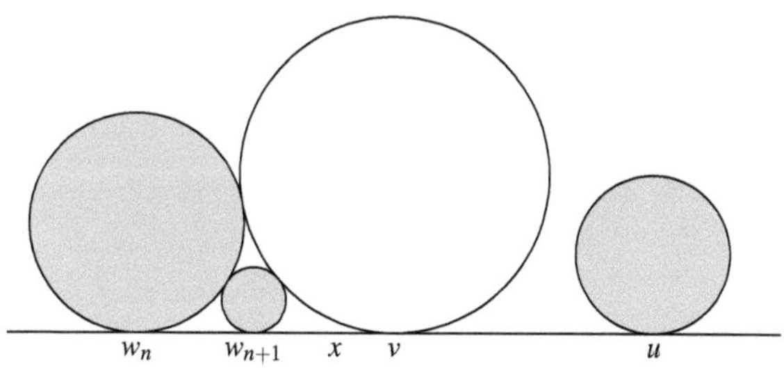

7 Concluding Remark

We have seen that a good deal of the theory of even-integer continued fractions can be understood by viewing such continued fractions as paths in the Farey tree. It may be of interest to study paths in other maps on surfaces, and investigate their relationship with continued fractions.

References

1. Beardon, A.F., Hockman, M., Short, I.: Geodesic continued fractions. Michigan Math. J. **61**, 133–150 (2012).
2. Ford, L.R.: A geometrical proof of a theorem of Hurwitz. Proc. Edinburgh Math. Soc. **35**, 59–65 (1917).
3. Ford, L.R.: Fractions. Amer. Math. Monthly **45**, 586–601 (1938).
4. Hardy, G.H., Wright, E.M.: An Introduction to the Theory of Numbers, 6th edn. Oxford Univ. Press, Oxford (2008).
5. Jones, G., Singerman, D.: Belyĭ functions, hypermaps and Galois groups. Bull. London Math. Soc. **28**, 561–590 (1996).
6. Khinchin, A.Ya.: Continued fractions, translated from the 3rd (1961) Russian edition, reprint of the 1964 translation. Dover, Mineola, NY (1997).
7. Knopp, M.I.: Modular functions in analytic number theory. Markham Publishing Co., Chicago, IL (1970).
8. Kraaikamp, C., Lopes, A.: The theta group and the continued fraction expansion with even partial quotients. Geom. Dedicata **59**, 293–333 (1996).
9. Schwartz, R.E.: Mostly surfaces. Student Mathematical Library, 60, Amer. Math. Soc., Providence, RI (2011).
10. Schweiger, F.: Continued fractions with odd and even partial quotients. Arbeitsber. Math. Inst. Univ. Salzburg **4**, 59–70 (1982).
11. Scott, W.T.: Approximation to real irrationals by certain classes of rational fractions. Bull. Amer. Math. Soc. **46**, 124–129 (1940).
12. Short, I.: Ford circles, continued fractions, and rational approximation. Amer. Math. Monthly **118**, 130–135 (2011).
13. Singerman, D.: Universal tessellations. Rev. Mat. Univ. Complut. Madrid **1**, 111–123 (1988).

Triangle Groups and Maps

David Singerman

Abstract We develop a Belyi type theory that applies to Klein surfaces, i.e. (possibly non-orientable) surfaces with boundary which carry a dianalytic structure. In particular we extend Belyi's famous theorem from Riemann surfaces to Klein surfaces.

1 Triangle Groups

Before we discuss maps we remind the reader about the basic facts about triangle groups. Let $T(l, m, n)$ be a triangle with angles $\pi/l, \pi/m, \pi/n$. Let

$$A = \frac{1}{l} + \frac{1}{m} + \frac{1}{n}.$$

Then $T(l, m, n)$ exists in the hyperbolic plane if $A < 1$, in the Euclidean plane if $A = 1$, and on the sphere if $A > 1$. Let $\Gamma(l, m, n)$ denote the group generated by the reflections a, b, c in the sides of $T(l, m, n)$ opposite the vertices with angles $\pi/m, \pi/n, \pi/l$, resp. Let $\mathscr{U}$ denote either the hyperbolic plane, the Euclidean plane or the sphere. Then the images of $T(l, m, n)$ form a triangular tessellation (or map) on $\mathscr{U}$.

A presentation of $\Gamma(l, m, n)$ is

$$\langle a, b, c \mid a^2 = b^2 = c^2 = (ab)^l = (bc)^m = (ca)^n = 1 \rangle.$$

We let $\Gamma[l, m, n]$ denote the subgroup of $\Gamma(l, m, n)$ consisting of those transformations that preserve orientation. Then $\Gamma[l, m, n]$ is generated by the rotations $x = ab, y = bc, z = ca$ and has presentation

$$\langle x, y, z \mid x^l = y^m = z^n = xyz = 1 \rangle.$$

D. Singermn (✉)
School of Mathematics, University of Southampton,
Highfield, Southampton SO17 1BJ, UK
e-mail: d.singerman@soton.ac.uk

J. Širáň and R. Jajcay (eds.), *Symmetries in Graphs, Maps, and Polytopes*,
Springer Proceedings in Mathematics & Statistics 159,
DOI 10.1007/978-3-319-30451-9_16

Groups of the form $\Gamma(l, m, n)$ are called *extended triangle groups* and those of the form $\Gamma[l, m, n]$ are *triangle groups*. Note that these groups are associated with triangular maps on $\mathcal{U}$.

2 Some Personal History

This paper is basically a personal survey of my thoughts about maps on Riemann surfaces. A lot of it is cooperative work with Gareth Jones, and later Bernhard Koeck, Jürgen Wolfart, and Milagros Izquierdo, and a lot comes from work that I did with my research students, Robin Bryant, David Corn, Robert Syddall, and Paul Watson. In 1974, I realised that associated to every map on a surface, there is a canonically defined complex structure on that surface which is associated to that map. For example every map automorphism is an automorphism of this Riemann surface. This realisation came about after listening to a seminar given by Norman Biggs, where he spoke about his result that every automorphism group of a map of genus g can be faithfully represented in the symplectic group $Sp(g, \mathbf{Z})$ [3]. As it is known that a group of automorphisms of a compact Riemann surface of genus g can be faithfully represented in $Sp(g, \mathbf{Z})$, it was natural to enquire whether the automorphism group of the map was also the automorphism group of some underlying Riemann surface of genus g. In [22] I showed that this was indeed the case. Soon after Gareth Jones and I explicitly constructed this Riemann surface in [15]. We found a subgroup M of a triangle group which uniformizes this Riemann surface and which has many nice properties related to this map. For example, the map is regular if and only if M is normal in the triangle group. Some years later Gareth and I were on the jury for a Ph.D. examination in Paris. The supervisor Tony Machì pointed out to us a recent paper "Drawing curves over number fields" by Shabat and Voevodsky which contained many of the ideas in [15]. This paper came from the Grothendieck Festschrift and was based on ideas first developed by Grothendieck in his Esquisse d'un Programme. There he had set out a theory of maps but added the vital new ingredient that a theorem of Belyi showed that the Riemann surfaces coming from maps are precisely those that come from complex algebraic curves defined over algebraic number fields.

3 Basic Concepts

In [11], Grothendieck called maps "dessin d'enfants" or just dessins. As we shall see these are basically the same as hypermaps, which are slight generalisations of maps. We prefer to use the terms map or hypermap. The observation about the relation between maps, Riemann surfaces and algebraic curves goes back to the nineteenth century to Felix Klein and his work on Klein's surface of genus 3 (see [17] and for an English translation [20]).

A map is basically a decomposition of a surface into simply-connected polygonal cells called faces. The most famous examples are the platonic solids on the sphere, but we could have an infinite set of maps without any symmetry. Another way of describing a map is as an embedding of a connected graph in a surface such that the complement of the graph in the surface consists of a collection of simply connected pieces, which are the faces. We could describe a hypermap in much the same way but now we embed a hypergraph. (A hypergraph is like a graph but an edge can have more than two vertices or only one vertex. We return to the concept of a hypermap in Sect. 7.)

Each map will have an automorphism group mapping vertices to vertices, edges to edges, and faces to faces, preserving incidence and preserving the orientation if the surface is orientable. The automorphism group of a platonic solid has the property that it acts transitively on directed edges (often these are called *darts*), and we use this property to define a regular map on any surface. The theory of regular maps was developed in the early 20th century. For a brief introduction to their history see the beginning of Chap. 8 of the book *Generators and Relations for Discrete Groups* [10], where it is stated that the theory began when Kepler in 1619 stellated a regular polyhedron to obtain the star polyhedron which is essentially a map of twelve pentagons on a surface of genus 4 (called the great dodecahedron; $\{5, \frac{5}{2}\}$ in Coxeter's notation). But one could also go back to the discovery of the regular polyhedra from ancient Greece. The book of Coxeter and Moser describes many of the regular maps of low genus. Even in this book the authors realised that there was a relationship between regular maps and Riemann surfaces and Sect. 8 is called "Maps on a two-sheeted Riemann surface" where they even associate an algebraic curve with a map.

The reason why regular maps are described in a book about group theory is that every regular map is associated with a finite group which has two generators R and S that obey the relations $R^m = S^n = (RS)^2 = 1$, and one can associate a regular map with such a group. Here R could be thought of as a rotation of the darts around a face of the map (with m being the face length of the faces of the map) and S the rotations of the darts around a vertex (with n denoting the degree of the map's vertices). Conversely, given a two-generator group, with one generator of order two, we can associate a regular map.

Another motivation for studying maps came from map colouring problems, the four-colour problem being the most notable example. This story will not be of interest to us here, but this problem did lead to a study of maps well away from the regular ones. A map can be defined by looking at the permutations of the *darts.* A dart is a directed edge. We usually draw a dart as an arrow along the edge pointing to a vertex. An edge will usually have two vertices and two darts. But we do allow edges with just one vertex. These may be loops (which still have two darts) or *free edges*. These are edges with just one vertex and one dart. For details see [15].

Let Ω denote the set of darts of a map. We define three permutations x, y, z of the darts as follows. The permutation x is the permutation that reverses the darts on each non-free edge, or fixes the dart on a free edge, and y cyclically permutes the darts directed towards each vertex v in an anticlockwise direction. The cycles of the permutation $z = y^{-1}x$ then describe the order of the darts around the faces following the orientation. (We are composing permutations on the right, so this means first do $y^{-1}$and then do x.) We then have the relations $x^2 = y^m = z^n = xyz = 1$ so that the group G generated by x, y, z is an image of the triangle group $\Gamma = \Gamma[2, m, n]$ which acts on one of the simply-connected Riemann surfaces $\mathscr{U}$ that is the Riemann sphere if $\frac{1}{m} + \frac{1}{n} > \frac{1}{2}$, the Euclidean plane if $\frac{1}{m} + \frac{1}{n} = \frac{1}{2}$, and the hyperbolic plane if $\frac{1}{m} + \frac{1}{n} < \frac{1}{2}$. We suppose that Γ is generated by X, Y, Z obeying the relations $X^2 = Y^m = Z^n = XYZ = 1$. Here m is the least common multiple of the vertex valencies and n is the lest common multiple of the face valencies. The ordered pair $\{n, m\}$ is called the *type* of the map.

Let G be the permutation group generated by x and y and z so that G is a transitive group acting on N points where $N = |\Omega|$, the number of darts. Transitivity follows from the connectedness of the map. There is then an epimorphism $\theta : \Gamma \longrightarrow G$ defined by $\theta(X) = x, \theta(Y) = y, \theta(Z) = z$. If G_α is the stabilizer of a dart in Ω, we let $M = \theta^{-1}(G_\alpha)$. Now M is a subgroup of index N in Γ, called a *map subgroup*, and thus a Fuchsian group provided $\mathscr{U}$ is the hyperbolic plane. The quotient space $R = R(\mathscr{M}) = \mathscr{U}/M$ is the Riemann surface associated with the map $\mathscr{M}$ and there is an embedding of the map $\mathscr{M}$ in R [15, 22].

The map subgroup turns out to tell us a lot about the map. For example, if $\mathscr{M}_1, \mathscr{M}_2$ are two maps with map subgroups M_1, M_2 respectively, then $\mathscr{M}_1$ covers $\mathscr{M}_2$ if and only if $M_1 \leq M_2$ (up to conjugacy). Two maps are isomorphic if and only if their map subgroups are conjugate in Γ and if $\mathscr{M}$ is a map with map subgroup M, then the $Aut(\mathscr{M})$, the automorphism group of $\mathscr{M}$, is isomorphic to $N_\Gamma(M)/M$ (where $N_\Gamma(M)$ is the normaliser of M in Γ) so that $\mathscr{M}$ is a regular map if and only if $M \lhd \Gamma$.

4 Belyi's Theorem

In the theory of dessin d'enfants a crucial role is played by Belyi's Theorem.

To state this we need the concept of a *critical value*. Let $f : R \longrightarrow \Sigma$ be a meromorphic function from a compact Riemann surface R to the Riemann sphere Σ. Then f is an n-sheeted branched cover of R over Σ. This means that every point p of Σ has at most n inverse images. If a point p has less than n inverse images then we call p a *critical value.*

An analytic function $w(z)$ is called an *algebraic function* if it satisfies a functional equation

$$A(z, w) = a_0(z)w^n + a_1(z)w^{n-1} + \cdots + a_n(z) = 0, \quad a_0(z) \neq 0. \qquad (1)$$

Here, $A(z, w)$ is an irreducible polynomial in z and w and the $a_i(z)$ are polynomials in z, with coefficients in some subfield F of the complex numbers. For example, F could be the field of complex numbers $\mathbb{C}$, the field of real numbers $\mathbb{R}$, the field $\overline{\mathbb{Q}}$ of algebraic numbers, or the field $\mathbb{Q}$ of rational numbers.

For each value of z there are at most n values of w. So we build an n-sheeted branched cover R of Σ such that w is a single valued function on X. We then call R the Riemann surface of w. This was Riemann's original approach to constructing Riemann surfaces. These days, Riemann surfaces are defined abstractly (using charts and atlases) as complex one-dimensional manifolds. It is a very deep result that these two approaches are equivalent. Thus every compact Riemann surface corresponds to a complex algebraic curve as in (1).

We say that R is defined over a field F if the polynomials $a_i(z)$ in (1) are defined over F.

A meromorphic function $\beta : R \longrightarrow \Sigma$ is called a *Belyi map* (or *Belyi function*) if it has at most three critical values. By composing this function with an element of $PSL(2, \mathbb{C})$ (the automorphism group of Σ), we can assume that these critical values lie in the set $\{0, 1, \infty\}$.

Theorem 1 (Belyi's Theorem [2]) *A compact Riemann surface R can be defined over the field $\overline{\mathbb{Q}}$ of algebraic numbers if and only if there exists a Belyi function $\beta : R \longrightarrow \Sigma$.*

If K is a subgroup of finite index in a triangle group $\Gamma = \Gamma[2, m, n]$, then the natural map from $\mathcal{U}/K \longrightarrow \mathcal{U}/\Gamma$ is a Belyi map. It can be shown that every Belyi function is of this form [6]. Thus we see that subgroups of the triangle group $\Gamma[2, m, n]$ play an important role in the theory of maps.

We sum up this section with the following theorem.

Theorem 2 *Let R be a compact Riemann surface. Then the following statements are equivalent.*

(i) R can be defined over $\overline{\mathbb{Q}}$,
(ii) there exists a Belyi function $\beta : R \longrightarrow \Sigma$,
(iii) $R = \mathcal{U}/M$ where M is a subgroup of a finite index in a triangle group $\Gamma[2, m, n]$.

The statement that R defined over $\overline{\mathbb{Q}}$ implies the existence of a Belyi function can be found in [13]. The converse is more difficult. In the early papers on the subject it was stated that this follows from Weil's irreducibility criterion and was sometimes called the "obvious" part of Belyi's theorem. This result turned out to be far from obvious. See [25] and also [19] for proofs.

5 Other Triangle Groups

We now play the same game after replacing $\Gamma[2, m, n]$ with other similarly defined groups. We start with the *extended triangle group* $\Gamma(2, m, n)$ defined in Sect. 1, generated by three reflections a, b, c in the sides of the triangle with angles $\pi/l, \pi/m, \pi/n$. Its presentation is

$$\{a, b, c | a^2 = b^2 = c^2 = (ab)^2 = (bc)^m = (ca)^n = 1\}. \quad (2)$$

Note that $\Gamma(2, m, n)$ contains $\Gamma[2, m, n]$ with index 2 as the subgroup generated by $x = ab$, $y = bc$.

This group is called the (non-oriented) *cartographic group* by Grothendieck in [11].

Interestingly, the idea of associating a map with a triple of involutions goes back to a paper of Tutte [23] in 1973.

We are now dealing with discrete groups that might contain orientation-reversing transformations. Such groups are called *non-Euclidean crystallographic groups* or *NEC groups*. If Λ is an NEC group then $\mathcal{U}/\Lambda$ might be non-orientable or might have boundary. (There is a non-empty boundary if and only if Λ contains reflections, that is conjugates of a, b or c.) Such a surface is called a Klein surface. Strictly speaking a Klein surface is one which has a dianalytic structure, which means that the change of coordinate maps are analytic or anti-analytic, whereas for a Riemann surface they are always analytic. Whereas compact Riemann surfaces correspond to complex algebraic curves, compact Klein surfaces correspond to real algebraic curves [1].

Again, we can associate a map $\mathcal{M}$ with a subgroup M of the extended triangle group $\Gamma(2, m, n)$. However as the group contains elements that reverse orientation, such as reflections and glide-reflections, the maps that we construct may now lie on non-orientable surfaces or may have boundary. We thus need to build a theory of maps which might be non-orientable or might lie on surfaces with boundary. We also want this theory to be related to the extended triangle group. This was first developed by Robin Bryant in a Southampton Ph.D. thesis in 1984. See [4, 5, 14]. Whereas the theory of maps on orientable surfaces uses darts, for non-orientable surfaces or surfaces with boundary we use *blades*. A blade is one of the halves of a dart. Figure 1 illustrates how an edge with two darts gives rise to four blades.

Note that each blade determines a unique vertex, edge and face and so can be regarded as a flag. Now on the set Ω^* of blades of $\mathcal{M}$ we can define three permutations τ, λ, ρ which we call the *transverse reflection*, the l*ongitudinal reflection* and the *rotary reflection* respectively. An interior edge will have four darts. If we draw the edge as a horizontal line then there will be two upper darts and two lower darts. The permutation τ interchanges the upper and lower halves of the dart. If the edge lies along the boundary we define τ to fix these darts. The longitudinal reflection λ interchanges both the upper darts and both the lower darts of an edge. Again, it could happen that the edge intersects the boundary but not at a vertex and then the

Fig. 1 Edge with four blades

edge could have only two blades. We then define λ to fix those blades. Finally, we have the rotary reflection ρ. Consider a vertex v of the map that does not lie on the boundary. At v there is a dart which consists of two blades, b and $\tau(b)$. The blade b determines a unique sense of orientation. (The blade $\tau(b)$ determines the opposite orientation.) We follow that orientation around until we meet the next edge of the map. This edge will contain two blades at v, one of which, b' say, determines the opposite orientation to that of b. We then define $\rho(b) = b'$, as in Fig. 2.

If v lies on the boundary it might be that following this orientation leads to the boundary in which case we define $\rho(b) = b$, as in Fig. 3, where the circle denotes a boundary component.

We note that $\lambda\tau = x$, $\rho\tau = y$ and $\lambda\rho = z$, so that we have the relations

$$\tau^2 = \rho^2 = \lambda^2 = (\tau\lambda)^2 = (\lambda\rho)^m = (\rho\tau)^n = 1.$$

As an example, Fig. 4 gives a map on the Möbius band. The corresponding permutations are

$$\tau = (1,4)(2,3), \quad \lambda = (1,3)(2,4), \quad \rho = (1)(2)(3,4).$$

Figure 5 gives a map on the disc with

$$\tau = (1)(2)(3,4)(5,6)(7,8), \quad \lambda = (1)(2)(3,4)(5,7)(6,8), \quad \rho = (1,5)(2,6)(3,7)(4,8).$$

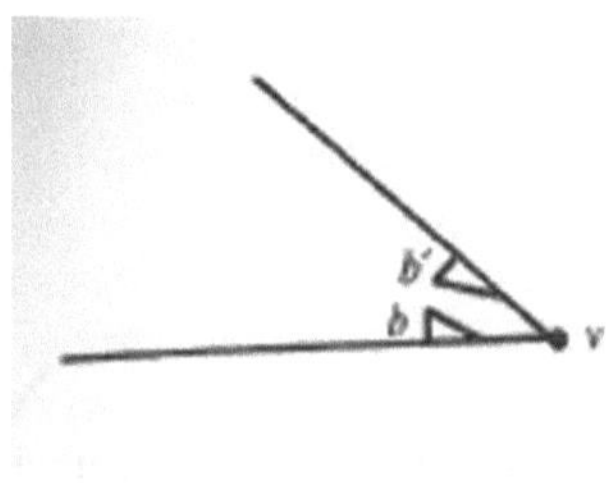

Fig. 2 Rotary reflection

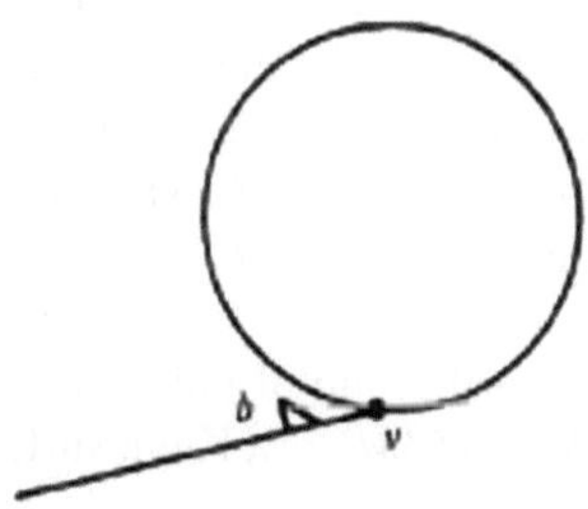

Fig. 3 Rotary reflection for v lying on the boundary

Fig. 4 Map on the Möbius band

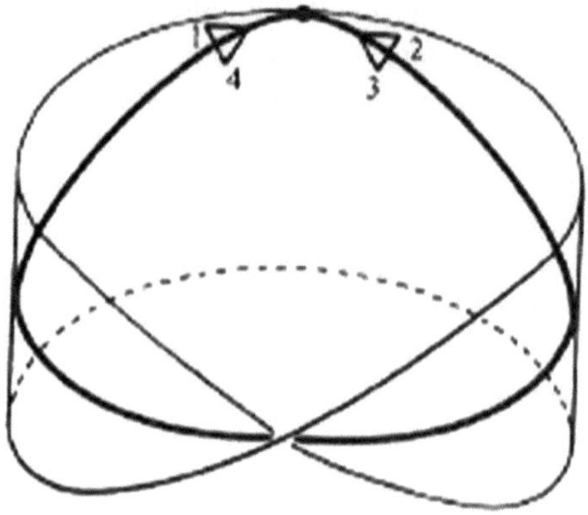

Fig. 5 Map on the disc

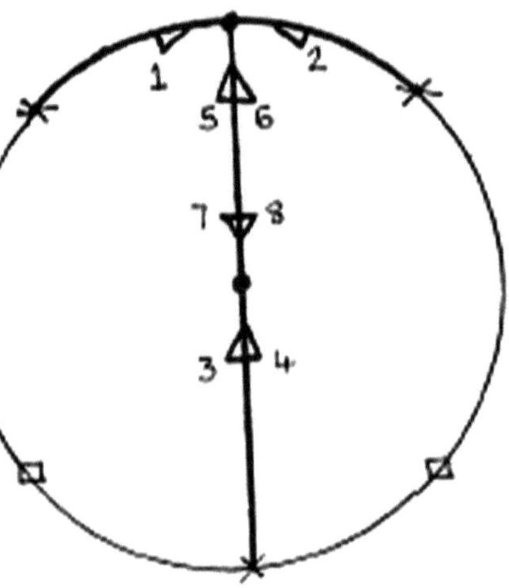

Here the dot represents a vertex, a cross a free edge, and an open square a face centre on the boundary. These last examples come from [5].

Now in the classical theory we showed that associated to a map or dessin $\mathcal{M}$ we have a canonical Riemann surface $R(\mathcal{M}) = \mathcal{U}/M$. Here M was a Fuchsian group provided $\mathcal{U}$ was the hyperbolic plane. Now let G^* be the permutation group generated by τ, λ, and ρ. Then there is an epimorphism $\theta : \Gamma(2, m, n) \longrightarrow G^*$, defined by $\theta(a) = \tau, \theta(b) = \lambda, \theta(c) = \rho$. If $|\Omega^*| = N$ then $M^* = \theta^{-1}(G^*_\alpha)$ is a subgroup of index N in $\Gamma(2, m, n)$ and is an NEC group. The quotient space $K = \mathcal{U}/M^*$ is a Klein surface and there is an embedding of $\mathcal{M}^*$ in K.

6 Klein Surfaces and Real Belyi Functions

We first define a real Belyi function. To do this we introduce the idea of the *complex double* of a Klein surface. This is done explicitly in [1], but it is easier to use NEC groups. It can be shown that every Klein surface S is of the form $\mathcal{U}/\Lambda$ where Λ is an NEC group without elliptic elements. Then the complex double is the Riemann surface $S^+ = \mathcal{U}/\Lambda^+$ where Λ^+ is the subgroup of Λ of index two consisting of those transformations that preserve orientation. Note that we have a projection map $\pi : S^+ \longrightarrow S$ defined by $\pi[z]_{\Lambda^+} = [z]_\Lambda$ (where $[z]_\Lambda$ denotes the Λ-orbit of z etc.). If S is a compact non-orientable surface without boundary then S^+ is the usual orientable two-sheeted cover. If S is a surface with boundary then S^+ is obtained by glueing the boundaries together and choosing the orientation so that S^+ is orientable.

For example, if S is the projective plane then S^+ is the sphere. If S is a Möbius band then S^+ is homeomorphic to a torus, and if S is an annulus then S^+ is also homeomorphic to a torus.

We let Δ denote the upper half of the complex plane including the equator which we regard as the great circle passing through $0, 1, \infty$. The *folding map* $\phi : \mathbb{C} \longrightarrow \mathbb{C}^* = \{a+ib|b \geq 0\}$ is defined by $\phi(a+ib) = a + i|b|$ and this can be extended to a map $\phi : \Sigma \longrightarrow \Delta$ by letting $\phi(\infty) = \infty$. Let S be a Klein surface and S^+ denote its complex double. A *real Belyi function* is a Belyi function $\beta : S \longrightarrow \Delta$ such that the following diagram commutes.

$$\begin{array}{ccc} S^+ & \xrightarrow{\beta^+} & \Sigma \\ \pi \downarrow & & \downarrow \phi \\ S & \xrightarrow[\beta]{} & \Delta \end{array}$$

Here, β^+ is a Belyi function defined on S^+.

A version of Belyi's theorem for Klein surfaces (the real Belyi theorem) was proved in [18].

Theorem 3 *Let K be a compact connected Klein surface. Then the following statements are equivalent:*

(i) K can be defined over $\overline{\mathbb{Q}} \cap \mathbb{R}$,
(ii) there exists a real Belyi function $\beta : K \longrightarrow \Delta$,
(iii) $K = \mathcal{U}/L$ where L is a subgroup of a finite index in an extended triangle group $\Gamma(2, m, n)$.

Real Belyi functions can be thought of as Belyi functions on Klein surfaces.

7 Hypermaps

We have been dealing with triangle groups of the form $\Gamma[2, m, n]$. Obviously, one should also investigate what happens if $\Gamma[l, m, n]$ was considered instead. When we developed a theory of maps based on subgroups of $\Gamma[2, m, n]$, we were aware of work being done in Bordeaux by Robert Cori (later joined by Antonio Machì) on *hypermaps* and their correspondence to subgroups of $\Gamma[l, m, n]$. In Cori's initial paper in 1975 [7], all hypermaps were planar (of genus 0) and the motivation was applications to computer science. Later, David Corn in his Southampton thesis [9] developed a theory of hypermaps based on the work of Jones and Singerman in [15] thus thinking of a hypermap as lying on a Riemann surface. Also see [8].

Whereas a map is an embedding of a graph in a surface, a hypermap is an embedding of a hypergraph in a surface X. A hypergraph is like a graph, except now the edges can have more than two vertices or just one vertex. The Fano plane is a well-known example. Here every (hyper)edge has three vertices. The definition of a hypermap after Cori [7] invokes two sets called S and A (S for sommets and A for arêtes). The elements of these sets are called the hypervertices and hyperedges respectively. We require the set $B = S \cap A$ of *brins* to be finite, on a compact surface. (Brin is sometimes translated to bit.) We also require $X \setminus (S \cup A)$ to be a union of simply connected regions called hyperfaces. Note that the hypervertices, hyperedges and hyperfaces are topological polygons with the brins as vertices. In this theory, the brins play the same role as the darts did in the theory of maps. Again, we can approach the theory in terms of permutations of the brins. We have three permutations x, y, z. Here x rotates the brins around a hypervertex, y rotates the brins around a hyperedge and then the cycles of $z = (xy)^{-1}$ describe the hyperfaces. If the orders of x, y, z are l, m, n respectively, then we say that the hypermap has *type* $\{l, m, n\}$. (If we perform xy we traverse two edges of a hyperface and hence the length of each cycle of xy is half the number of sides of the corresponding hyperface.) For example, in Fig. 6 we have three hypervertices of length 3, three hyperedges of length 3, and three hyperfaces of length 6.

Again, as with maps, we can approach all this algebraically just using permutations. So now a hypermap is just a transitive group on a set B, with two generators x, y.

We now describe the Walsh representation of a hypermap. In the interior of each hypervertex we place a black vertex and in the interior of a each hyperedge we place a white vertex. If the hyperedge intersects the hypervertex in a brin we join this black vertex to the white vertex by an edge through the brin. We end up with a map and the number of edges of this map is the number of brins of the hypermap. This is called the *Walsh* map $W(\mathscr{H})$ of the hypermap $\mathscr{H}$ [24]. This gives another way of describing hypermaps. They are just bipartite maps with vertices coloured black or white. The permutations x and y are now the cyclic rotations of the edges emanating out of a black or white vertex.

In the Cori definition these permutations just become the anticlockwise permutations of the brins around a hypervertex or hyperedge. Details about hypermaps as bipartite maps may be found in a paper of Gareth Jones [16].

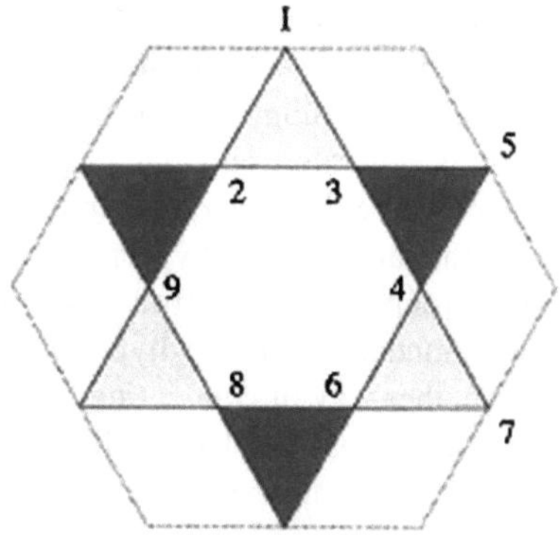

Fig. 6 Hypermap of type $\{3, 3, 3\}$ on the torus (opposite sides are identified)

Example. Figure 6 gives a hypermap of type $\{3, 3, 3\}$ on the torus formed by identifying the opposite edges of the hexagon. The permutations are

$$x = (1, 2, 3)(9, 5, 8)(4, 6, 7), \quad y = (2, 7, 9)(8, 1, 6)(3, 4, 5),$$

and then

$$z = (1, 5, 7)(2, 8, 4)(3, 9, 6).$$

8 Triangle Groups with Infinite Periods

An element of finite order n in a triangle group (and more generally in any Fuchsian group) is called an *elliptic element*. It represents a rotation of order n. In a fundamental region for the triangle group this rotation is about a vertex whose angle is $2\pi/n$. We also have limit rotations. These are rotations about points on the real axis which is the line at infinity in the hyperbolic plane. At these points the angle is $0 = 2\pi/\infty$ so we then have an element of infinite order which is usually referred to as a *parabolic* element. (For example $z \mapsto z+1$ is a parabolic element which is a rotation about the point at infinity.) In the presentation of the triangle group we consider the relation $z^{\infty} = 1$ as being the empty relation, so for example the triangle group $\Gamma[l, m, \infty]$ has a presentation $\langle x, y | x^l = y^m = 1 \rangle$ and so is isomorphic to a free product $C_l * C_m$.

Note that there is a homomorphism from $\Gamma[l, m, \infty]$ to $\Gamma[l, m, n]$ for any n so that any map subgroup in $\Gamma[2, m, n]$ can be pulled back to $\Gamma[2, m, \infty]$. The most well known example is $\Gamma[2, 3, \infty]$ which is known to be isomorphic to the classical modular group $\mathrm{PSL}(2, \mathbb{Z})$. Let $\mathcal{M}$ be a triangular map on an oriented surface. If the least common multiple of the vertex valencies is equal to N, then this map can be represented by a map subgroup M lying in the triangle group $\Gamma[2, 3, N]$. There is a homomorphism $\theta : PSL(2,\mathbb{Z}) \longrightarrow \Gamma[2,3,N]$ and then M can be pulled back to a subgroup $\theta^{-1}(M)$ inside the classical modular group. Thus all triangular maps correspond to subgroups of the classical modular group. The converse is true if we allow degenerate triangular maps, where triangles can degenerate to points. (This occurs where the image of y in the permutation group has a fixed point.)

In the same way, all hypermaps can be represented by subgroups of $\Gamma[\infty,\infty,\infty]$. This group is just the free group F_2 on two generators so there is a ono-to-one correspondence between hypermaps and the conjugacy classes of subgroups of F_2.

Another way of connecting together maps and hypermaps is to use inclusions of triangle groups. The list of all possible inclusions between triangle groups is given in [21]. The simplest of these inclusions is $\Gamma[m,m,n] < \Gamma[2,m,2n]$ with index 2. Hence $\Gamma[\infty,\infty,\infty] < \Gamma[2,\infty,\infty]$ with index 2 and so if we take a hypermap subgroup H of index N in $\Gamma(\infty,\infty,\infty)$ it becomes a map subgroup of index $2N$ in $\Gamma[2,\infty,\infty]$. This corresponds to the Walsh representation of a hypermap where we go from a hypermap with N brins to a map with $2N$ darts, see [8]. A consequence is that there are no more Riemann surfaces, or algebraic curves, that we get from hypermaps than we do from maps.

One should also consider subgroups of $\Gamma(l,m,n)$ where none of l,m,n are equal to 2. These should correspond to hypermaps on non-orientable surfaces or surfaces with boundary. A basic theory was developed in [12].

9 Away from Triangle Groups

Triangle groups are just Fuchsian groups with three periods. It is possible to generalise the above theories by considering Fuchsian groups with more than three periods. This leads to the work of Zvonkin [26]. He defines a *k-constellation* to be a sequence of permutations $\sigma_1,\sigma_2\ldots,\sigma_k$ with $\sigma_1\sigma_2\ldots\sigma_k = 1$, acting transitively on a finite set. This idea was also mentioned in [22] where it was given the rather ugly name of marked finite permutation group. The theory of constellations is still in its infancy. One thing is worth mentioning here. Triangle groups are *rigid*. Any two isomorphic triangle groups are conjugate in the group of all Möbius transformations. Hence their subgroups correspond to the same Riemann surface or algebraic curve. As soon as $k > 3$, there is a whole continuum of Riemann surfaces. This leads to Teichmüller theory and well away from discrete mathematics.

References

1. Alling N.L., Greenleaf N.: Foundations of the Theory of Klein Surfaces. Springer Verlag, Berlin (1971).
2. Belyĭ G.V.: On Galois extensions of a maximal cyclotomic field. Izv. Akad. Nauk SSSR Ser. Mat. **43**, 267–276, 479 (1979).

3. Biggs N.: The symplectic representation of map automorphisms. Bull. London Math. Soc. **4**, 303–306 (1972).
4. Bryant R.P., Singerman D.: Foundations of the theory of maps on surfaces with boundary. Q. J. Math. Oxford II. **36**, 17–41 (1985).
5. Bryant R.P.: Maps on surfaces with boundary. Ph.D. thesis, University of Southampton (1984).
6. Cohen P.B., Itzykson C., Wolfart J.: Fuchsian triangle groups and Grothendieck dessins. Variations on a theme of Belyi. Commun. Math. Phys. **163**, No. 3, 605–627 (1994).
7. Cori R.: Un Code pour les graphes planaires et ses applications. Astérisque **27**, 1–169, Soc. Math. France, Paris (1975).
8. Corn D., Singerman D.: Regular hypermaps. Eur. J. Comb. **9**, No.4, 337–351 (1988).
9. Corn D.: Regular hypermaps. Ph.D.thesis, University of Southampton (1989).
10. Coxeter H.S.M., Moser W.O.J.: Generators and Relations for Discrete Groups, 4th ed., Springer-Verlag, Berlin / Heidelberg / New York (1980).
11. Grothendieck A.: Esquisse d'un Programme. Geometric Galois Actions I, Around Grothendieck's Esquisse d'un Programme, Lochak P., Schneps L. (eds.) London Math. Soc. Lecture Note Ser. **242**, 5–48, Cambridge University Press, Cambridge (1997).
12. Izquierdo M., Singerman D.: Hypermaps on surfaces with boundary. Eur. J. Comb. **15**, No.2, 159–172 (1994).
13. Jones G.A., Singerman D.: Belyĭ functions, hypermaps and Galois groups. Bull. London Math. Soc. **28**, 561–590 (1996).
14. Jones G. A., Singerman D.: Maps, hypermaps and triangle groups. The Grothendieck Theory of Dessins d'Enfants, Schneps L. (ed.), London Math. Soc. Lecture Note Ser. **200**, 115–145, Cambridge University Press, Cambridge (1994).
15. Jones G.A., Singerman D.: Theory of maps on orientable surfaces. Proc. London Math. Soc. (3) **37**, 273–307 (1978).
16. Jones G.A.: Maps on surfaces and Galois groups. Math.Slovaca **47**,1–33 (1997).
17. Klein F.: Über die Transformationen siebenter Ordnung der elliptischen Funktionen. Math. Ann. **14**, 428–471 (1879).
18. Köck B., Singerman D.: Real Belyi theory. Q. J. Math. **58**, 463–478 (2007).
19. Köck B.: Belyi's theorem revisited. Beitr. Algebra Geom. **45**, No. 1, 253–265 (2004).
20. Levy S.: The eightfold way. The beauty of Kleins quartic curve. Mathematical Sciences Research Institute Publications **35**, Cambridge University Press, Cambridge (2001).
21. Singerman, D.: Finitely maximal Fuchsian groups. J. London Math. Soc. (2) **6**, 29–38 (1972).
22. Singerman D.: Automorphisms of maps, permutation groups and Riemann surfaces. Bull. Lond. Math. Soc. **8**, 65–68 (1976).
23. Tutte W.T.: What is a map? New directions in graph theory, Harary F. (ed.), 309–325, Academic Press (1973).
24. Walsh T.R.S.: Hypermaps versus bipartite maps. J. Combin. Theory Ser. B **18**, 155–163 (1975).
25. Wolfart J.: The 'obvious' part of Belyis theorem and Riemann surfaces with many automorphisms. Schneps L. (ed.), Geometric Galois actions. 1. Around Grothendiecks "Esquisse dun programme". Proceedings of the conference on geometry and arithmetic of moduli spaces, Luminy, France, August 1995. Lond. Math. Soc. Lect. Note Ser. **242**, 97–112, Cambridge University Press, Cambridge (1997).
26. Zvonkin, A.: Megamaps: Construction and examples. Discrete models: combinatorics, computation, and geometry. Proceedings of the 1st international conference (DM-CCG), Paris, France, July 2-5, 2001. Discrete Math. Theor. Comput. Sci., Proc. AA, 329–340, Maison de l'Informatique et des Mathématiques Discrètes, Paris (2001).

Nilpotent Symmetric Dessins of Class Two

Na-Er Wang, Roman Nedela and Kan Hu

Abstract A dessin is a cellular embedding of a connected bipartite graph into an orientable closed surface with a fixed colouring of vertices and prescribed global orientation. A dessin is regular if its group of colour- and orientation-preserving automorphisms acts regularly on the set of edges, and a regular dessin is symmetric if it admits an external symmetry transposing the vertex colours. The symmetric dessins whose automorphism groups are nilpotent of class two are classified.

1 Introduction

A *map* $\mathscr{M}$ is a 2-cell embedding of a connected graph into an orientable closed surface. A *dessin* is a bipartite map with a fixed colouring of vertices and prescribed global orientation. A dessin is often defined by a 2-cell embedding of a bipartite bicoloured graph. An automorphism of a dessin $\mathscr{D}$ is a permutation of the edges which preserves the graph incidence and vertex colourings, and extends to an orientation-preserving self-homeomorphism of the supporting surface. The set of automorphisms of $\mathscr{D}$ forms the automorphism group $\mathrm{Aut}(\mathscr{D})$ under composition. It is well known that $\mathrm{Aut}(\mathscr{D})$ acts semi-regularly on the edges. If this action is transitive, and hence regular, then the dessin is called *regular* as well.

Dessins were introduced by Grothendieck as a tool to investigate the absolute Galois group. By Belyi theorem each dessin determines an algebraic curve defined

N.-E. Wang · K. Hu
School of Mathematics, Physics and Information Science,
Zhejiang Ocean University, Zhoushan, Zhejiang 316000, People's Republic of China
e-mail: wangnaer@zjou.edu.cn

K. Hu
e-mail: hukan@zjou.edu.cn

R. Nedela (✉)
Faculty of Natural Sciences, Matej Bel University, Tajovského 40,
974 01 Banská Bystrica, Slovakia
e-mail: nedela@savbb.sk

J. Širáň and R. Jajcay (eds.), *Symmetries in Graphs, Maps, and Polytopes*,
Springer Proceedings in Mathematics & Statistics 159,
DOI 10.1007/978-3-319-30451-9_17

over the field of algebraic numbers. Through this correspondence a faithful action of the absolute Galois group on dessins is defined. It is known that this action remains faithful if we are restricted to plane trees, or to regular dessins. For more details on this interesting correspondence the reader is referred to [24, 26].

For several reasons classification of regular dessins became important in the development of this area of research. The regular dessins have been studied by imposing certain constraints on the supporting surfaces, on the embedded graphs or on their automorphism groups; see [4, 6, 9, 10, 13, 17, 18, 20, 25] and references therein.

In the present paper we follow the third approach and investigate regular dessins with a nilpotent automorphism group, focusing on those with an external symmetry that transposes the vertex colours. These will be referred to as *nilpotent symmetric* dessins. Symmetric dessins have appeared in the classification of regular embeddings of complete bipartite graphs $K_{n,n}$ [9, 10, 17, 19, 20]. In particular, if n is prime power, then the automorphism group of the respective dessin is a p-group, and therefore the dessin is nilpotent symmetric. Nilpotent regular dessins of class one, namely the *abelian dessins*, were investigated by several authors [11, 12, 23]. A complete classification of the abelian dessins can be found in [13]. The curves associated to the abelian dessins include some popular families of curves such as the curves of Fermat and Lefschetz type.

In this paper we classify the symmetric dessins whose automorphism groups are nilpotent of class two. Similarly as in the theory of nilpotent groups, we first show that nilpotent dessins decompose into a parallel product of dessins whose automorphism groups are p-groups. This allows us to reduce the classification problem to the classification of symmetric dessins whose automorphism groups are p-groups of class two. This is done in two steps. We first classify the symmetric p-dessins of class two with simple underlying graphs; see Theorem 1. Then we extend the classification to the general case; see Theorems 2 and 3.

2 Algebraic Dessins

In this section, we briefly outline the algebraic theory of regular dessins; see [7, 18] for more details.

Let $\mathscr{D}$ be a dessin on an orientable surface $\mathscr{S}$. The fixed global orientation of $\mathscr{S}$ induces two permutations ρ and λ which cyclically permute the edges around the black and white vertices, respectively. Due to the connectivity of the underlying graph, the group $\mathrm{Mon}(\mathscr{D}) = \langle \rho, \lambda \rangle$ acts transitively on the edges of $\mathscr{D}$. Conversely, given a triple $(E; \rho, \lambda)$, where E is a non-empty finite set and $\rho, \lambda \in \mathrm{Sym}(E)$ generate a transitive permutation group on E, a dessin $\mathscr{D}$ is defined as follows: We identify the elements of E with the edges, and the cycles of ρ and λ with the black and white vertices of $\mathscr{D}$, with the incidence given by containment. In this way we obtain a bipartite graph $\mathscr{G}$. The cyclic order of every cycle of the permutations ρ and λ determine the local rotation of edges around the respective vertex. In this way an embedding of $\mathscr{G}$ into an oriented surface is determined.

Let $\mathscr{D}_i = (E_i; \rho_i, \lambda_i)$ $(i = 1, 2)$ be two dessins. A *homomorphism* from $\mathscr{D}_1$ to $\mathscr{D}_2$ is a mapping $\phi : E_1 \to E_2$ such that

$$\rho_1\phi = \phi\rho_2 \quad \text{and} \quad \lambda_1\phi = \phi\lambda_2,$$

where the composition is from the left to the right. In particular, if ϕ is a bijection, then it is called an *isomorphism* from $\mathscr{D}_1$ to $\mathscr{D}_2$ and is denoted by $\mathscr{D}_1 \cong \mathscr{D}_2$. An isomorphism of a dessin $\mathscr{D}$ onto itself is called an *automorphism* of $\mathscr{D}$. The set of automorphisms of $\mathscr{D}$ forms the automorphism group $\mathrm{Aut}(\mathscr{D})$ of $\mathscr{D}$ under composition. By the above commuting rule we have $\mathrm{Aut}(\mathscr{D}) = C_{\mathrm{Sym}(E)}(\mathrm{Mon}(\mathscr{D}))$, the centralizer of $\mathrm{Mon}(\mathscr{D})$ in the symmetric group $\mathrm{Sym}(E)$. Since $\mathrm{Mon}(\mathscr{D})$ is transitive, $\mathrm{Aut}(\mathscr{D})$ is semiregular on E. If the action is transitive, and hence regular, then the dessin is called a regular dessin.

Every regular dessin $\mathscr{D}$ can be identified with a triple (G, x, y) called an *an algebraic dessin* where $G = \mathrm{Aut}(\mathscr{D})$, and x and y generate, respectively, the cyclic stablizers of a black vertex and an adjacent white vertex. The monodromy group and the automorphism group of $\mathscr{D}$ are identified with the right and the left regular representations of G, respectively. Under the identification two regular dessins (G_i, x_i, y_i) $(i = 1, 2)$ are isomorphic if the assignment $x_1 \mapsto x_2, y_1 \mapsto y_2$ extends to an isomorphism from G_1 onto G_2.

For a regular dessin $\mathscr{D} = (G, x, y)$, the triple (l, m, n) is called the *type* of $\mathscr{D}$ where $l = o(x)$, $m = o(y)$ and $n = o(xy)$. The *genus* g of $\mathscr{D}$ is the genus of its supporting surface. It is determined by the Euler-Poincaré formula

$$2 - 2g = |G|\left(\frac{1}{l} + \frac{1}{m} + \frac{1}{n} - 1\right).$$

3 External Symmetries

Every dessin $\mathscr{D} = (E; \rho, \lambda)$ determines a transitive permutation representation $F_2 \to \mathrm{Mon}(\mathscr{D}), X \mapsto \rho, Y \mapsto \lambda$ where $F_2 = \langle X, Y \mid - \rangle$ is the free group of rank two. The stabilizer N in F_2 of an element $e \in E$ is a subgroup of finite index in F_2. This subgroup is uniquely determined up to conjugacy, and will be referred to as the *dessin subgroup* associated with $\mathscr{D}$. In particular, a regular dessin $\mathscr{D}$ corresponds to a normal subgroup N, in which case $\mathrm{Aut}(\mathscr{D}) \cong F_2/N$.

Let $\mathscr{D}$ be a dessin, and let N be the associated dessin subgroup. An automorphism σ of F_2 sends N to N^σ, and hence transforms $\mathscr{D}$ to a dessin $\mathscr{D}^\sigma$. In particular if σ is an inner automorphism of F_2, then N is conjugate to N^σ, and hence $\mathscr{D}$ is isomorphic to $\mathscr{D}^\sigma$. It follows that the outer automorphism group $\Omega := \mathrm{Out}(F_2) = \mathrm{Aut}(F_2)/\mathrm{Inn}(F_2)$ acts as the *group of dessin operations* on the isomorphism classes of dessins.

For example, the operation ω_τ induced by the automorphism $\tau : X \mapsto Y, Y \mapsto X$ transposes black and white vertices while preserving faces and orientation. This is one of the six duality operations studied in [21]. The operation ω_π induced by the automorphism $\pi : X \mapsto X^{-1}, Y \mapsto Y$ reverses the orientation around the black vertices but preserves it around the white vertices. Note that the automorphism $\pi_1 = \tau\pi\tau : X \mapsto X, Y \mapsto Y^{-1}$ induces operation which reverses the orientation around the white vertices while preserves it around the black ones. These are sometimes called Petrie operations, since they transpose faces and Petrie polygons (the zig-zag walks). Finally the automorphism $\iota = \pi\pi_1 : X \mapsto X^{-1}, Y \mapsto Y^{-1}$ induces an operation ω_ι which transforms a dessin to its mirror image. The two operations ω_τ and ω_π generate a maximal finite subgroup $\Omega_1 := \langle\omega_\tau, \omega_\pi\rangle \cong D_8$ of Ω [21].

Regular dessins which are invariant under a certain operation are said to possess a corresponding external symmetry. More specifically, a regular dessin $\mathscr{D}$ is *symmetric* if $\omega_\tau(\mathscr{D}) \cong \mathscr{D}$, *self-Petrie-dual* if $\omega_\pi(\mathscr{D}) \cong \mathscr{D}$, and *reflexible* if $\omega_\iota(\mathscr{D}) \cong \mathscr{D}$. Moreover, a regular dessin will be called Ω_1-*invariant* if it is invariant under all operations in Ω_1, and Ω-*invariant* (or totally symmetric) if it is invariant under all dessin operations.

Given two regular dessins $\mathscr{D}_i = (G_i, x_i, y_i)$ $(i = 1, 2)$, let N_i be the associated normal dessin subgroups. Then $N_1 \cap N_2$, being the intersection of two normal subgroups of finite index in F_2, is also a normal subgroup of finite index in F_2. The corresponding regular dessin is called the *parallel product* of $\mathscr{D}_1$ and $\mathscr{D}_2$ and is denoted by $\mathscr{D}_1 \vee \mathscr{D}_2$. In particular, if $\gcd(|G_1|, |G_2|) = 1$ and $\mathscr{D}_i$ $(i = 1, 2)$ are symmetric (resp. self-Petrie-dual, reflexible), then so is $\mathscr{D}_1 \vee \mathscr{D}_2$ [13, Proposition 10].

The existence of multiple edges of a regular dessin $\mathscr{D} = (G, x, y)$ depends on the non-triviality of the central subgroup $K = \langle x\rangle \cap \langle y\rangle$ in G. The subgroup $K \trianglelefteq G$ determines a unique regular dessin $\bar{\mathscr{D}} = (\bar{G}, \bar{x}, \bar{y})$ with the simple underlying graph, where $\bar{G} = G/K$, $\bar{x} = xK$ and $\bar{y} = yK$. Such a quotient dessin will be referred to as the *shadow dessin* of $\mathscr{D}$. Regular dessins with simple underlying graphs will be called *simple* regular dessins.

Proposition 1 [13, Corollary 5] *The shadow dessin of a symmetric dessin is symmetric, and the shadow dessin of a reflexible dessin is reflexible.*

Proposition 2 *Let $\mathscr{D} = (G, x, y)$ be a regular dessin. If $\mathscr{D}$ is self-Petrie-dual, then the underlying graph of $\mathscr{D}$ has multiplicity at most two.*

Proof Assume that the underlying graph of $\mathscr{D}$ has multiplicity m. Then $o(x) = mr$ and $o(y) = ms$ for some positive integers r, s, and $\langle x\rangle \cap \langle y\rangle = \langle x^r\rangle = \langle y^s\rangle$. It follows that $y^s = x^{r\varepsilon}$ for some ε coprime to m. Since $\mathscr{D}$ is self-Petrie-dual, the assignment $\pi : x \mapsto x^{-1}, y \mapsto y$ extends to an automorphism of G. Hence

$$y^s = \pi(y^s) = \pi(x^{r\varepsilon}) = x^{-r\varepsilon}.$$

Combining this with the preceding relation we get $y^{2s} = 1$. Therefore $m \mid 2$. □

Let $\mathscr{D} = (G, x, y)$ be a symmetric dessin of type (m, m, n). Another set of operations on dessins preserving the automorphism group was introduced by Wilson [28]. The *jth Wilson's operation* H_j transforms $\mathscr{D}$ to a symmetric dessin $H_j(\mathscr{D}) = (G, x^j, y^j)$ where j is coprime to m.

Proposition 3 *Let $\mathscr{D}_i$ be two symmetric dessins of type (m_i, m_i, n_i) $(i = 1, 2)$, and let H_j be the jth Wilson's operation.*

1. *If $m_2 \mid m_1$, then for each number j such that $\gcd(j, m_2) = 1$ there is a number j' such that $j' \equiv j \pmod{m_1}$ and $\gcd(j', m_1) = 1$.*
2. *$\mathscr{D}_1$ covers $\mathscr{D}_2$ if and only if $H_j(\mathscr{D}_1)$ covers $H_j(\mathscr{D}_2)$ where j is coprime to m_1.*

Proof The first part follows from [14, Lemma 1]. To prove the second we let $\mathscr{D}_i = (G_i, x_i, y_i)$ $(i = 1, 2)$. If $\mathscr{D}_1$ covers $\mathscr{D}_2$, then the assignment $\phi : x_1 \mapsto x_2, y_1 \mapsto y_2$ extends to an epimorphism from G_1 onto G_2. Since $o(x_1) = o(y_1) = m_1$ and $o(x_2) = o(y_2) = m_2$, we have $m_2 \mid m_1$. Since j is coprime to m_1, j is also coprime to m_2. So we have $H_j(\mathscr{D}_i) = (G_i, x_i^j, y_i^j)$ $(i = 1, 2)$. Observe that $\phi(x_1^j) = (\phi(x_1))^j = x_2^j$ and $\phi(y_1^j) = (\phi(y_1))^j = y_2^j$. It follows that the assignment $x_1^j \mapsto x_2^j, y_1^j \mapsto y_2^j$ determines the same epimorphism from G_1 onto G_2 as ϕ, and hence $H_j(\mathscr{D}_1)$ covers $H_j(\mathscr{D}_2)$. Conversely, if $H_j(\mathscr{D}_1)$ covers $H_j(\mathscr{D}_2)$, then $\mathscr{D}_1 = H_k(H_j(\mathscr{D}_1))$ covers $\mathscr{D}_2 = H_k(H_j(\mathscr{D}_2))$ where $kj \equiv 1 \pmod{m_1}$. □

4 Nilpotent Regular Dessins

In this section we recall some known facts on nilpotent groups and nilpotent regular dessins. Let G be a finite group. The upper central series for G is the series

$$1 = Z_0 \leq Z_1(G) \leq Z_2(G) \leq \cdots \leq Z_i(G) \leq Z_{i+1}(G) \leq \cdots,$$

where $Z_i(G)$ $(i \geq 1)$ is defined by the rule: $Z_i(G)/Z_{i-1}(G)$ is the center of $G/Z_{i-1}(G)$. A group G is *nilpotent* if its upper central series contains G. It is well known that in a finite nilpotent group G, the upper central series has finite length c. The number c is called the *class* of G, and is denoted by $c(G)$.

Lemma 1 [15, Chap. III, Lemma 1.11] *Let $G = \langle x, y\rangle$ be a group. Then $G' = \langle [x, y]^g \mid g \in G\rangle$.*

Lemma 2 [15, Chap. III, Lemma 1.3] *Let G be a nilpotent group of class two. Then, for any $x, y \in G$,*

$$[x^n, y] = [x, y^n] = [x, y]^n \quad \textit{and} \quad (xy)^n = x^n y^n [y, x]^{\binom{n}{2}},$$

where $n \geq 1$ is a positive integer.

A regular dessin whose automorphism group is a p-group will be called a *regular p-dessin*. It is well known that every nilpotent group is a direct product of its Sylow subgroups. So if $\mathscr{D} = (G, x, y)$ is a nilpotent regular dessin, then for each prime factor p of $|G|$, $G = G_p \times K$ where G_p denotes the Sylow p-subgroup of G, supplemented by K. The quotient $\mathscr{D}_p = (G/K, xK, yK)$ is a regular p-dessin with $\mathrm{Aut}(\mathscr{D}_p) \cong G_p$. It will be the called the *Sylow p-dessin* of $\mathscr{D}$.

Lemma 3 [13, Theorem 12] *Every nilpotent regular dessin $\mathcal{D} = (G, x, y)$ is uniquely decomposed into a parallel product of its Sylow p-dessins $\mathcal{D}_p$ where p ranges over all distinct prime factors of $|G|$. Moreover, $\mathcal{D}$ possesses an external symmetry if and only if so does every Sylow p-dessin of $\mathcal{D}$.*

For nilpotent regular dessins with multiple edges we have

Lemma 4 [13, Proposition 11] *Let $\mathcal{D} = (G, x, y)$ be a nilpotent regular dessin, and let $\bar{\mathcal{D}} = (\bar{G}, \bar{x}, \bar{y})$ be its shadow dessin. If $c(G) = c \geq 2$, then $c - 1 \leq c(\bar{G}) \leq c$. Conversely, if $c(\bar{G}) = c$, then $c \leq c(G) \leq c + 1$.*

We end this section by a classification of abelian p-dessins proved in [13].

Lemma 5 [13, Theorem 19, Corollary 21] *Let p be a prime, and let a and b be integers, $0 \leq b \leq a$. Then each regular dessin $\mathcal{D} = (G, x, y)$ with $G \cong \mathbb{Z}_{p^a} \times \mathbb{Z}_{p^b}$ is determined by the presentation*

$$G = \langle x, y \mid x^{p^a} = y^{p^{b+c}} = [x, y] = 1, y^{p^b} = x^{ep^{a-c}} \rangle, \tag{1}$$

where

$$0 \leq c \leq a - b \quad and \quad e \in \mathbb{Z}_{p^c}^*. \tag{2}$$

Moreover, up to the duality swapping the black and white vertices, the isomorphism classes of regular dessins $\mathcal{D}$ with $\mathrm{Aut}(\mathcal{D}) \cong \mathbb{Z}_{p^a} \times \mathbb{Z}_{p^b}$ *are in one-to-one correspondence with the integer pairs (c, e) satisfying* (2).

Finally, the dessin $\mathcal{D}$ is symmetric if and only if $c = a - b$ and $e^2 \equiv 1 \pmod{p^c}$, and it is simple if and only if $a = b$.

5 Simple Symmetric *p*-Dessins of Class Two

Recall that a simple regular dessin is a regular dessin with simple underlying graph. In this section we classify the simple symmetric p-dessins of class two. In order to state the result we let $G(p; a, b, c)$ be a p-group with a presentation

$$\langle x, y \mid x^{p^a} = y^{p^a} = z^{p^{a+b-c}} = [x, z] = [y, z] = 1, z^{p^b} = x^{-p^c} y^{p^c}, z = [x, y] \rangle. \tag{3}$$

Theorem 1 *Let p be a prime, and let $\mathcal{D}$ be a simple symmetric p-dessin of class two, then $\mathcal{D} \cong (G(p; a, b, c), x^\delta, y^\delta)$ where*

$$\max(0, 1 + c - a) \leq b \leq c \leq a \leq 2c - b \quad and \quad \delta \in \mathbb{Z}_{p^{a-c}}^*. \tag{4}$$

Proof Let $\mathcal{D} = (G, x_1, y_1)$ be a symmetric p-dessin of class two. Then the mapping $\tau : x_1 \mapsto y_1, y_1 \mapsto x_1$ extends to an automorphism of G. Assume that $o(x_1) = p^a$

where $a \geq 0$. Then $o(y_1) = o(\tau(x_1)) = o(x_1) = p^a$. Define $z_1 = [x_1, y_1]$ and assume that $o(z_1) = p^d$. Since $c(G) = 2$, we have $G' \leq Z(G)$, and hence by Lemma 1 $G' = \langle z_1 \rangle \cong \mathbb{Z}_{p^d}$ where $d \geq 1$.

Note that $\tau(z_1) = [\tau(x_1), \tau(y_1)] = z_1^{-1}$. We have $\tau(\langle x_1 \rangle \cap \langle z_1 \rangle) = \langle y_1 \rangle \cap \langle z_1 \rangle$. So $\langle x_1 \rangle \cap \langle z_1 \rangle$ and $\langle y_1 \rangle \cap \langle z_1 \rangle$, being subgroups of the same order in the cyclic group $G' = \langle z_1 \rangle$, are identical. It follows that

$$\langle x_1 \rangle \cap \langle z_1 \rangle = (\langle x_1 \rangle \cap \langle z_1 \rangle) \cap (\langle y_1 \rangle \cap \langle z_1 \rangle) = \langle x_1 \rangle \cap \langle y_1 \rangle \cap \langle z_1 \rangle \subseteq \langle x_1 \rangle \cap \langle y_1 \rangle.$$

By our assumption $\mathscr{D}$ is simple, so $\langle x_1 \rangle \cap \langle y_1 \rangle = 1$. Therefore $\langle x_1 \rangle \cap \langle z_1 \rangle = 1$. By Lemma 2, $z_1^{p^{p^a}} = [x_1^{p^{p^a}}, y_1] = 1$, we get $d \leq a$. Let $N = \langle x_1, z_1 \rangle$. Since $z_1^{y_1} = z_1$ and $x_1^{y_1} = x_1 z_1$, we have $N \trianglelefteq G$ and $G/N = \langle y_1 N \rangle$. Assume that $G/N \cong \mathbb{Z}_{p^c}$ where $0 \leq c \leq a$. Then $y_1^{p^c} = x_1^{\kappa p^l} z_1^{\mu p^b}$ for some integers l, b, μ, κ where $0 \leq l \leq a$, $0 \leq b \leq d$, $\mu \in \mathbb{Z}^*_{p^{d-b}}$ and $\kappa \in \mathbb{Z}^*_{p^{a-l}}$. Since $[x_1, y_1^{p^c}] = [x_1, x_1^{\kappa p^l} z_1^{\mu p^b}] = 1$, we have $y_1^{p^c} \in Z(G)$. Note that $Z(G)$ char G, we get $x_1^{p^c} = \tau(y_1^{p^c}) \in Z(G)$.

We proceed to prove that $c = l$. Without loss of generality we suppose to the contrary that $c < l$. The relation $y_1^{p^c} = x_1^{\kappa p^l} z_1^{\mu p^b}$ can be rewritten as the form $z_1^{\mu p^b} = x_1^{-\kappa p^l} y_1^{p^c}$. It follows that

$$z_1^{-\mu p^b} = (z_1^{\mu p^b})^\tau = (x_1^{-\kappa p^l} y_1^{p^c})^\tau = y_1^{-\kappa p^l} x_1^{p^c}.$$

By equating these two relations, we get $x_1^{-\kappa p^l} y_1^{p^c} = x_1^{-p^c} y_1^{\kappa p^l}$, or equivalently,

$$x_1^{(\kappa p^{l-c}-1)p^c} = y_1^{(1-\kappa p^{l-c})p^c}. \tag{5}$$

Recall that $\langle x_1 \rangle \cap \langle y_1 \rangle = 1$. So by the identity (5) we have $c = a$. This contradicts the assumption that $c < l \leq a$. Therefore $z_1^{\mu p^b} = x_1^{-\kappa p^c} y_1^{p^c}$, and the identity (5) is of the form $x_1^{(\kappa-1)p^c} = y_1^{(1-\kappa)p^c}$. Since $\langle x_1 \rangle \cap \langle y_1 \rangle = 1$, we get $k \equiv 1 \pmod{p^{a-c}}$. Thus,

$$z_1^{\mu p^b} = x_1^{-p^c} y_1^{p^c}. \tag{6}$$

The identity implies that $o(z_1^{\mu p^b}) = o(x_1^{-p^c} y_1^{p^c}) = p^{a-c}$. So $o(z) = p^{a+b-c}$, and hence $d = a + b - c$. Therefore G has a presentation

$$\langle x_1, y_1 | x_1^{p^a} = y_1^{p^a} = z_1^{p^{a+b-c}} = [x_1, z_1] = [y_1, z_1] = 1, z_1^{\mu p^b} = x_1^{-p^c} y_1^{p^c}, z_1 = [x_1, y_1] \rangle.$$

We show that the parameters a, b and c satisfy condition (4). First we deduce from the relation $x_1^{y_1} = x_1 z_1$ that $x_1 = x_1^{x_1^{p^c} z_1^{\mu p^b}} \overset{(6)}{=} x_1^{y_1^{p^c}} = x_1 z_1^{p^c}$. So $z_1^{p^c} = 1$, and hence $a + b - c \leq c$. This is equivalent to $b - c \leq c - a$ or $a \leq 2c - b$. Since $c \leq a$, we get $b \leq c \leq a \leq 2c - b$. Recall that $1 \leq d = b + a - c$, we get $1 + c - a \leq b$. Combining this with the inequality $0 \leq b$ yields $\max(0, 1 + c - a) \leq b$.

Finally, let δ be the modular inverse of μ in $\mathbb{Z}_{p^{a-c}}$, that is, $\delta\mu \equiv 1 \pmod{p^{a-c}}$. It is straightforward to verify that the assignment $x_1 \mapsto x^\delta$, $y_1 \mapsto y^\delta$ extends to a group isomorphism from G onto $G(p; a, b, c)$. Hence $\mathcal{D} \cong (G(p; a, b, c), x^\delta, y^\delta)$. □

Corollary 1 *Let G(p; a, b, c) be the group as above, then the following statements hold true:*

1. *The group G(p; a, b, c) is a p-group of class two and order* p^{2a+b}.
2. *Two groups* $G(p_i; a_i, b_i, c_i)$ $(i = 1, 2)$ *are isomorphic if and only if* $p_1 = p_2$, $a_1 = a_2$, $b_1 = b_2$ *and* $c_1 = c_2$.
3. *The group G(p; a, b, c) underlies* $\varphi(p^{a-c})$ *simple symmetric p-dessins where* φ *is the Euler's totient function, and these dessins form a single orbit under the Wilson's operation.*

Proof Let $G = G(p; a, b, c)$. In the proof of Theorem 1 we have seen that $N = \langle x_1, z_1 \rangle \trianglelefteq G$ and $G/N = \langle y_1 N \rangle$. So $|G| = |N||\langle y_1 N \rangle| = p^{a+d+c} = p^{2a+b}$. By the presentation of G it is clear that G has class two.

Moreover, since $G' = \langle z \rangle \cong \mathbb{Z}_{p^{a+b-c}}$ and $G/G' \cong \mathbb{Z}_{p^a} \times \mathbb{Z}_{p^c}$, distinct parameters correspond to non-isomorphic groups.

Finally, we have also shown that if $\mathcal{D}$ is a simple symmetric p-dessin of class two, then $\mathcal{D} \cong (G; x^\delta, y^\delta)$ where $\delta \in \mathbb{Z}^*_{p^{a-c}}$. It is clear that distinct parameters $\delta \in \mathbb{Z}^*_{p^{a-c}}$ give rise to non-isomorphic regular dessins. Hence there are precisely $\varphi(p^{a-c})$ simple symmetric p-dessins $\mathcal{D}$ with $\mathrm{Aut}(\mathcal{D}) \cong G$. Let H_δ be the δth Wilson's operation. Then H_δ transforms the regular dessin (G, x, y) to (G, x^δ, y^δ), and hence the regular dessins (G, x^δ, y^δ) $(\delta \in \mathbb{Z}^*_{p^{a-c}})$ form a single orbit under the Wilson's operation. □

Corollary 2 *The type and genus of the simple symmetric p-dessins* $\mathcal{D}$ *of class two with* $\mathrm{Aut}(\mathcal{D}) \cong G(p; a, b, c)$ *determined in Theorem 1 are given in Table 1.*

Proof Let $G = G(p; a, b, c)$. In the proof of Theorem 1 we have shown that if $\mathcal{D}$ is a simple symmetric p-dessin of class two, then $\mathcal{D} \cong (G, x_1, y_1)$ where $x_1 = x^\delta$ and $y_1 = y^\delta$. To derive the type and genus of the dessin it suffices to evaluate the order of $x_1 y_1$.

Assume that $o(x_1 y_1) = p^n$. Then by Lemma 2 we have $(x_1 y_1)^{p^n} = x_1^{p^n} y_1^{p^n} z_1^{-\binom{p^n}{2}}$, so $x_1^{p^n} y_1^{p^n} z_1^{-\binom{p^n}{2}} = 1$, and hence $y_1^{p^n} = x_1^{-p^n} z_1^{\binom{p^n}{2}} \in N = \langle x_1, z_1 \rangle$. By the minimality

Table 1 Type and genus of the simple symmetric dessins

Classes	Subclass	Type	Genus
$p > 2$	$1 \le b \le c \le a \le 2c - b$	(p^a, p^a, p^a)	$1 + p^{a+b}(p^a - 3)/2$
$p = 2$	$1 \le b = c = a$	$(2^a, a^a, a^{a+1})$	$1 + 2^{a+b-2}(2^{a+1} - 5)$
$p = 2$	$2 \le b + 1 \le c = a$	$(2^a, 2^a, 2^a)$	$1 + 2^{a+b-1}(2^a - 3)$
$p = 2$	$1 \le b + 1 = c = a - 1$	$(2^a, 2^a, 2^{a-1})$	$1 + 2^{a+b+1}(2^{a-2} - 1)$
$p = 2$	$2 \le b + 2 \le c < a \le 2c - b$	$(2^a, 2^a, 2^a)$	$1 + 2^{a+b-1}(2^a - 3)$

of c (see the proof of Theorem 1) we have $c \leq n$. Recall that $z_1^{\mu p^b} = x_1^{-p^c} y_1^{p^c}$, so $x_1^{p^c} = y_1^{p^c} z_1^{-\mu p^b}$. Then by Lemma 2 we have

$$(x_1 y_1)^{p^c} = x_1^{p^c} y_1^{p^c} z_1^{-\binom{p^c}{2}} = (y_1^{p^c} z_1^{-\mu p^b}) y_1^{p^c} z_1^{-\binom{p^c}{2}} = y_1^{2p^c} z_1^{-(\binom{p^c}{2}+\mu p^b)}. \tag{7}$$

Let $o(y_1^{2p^c}) = p^s$ and $o(z_1^{\binom{p^c}{2}+\mu p^b}) = p^t$. Since $\langle y_1 \rangle \cap \langle z_1 \rangle = 1$ and y_1 commutes with z_1, we have $n = c + \max(s, t)$. Recall that $o(y_1) = p^a$ and $o(z_1) = p^{a+b-c}$ where $b \leq c \leq a$.

If p is an odd prime, then $s = a - c$ and $t \leq a - c$. So in this case $n = c + s = a$.

If $p = 2$, then $z_1^{\binom{2^c}{2}+\mu 2^b} = z^{2^{c-1}(2^c-1)+\mu 2^b}$. By the condition (4) we distinguish five subcases. (i) If $b = c = a$, then $s = 0$ and $t = 1$, and hence $n = a + 1$. (ii) If $b + 1 \leq c = a$, then $s = 0$ and $t = 0$, and hence $n = a$. (iii) If $b = c < a$, then by (4) we have $a + b \leq 2c = 2b$, so $a \leq b$; since $b \leq c \leq a$, we have $a = b = c$, a contradiction. Hence this case cannot happen. (iv) If $b + 1 = c < a$, then by (4) we have $b + a - c \leq c = b + 1$, and hence $a \leq c + 1$. Since $a > c$, $b + 1 = c = a - 1$. Therefore $s = 0$ and $t = 0$. Consequently $n = a - 1$. (v) If $b + 2 \leq c < a$, then $s = a - c - 1$ and $t = a - c$. Hence $n = a$. □

Corollary 3 *Let $\mathscr{D}$ be a simple symmetric p-dessin from Theorem 1. If $\mathscr{D}$ is reflexible, then it isomorphic to one of the following regular dessins:*

1. *$\mathscr{D}_1(p, a, b) = (G, x, y)$ where p is an odd prime, $1 \leq b \leq a$, and*

$$G = \langle x, y \mid x^{p^a} = y^{p^a} = z^{p^b} = [x, z] = [y, z] = 1, z = [x, y]\rangle.$$

2. *$\mathscr{D}_2(a, b) = (G, x, y)$ where $1 \leq b \leq a$ and*

$$G = \langle x, y \mid x^{2^a} = y^{2^a} = z^{2^b} = [x, z] = [y, z] = 1, z = [x, y]\rangle.$$

3. *$\mathscr{D}_3(a, b) = (G, x, y)$ where $0 \leq b \leq a - 2$, and*

$$G = \langle x, y \mid x^{2^a} = y^{2^a} = z^{p^{b+1}} = [x, z] = [y, z] = 1, z^{2^b} = x^{2^{a-1}} y^{2^{a-1}}, z = [x, y]\rangle.$$

Moreover, the reflexible and symmetric simple p-dessins of class two are all self-Petrie-dual.

Proof Let $\mathscr{D} = (G, x_1, y_1)$ where $G = G(p; a, b, c)$ and $x_1 = x^\delta, y_1 = y^\delta$. If $\mathscr{D}$ is reflexible, then the assignment $\iota : x_1 \mapsto x_1^{-1}, y_1 \mapsto y_1^{-1}$ extends to an automorphism of G. We have $\iota(z_1) = [\iota(x_1), \iota(y_1)] = [x_1^{-1}, y_1^{-1}] = z_1$. Recall that $z_1^{\mu p^b} = x_1^{-p^c} y_1^{p^c}$, so $z_1^{\mu p^b} = \iota(z_1^{\mu p^b}) = \iota(x_1^{-p^c} y_1^{p^c}) = x_1^{p^c} y_1^{-p^c}$. Equating these two relations yields $x_1^{2p^c} = y_1^{-2p^c}$. Recall that $\langle x_1 \rangle \cap \langle y_1 \rangle = 1$. Therefore, if $p > 2$ then $a = c$, and if $p = 2$ then either $c = a$ or $c = a - 1$, giving the three families of dessins listed above. Conversely, it is straightforward to verify that the three families of regular dessins are all reflexible.

Moreover, if the symmetric p-dessin (G, x_1, y_1) is self-Petrie-dual, then it is also reflexible. Checking the reflexible dessins it is easy to verify that they are all self-Petrie-dual. The details are left as an exercise to the reader. □

Corollary 4 *Let p be a prime and $a \geq 1$, and let ξ denote the number of isomorphism classes of p^a-valent simple symmetric p-dessins of class two. Then*

$$\xi = \begin{cases} \frac{p+1}{p-1}(p^{\frac{a}{2}} - 1), & \text{if } a \text{ is even,} \\ \frac{2}{p-1}(p^{\frac{a+1}{2}} - 1) - 1, & \text{if } a \text{ is odd.} \end{cases}$$

Proof By Theorem 1 and Corollary 1, the number ξ is equal to the number of triples (b, c, δ) where b and c satisfy (4) and $\delta \in \mathbb{Z}^*_{p^{a-c}}$. If $a = c$, then these conditions reduce to $1 \leq b \leq c = a$ and $\delta = 1$. If $a > c$, then by the relations $a \leq 2c - b$ and $b \geq \max(0, 1 + c - a) = 0$ we have $2c \geq a + b \geq a$, and hence (4) reduces to $a/2 \leq c \leq a$ and $0 \leq b \leq 2c - a$. Let $l = \lceil a/2 \rceil$, we have $l = a/2$ if a is even, and $l = (a + 1)/2$ if a is odd. Then

$$\begin{aligned} \xi &= a + \sum_{c=l}^{a-1} (2c - a + 1)\varphi(p^{a-c}) \\ &= a + 2\sum_{c=l}^{a-1} c\varphi(p^{a-c}) - (a - 1)\sum_{c=l}^{a-1} \varphi(p^{a-c}) \\ &= a + 2(p - 1)\sum_{c=l}^{a-1} cp^{a-c-1} - (a - 1)(p^{a-l} - 1), \\ &= \frac{2(p^{a-l} - 1)}{p - 1} + (2l - a + 1)p^{a-l} - 1. \end{aligned}$$

The value of ξ is obtained by substitution for l. □

Remark 1 A triple (G, x, y) is called an *n-isobicyclic triple* if G is a finite group, $G = \langle x \rangle \langle y \rangle$ where $o(x) = o(y) = n$, and the assignment $\tau : x \mapsto y, y \mapsto x$ extends to an automorphism of G. The notion of n-isobicyclic triples (G, x, y) was employed by Jones et al. to classify regular embeddings of the complete bipartite graphs $K_{n,n}$, where G is the index-two subgroup of colour- and orientation-preserving automorphisms, see [20]. Each such a map defines a simple symmetric dessin $\mathscr{D} = (G, x, y)$. If $n = p^a$ is a prime power, then the dessin is a simple symmetric p-dessin. In what follows we determine which of these dessins are of class two.

Let $\mathscr{D} = (G, x^\delta, y^\delta)$ be the simple symmetric p-dessins classified in Theorem 1 where $G = G(p; a, b, c)$. Since $o(x^\delta) = o(y^\delta) = p^a$ and $\langle x^\delta \rangle \cap \langle y^\delta \rangle = 1$, $|\langle x^\delta \rangle \langle y^\delta \rangle| = |\langle x^\delta \rangle||\langle y^\delta \rangle| = p^{2a}$. By Corollary 1 $|G| = p^{2a+b}$. It follows that (G, x^δ, y^δ) is a p^a-isobicyclic triple if and only if $b = 0$, in which case it corresponds to a regular embedding of the complete bipartite graph K_{p^a,p^a}. To give the identification, we recall that Wilson's operation preserves the underlying bipartite graph, so by Corollary 1 it suffices to verify the symmetric dessin (G, x, y). Since $b = 0$, by the presentation we have $z = x^{-p^c} y^{p^c}$. By (4) we have $1 \le c+1 \le a \le 2c$. Let $h = x^{-1}y$ and $g = x$. Then

$$h^g = hx^{p^c} y^{-p^c}.$$

If p is odd, then $h^{p^c} = (x^{-1}y)^{p^c} = x^{-p^c} y^{p^c} z^{\binom{p^c}{2}} = x^{-p^c} y^{p^c}$. So $h^g = h^{1-p^c}$, and hence G is metacyclic. The associated symmetric p-dessins correspond to the subclass $\mathcal{M}(p; e, f)$ $(e/2 \le f < e)$ of regular embeddings of K_{p^e,p^e} [19].

On the other hand, if $p = 2$ and $a \le 2c-1$, then $h^{2^c} = x^{-2^c} y^{2^c} z^{\binom{2^c}{2}} = x^{-2^c} y^{2^c}$. So we have $h^g = h^{1-2^c}$, and hence G is metacyclic. While if $p = 2$ and $c+2 \le a = 2c$, then

$$z^{2^{c-1}} = (x^{-2^c} y^{2^c})^{2^{c-1}} = x^{-2^{2c-1}} y^{2^{2c-1}} = (x^{-2^{c+1}} y^{2^{c+1}})^{2^{c-2}} = h^{2^{2c-1}},$$

and hence

$$h^{2^c} = (x^{-1}y)^{2^c} = x^{-2^c} y^{2^c} z^{2^{c-1}(2^c-1)} = x^{-2^c} y^{2^c} h^{2^{2c-1}(2^c-1)} = x^{-2^c} y^{2^c} h^{-2^{2c-1}}.$$

Consequently $h^g = hx^{2^c} y^{-2^c} = h^{1-2^c(1+2^{c-1})}$. Therefore G is metacyclic as well. These dessins cover the subclass $\mathcal{M}(2; e, f)$ $(e/2 \le f < e)$ of regular embeddings of $K_{2^e,2^e}$ with a metacyclic subgroup of colour- and orientation-preserving automorphisms classified in [9].

Finally, if $p = 2$ and $c+1 = a = 2c$, then $c = 1$ and $a = 2$. In this case, the group G is non-metacyclic, and the dessin corresponds to a regular embedding of $K_{4,4}$ into the torus, denoted $\mathcal{N}(4; 0, 0)$ in [10, Theorem1.1].

6 Symmetric p-dessins of Class Two with Multiple Edges

By Lemma 4 the shadow dessin $\bar{\mathcal{D}}$ of a symmetric p-dessin $\mathcal{D}$ of class two is a simple symmetric p-dessin $\mathcal{D}$ of class one or two. If $\bar{\mathcal{D}}$ is of class one, then it is an abelian simple symmetric p-dessin from Lemma 5. In the second case, by Theorem 1 we have $\bar{\mathcal{D}} \cong (G, x^\delta, y^\delta)$ where $G = G(p; a, b, c)$ and a, b and c satisfy (4) and $\delta \in \mathbb{Z}^*_{p^{a-c}}$. In what follows we shall distinguish these two cases.

In order to formulate our first result in this section, we let $L(p; a, b, m)$ be a group with a presentation

$$L(p; a, b, m) = \langle x, y \mid x^{p^{a+m}} = y^{p^{a+m}} = 1, y^{p^a} = x^{p^a}, [x, y] = x^{p^{a+b}} \rangle. \tag{8}$$

Theorem 2 *Let p be a prime, $a \ge 0$ and $m \ge 1$, and let $\mathcal{D}$ be a class-two symmetric p-dessin of multiplicity p^m. If the shadow dessin of $\mathcal{D}$ is the p^a-valent abelian simple symmetric dessin, then $\mathcal{D} \cong (L(p; a, b, m), x^{\xi\gamma}, y^\xi)$ where*

$$0 \le b < m \le a + b, \tag{9}$$

and $\gamma \in \mathbb{Z}^*_{p^m}$, $\zeta \in \mathbb{Z}^*_{p^{m-b}}$ *satisfy the following conditions*

$$\gamma^2 \equiv 1 \pmod{p^m} \quad and \quad \gamma \equiv -1 \pmod{p^{m-b}}. \tag{10}$$

Proof Let $\mathscr{D} = (G, x_1, y_1)$ and $K = \langle x_1 \rangle \cap \langle y_1 \rangle$. Then by the assumption $\bar{\mathscr{D}} = (\bar{G}, \bar{x}, \bar{y})$ is a p^a-valent abelian simple symmetric p-dessin where $G = G/K$, $\bar{x} = xK$ and $\bar{y} = yK$. By Lemma 5 we have

$$\bar{G} = \langle \bar{x}_1, \bar{y}_1 \mid \bar{x}_1^{p^a} = \bar{y}_1^{p^a} = [\bar{x}_1, \bar{y}_1] = \bar{1} \rangle.$$

So G, being a cyclic extension of $\bar{G}$ by $K = \langle x_1 \rangle \cap \langle y_1 \rangle \cong \mathbb{Z}_{p^m}$, has a presentation

$$\langle x_1, y_1 \mid x_1^{p^{a+m}} = y_1^{p^{a+m}} = 1, y_1^{\gamma p^a} = x_1^{p^a}, [x_1, y_1] = x_1^{\zeta p^{a+b}} \rangle,$$

where $0 \le b \le m$, $\gamma \in \mathbb{Z}^*_{p^m}$ and $\zeta \in \mathbb{Z}^*_{p^{m-b}}$.

By the assumption the group G is of class two, so $[x_1, y_1] \ne 1$, and hence $b < m$. Clearly $[x_1, y_1] \in Z(G)$. So by Lemma 2 we have $1 = [x_1, y_1^{p^a}] = [x_1, y_1]^{p^a} = x_1^{\zeta p^{2a+b}}$, which implies $m \le a + b$.

Since $\mathscr{D}$ is symmetric, the assignment $\tau : x_1 \mapsto y_1, y_1 \mapsto x_1$ extends to an automorphism of G. So we have $x_1^{\gamma p^a} = \tau(y_1^{\gamma p^a}) = \tau(x_1^{p^a}) = y_1^{p^a}$. Combining this relation with $y_1^{\gamma p^a} = x_1^{p^a}$ yields $x_1^{p^a} = x_1^{\gamma^2 p^a}$, which implies that $\gamma^2 \equiv 1 \pmod{p^m}$. Similarly we have $[y_1, x_1] = \tau([x_1, y_1]) = \tau(x_1^{\zeta p^{a+b}}) = y_1^{\zeta p^{a+b}}$. It follows that $y_1^{\zeta p^{a+b}} = [y_1, x_1] = x_1^{-\zeta p^{a+b}} = y_1^{-\zeta \gamma p^{a+b}}$, or equivalently $y_1^{\zeta(\gamma+1)p^{a+b}} = 1$. Since $b \le m$ and $\zeta \in \mathbb{Z}^*_{p^{m-b}}$ we get $\gamma \equiv -1 \pmod{p^{m-b}}$.

Finally, it is straightforward to verify that the assignment $G \to L(p; a, b, m)$, $x_1 \mapsto x^{\zeta\gamma}$, $y_1 \mapsto y^{\zeta}$ extends to a group isomorphism. So $\mathscr{D} \cong (L(p; a, b, m), x^{\zeta\gamma}, y^{\zeta})$. □

Corollary 5 *Let* $L(p; a, b, m)$ *be the group defined as above, then the following holds true:*

1. *The group* $L(p; a, b, m)$ *is a metacylcic* p*-group of class two and order* p^{2a+m}.
2. *Two groups* $L(p_i; a_i, b_i, m_i)$ $(i = 1, 2)$ *are isomorphic if and only if* $p_1 = p_2$, $a_1 = a_2$, $b_1 = b_2$ *and* $m_1 = m_2$.
3. *Let* ξ *be the number of symmetric* p*-dessins* $\mathscr{D}$ *with multiplicity* p^m *such that* $\mathrm{Aut}(\mathscr{D}) \cong L(p; a, b, m)$, *then*

$$\xi = \begin{cases} \varphi(p^{m-b}), & \text{if } p \text{ is odd}, \\ 1, & \text{if } p = 2 \text{ and } m = 1, \\ 2, & \text{if } p = 2 \text{ and } m = 2, \\ 4\varphi(2^m), & \text{if } p = 2,\ m \ge 3 \text{and } b = 0, \\ 2\varphi(2^{m-b}), & \text{if } p = 2 \text{ and } 1 \le b \le m-2, \\ 4, & \text{if } p = 2 \text{ and } b + 1 = m \ge 3. \end{cases}$$

Proof Let $G = L(p; a, b, m)$. Since $[x, y] \in Z(G)$, G has class two. Since $x^y = x^{1+p^{a+b}}$ and $G = \langle x, y\rangle$, G is metacyclic.

Observe that $G' \cong \mathbb{Z}_{p^{m-b}}$ and $G/G' \cong \mathbb{Z}_{p^a} \times \mathbb{Z}_{p^{a+b}}$. So distinct triples (a, b, m) correspond to non-isomorphic groups.

Moreover, for fixed a, b and m it is straightforward to verify that two regular dessins $(G, x^{\zeta_i\gamma_i}, y^{\zeta_i})$ $(i = 1, 2)$ are isomorphic if and only if $\gamma_1 \equiv \gamma_2 \pmod{p^m}$ and $\zeta_1 \equiv \zeta_2 \pmod{p^{m-b}}$. So the number ξ is equal to the number of pairs (γ, ζ) satisfying (10) where $\gamma \in \mathbb{Z}^*_{p^m}$ and $\zeta \in \mathbb{Z}^*_{p^{m-b}}$. The value of ξ is obtained by solving the congruences in (10).

The type and genus of the dessins from Theorem 2 are summarized in the following corollary. The proof is easy and we leave it as an exercise to the reader.

Corollary 6 *Let $\mathcal{D} = (L(p; a, b, m), x^{\zeta\gamma}, y^{\zeta})$. If p is odd, then $\mathcal{D}$ is of type (p^{a+m}, p^{a+m}, p^a) and genus $\frac{1}{2}p^a(p^{a+m} - p^m - 2) + 1$, while if $p = 2$, then it has type $(2^{a+m}, 2^{a+m}, 2^{a+m}/d)$ and genus $1 + 2^{a-1}(2^{a+m} - d - 2)$ where $d = \gcd(1 + \gamma + \zeta 2^{a+b-1}(1 - 2^a), 2^m)$.*

Corollary 7 *Let $\mathcal{D} = (L(p; a, b, m), x^{\zeta\gamma}, y^{\zeta})$.*

1. *If p is odd, then $\mathcal{D}$ is chiral,*
2. *If $p = 2$ and $\mathcal{D}$ is reflexible, then $\mathcal{D} \cong (L(2; a, m-1, m), x^{\gamma}, y)$ where $m \geq 1$, $a \geq 1$ and $\gamma^2 \equiv 1 \pmod{2^m}$.*
3. *If $\mathcal{D}$ is self-Petrie-dual, then $\mathcal{D} \cong (L(2; a, 0, 1), x, y)$.*

Proof Let $L = L(p; a, b, m)$ and $x_1 = x^{\zeta\gamma}$, $y_1 = y^{\zeta}$. If $\mathcal{D}$ is reflexible, then the assignment $\iota : x_1 \mapsto x_1^{-1}, y_1 \mapsto y_1^{-1}$ extends to an automorphism of L. So by the relation $[x_1, y_1] = x_1^{\zeta p^{a+b}}$ we have $[x_1, y_1] = \iota([x_1, y_1]) = \iota(x_1^{\zeta p^{a+b}}) = x_1^{-\zeta p^{a+b}}$. Hence $x_1^{2\zeta p^{a+b}} = 1$.

If p is odd, then we have $b = m$, it contradicts to (9). So in the case that p is odd the dessins $\mathcal{D}$ are all chiral.

If $p = 2$, then $x_1^{\zeta 2^{a+b+1}} = 1$. So $b = m - 1$ and hence $\zeta = 1$. Therefore $\mathcal{D} \cong (L(2; a, m-1, m), x^{\gamma}, y)$. Conversely it is straightforward to verify that the dessin $(L(2; a, m-1, m), x^{\gamma}, y)$ is reflexible.

Finally, if the symmetric dessin $\mathcal{D}$ is self-Petrie-dual, then it must be reflexible. By Proposition 2 we have $m = 1$. Therefore $b = 0$ and $\gamma = \zeta = 1$. □

In the remainder of the paper we classify the symmetric p-dessins of class two with multiple edges whose shadow dessins are the simple symmetric p-dessins $\mathcal{D}$ of class two classified in Theorem 1. Recall that $\mathcal{D}(\delta) = (G, x^{\delta}, y^{\delta})$ where $G = G(p; a, b, c)$ and $\delta \in \mathbb{Z}^*_{p^{a-c}}$. By Corollary 1, $\mathcal{D}(\delta) = H_{\delta}(\mathcal{D}(1))$ where H_{δ} is the δth Wilson's operation. So by Proposition 3, it suffices to classify the symmetric p-dessins of class two whose shadow dessin is $\mathcal{D}(1) = (G, x, y)$.

Theorem 3 *Let p be a prime and $m \geq 1$. If $\mathcal{D}$ is a class-two symmetric p-dessin of multiplicity p^m and its shadow dessin is the simple symmetric p-dessin $\mathcal{D}(1)$ from*

Theorem 1, then $\mathscr{D} \cong \mathscr{D}(p; a, b, c; m, d, f, \alpha, \beta, \gamma) = (U, x, y)$ *where* U *has a presentation*

$$\begin{aligned} \langle x, y \mid & x^{p^{a+m}} = y^{p^{a+m}} = [x, z] = [y, z] = 1, y^{p^a} = x^{\alpha p^a}, z^{p^{a+b-c}} = x^{\beta p^{a+d}}, \\ & z^{p^b} = x^{p^c(\gamma p^{a+f-c}-1)} y^{p^c}, z = [x, y] \rangle, \end{aligned} \tag{11}$$

where a, b, c *are given by* (4), *the integer parameters* d *and* f *satisfy*

$$\max(0, m + a + b - 2c) \le d \le m \quad and \quad 0 \le f \le m,$$

and $\alpha \in \mathbb{Z}^*_{p^m}$, $\beta \in \mathbb{Z}^*_{p^{m-d}}$ *and* $\gamma \in \mathbb{Z}^*_{p^{m-f}}$ *fulfil the following conditions*

$$\alpha^2 \equiv 1 \pmod{p^m}, \tag{12}$$

$$\gamma p^{a+f-c} + \alpha - 1 \equiv \beta p^d \pmod{p^m}, \tag{13}$$

$$\alpha \equiv -1 \pmod{p^{m-d}}, \tag{14}$$

$$\alpha \equiv -1 \pmod{p^{m-f}}. \tag{15}$$

Proof By the assumption the symmetric dessin $\mathscr{D} = (U, x, y)$ has multiplicity p^m and valency p^{a+m}, so $K = \langle x \rangle \cap \langle y \rangle = \langle x^{p^a} \rangle = \langle y^{p^a} \rangle \cong \mathbb{Z}_{p^m}$. It follows that the p-group U of class two, being a cyclic extension of $G(p; a, b, c)$ by K, has the presentation (11), where $0 \le d \le m$, $0 \le f \le m$, $\alpha \in \mathbb{Z}^*_{p^m}$, $\beta \in \mathbb{Z}^*_{p^{m-d}}$ and $\gamma \in \mathbb{Z}^*_{p^{m-f}}$.

Since $z^{p^{a+b-c}} = x^{\beta p^{a+d}}$ and $o(x) = p^{a+m}$, we have $o(z) = p^{a+b-c+m-d}$. Note that the relations $z^{p^b} = x^{(\gamma p^{a+f-c}-1)p^c} y^{p^c}$ and $z = [x, y]$ can be rewritten as the form

$$y^{p^c} = x^{(1-\gamma p^{a+f-c})p^c} z^{p^b} \quad \text{and} \quad x^y = xz.$$

It follows that $x = x^{x^{(1-\gamma p^{a+f-c})p^c} z^{p^b}} = x^{y^{p^c}} = xz^{p^c}$. By cancellation we get $z^{p^c} = 1$, and hence $m + a + b - 2c \le d$. Further, we deduce from the relations $y^{p^a} = x^{\alpha p^a}$, $z^{p^{a+b-c}} = x^{\beta p^{a+d}}$ and $z^{p^b} = x^{(\gamma p^{a+f-c}-1)p^c} y^{p^c}$ that

$$x^{\beta p^{a+d}} = z^{p^{a+b-c}} = (z^{p^b})^{p^{a-c}} = x^{(\gamma p^{a+f-c}-1)p^a} y^{p^a} = x^{(\gamma p^{a+f-c}+\alpha-1)p^a}.$$

Hence $\gamma p^{a+f-c} + \alpha - 1 \equiv \beta p^d \pmod{p^m}$.

Since (U, x, y) is symmetric, the assignment $\tau : x \mapsto y, y \mapsto x$ extends to an automorphism of U. Applying τ to the relation $y^{p^a} = x^{\alpha p^a}$ we get $x^{p^a} = \tau(y^{p^a}) = \tau(x^{\alpha p^a}) = y^{\alpha p^a}$. So $x^{p^a} = x^{\alpha^2 p^a}$, and hence $\alpha^2 \equiv 1 \pmod{p^m}$. Applying τ to the relation $z^{p^{a+b-c}} = x^{\beta p^{a+d}}$ we get

$$x^{-\beta p^{a+d}} = z^{-p^{a+b-c}} = \tau(z^{p^{a+b-c}}) = \tau(x^{\beta p^{a+d}}) = y^{\beta p^{a+d}} = x^{\alpha\beta p^{a+d}},$$

and hence $\alpha \equiv -1 \pmod{p^{m-d}}$. Similarly, we deduce from $z^{p^b} = x^{(\gamma p^{a+f-c}-1)p^c} y^{p^c}$ that

$$y^{-p^c} x^{(-\gamma p^{a+f-c}+1)p^c} = z^{-p^b} = \tau(z^{p^b}) = \tau(x^{(\gamma p^{a+f-c}-1)p^c} y^{p^c}) = y^{(\gamma p^{a+f-c}-1)p^c} x^{p^c}.$$

By collecting similar terms we obtain that $x^{\gamma p^{a+f}} = y^{-\gamma p^{a+f}}$. Since $y^{p^a} = x^{\alpha p^a}$, using substitution we get $x^{\gamma p^{a+f}} = y^{-\gamma p^{a+f}} = x^{-\alpha\gamma p^{a+f}}$, and hence $\alpha \equiv -1 \pmod{p^{m-f}}$.

Corollary 8 *If the dessin $\mathscr{D}(p; a, b, c; m, d, f, \alpha, \beta, \gamma)$ from Theorem 3 is reflexible, then*

1. *if $p > 2$, then $1 \le b \le a$, $d = f = m \ge 1$ and $\alpha = \beta = 1$.*
2. *if $p = 2$, then the values of parameters in $\mathscr{D}(p, a, b, c; m, d, f, \alpha, \beta, \gamma)$ are determined by Table 2.*

Table 2 Reflexible symmetric dessins with multiple edges

Classes	d	m	f	α	β	γ
$1 \le b \le c = a$	$d = m - 1$	1	0	1	1	1
$1 \le b \le c = a$	$d = m - 1$	2	1	1	1	1
$1 \le b \le c = a$	$d = m - 1$	2	2	3	1	1
$1 \le b \le c = a$	$d = m - 1$	≥ 3	$m - 1$	1	1	odd
$1 \le b \le c = a$	$d = m - 1$	≥ 3	m	$2^{m-1}+1$	1	odd
$1 \le b \le c = a$	$d = m - 1$	≥ 3	1	$2^{m-1}-1$	1	$1, 2^{m-1}+1$
$1 \le b \le c = a$	$d = m - 1$	≥ 3	1	$2^m - 1$	1	$2^{m-2}+1$
$1 \le b \le c = a$	$d = m$	1	1	1	1	1
$1 \le b \le c = a$	$d = m$	2	2	1	1	1, 3
$1 \le b \le c = a$	$d = m$	2	1	3	1	1, 3
$1 \le b \le c = a$	$d = m$	≥ 3	m	1	1	odd
$1 \le b \le c = a$	$d = m$	≥ 3	1	$2^m - 1$	1	$1, 2^{m-1}+1$
$1 \le b \le c = a$	$d = m$	≥ 3	$m - 1$	$2^{m-1}+1$	1	odd
$1 \le b \le c = a$	$d = m$	≥ 3	1	$2^{m-1}-1$	1	$2^{m-2}+1$
$2 \le b+2 \le a = c+1$	$d = m$	1	0	1	1	1
$2 \le b+2 \le a = c+1$	$d = m$	1	1	1	1	1
$2 \le b+2 \le a = c+1$	$d = m$	2	1	1	1	1, 3
$2 \le b+2 \le a = c+1$	$d = m$	2	2	1	1	1, 3
$2 \le b+2 \le a = c+1$	$d = m$	2	0	1	1	1, 3
$2 \le b+2 \le a = c+1$	$d = m$	≥ 3	$m - 1$	1	1	odd
$2 \le b+2 \le a = c+1$	$d = m$	≥ 3	m	1	1	odd
$2 \le b+2 \le a = c+1$	$d = m$	≥ 3	0	$2^m - 1$	1	$1, 2^{m-1}+1$
$2 \le b+2 \le a = c+1$	$d = m$	≥ 3	$m - 2$	$2^{m-1}+1$	1	odd
$2 \le b+2 \le a = c+1$	$d = m$	≥ 3	0	$2^{m-1}-1$	1	$2^{m-2}+1$

Proof If (U, x, y) is reflexible, then the assignment $\iota : x \mapsto x^{-1}, y \mapsto y^{-1}$ extends to an automorphism of U. We have $\iota(z) = [\iota(x), \iota(y)] = z$. Applying ι to the defining relations of U we get $z^{p^{a+b-c}} = x^{-\beta p^{a+d}}$ and $z^{p^b} = x^{-p^c(\gamma p^{a+f-c}-1)} y^{-p^c}$. Combining these with the defining relations of U we deduce that

$$y^{2p^c} = x^{-2p^c(\gamma p^{a+f-c}-1)} \quad \text{and} \quad x^{2\beta p^{a+d}} = 1.$$

Recall that $\langle x\rangle \cap \langle y\rangle = \langle x^{p^a}\rangle = \langle y^{p^a}\rangle$. So $2p^c \equiv 0 \pmod{p^a}$ and $2\beta \equiv 0 \pmod{p^{m-d}}$.

If p is odd, then $a = c$ and $d = m$. By (13) we also have $m = f$. Hence $\alpha = \beta = \gamma = 1$.

If $p = 2$, then either $c = a$ or $c = a - 1$. We distinguish two cases.

Case 1: $c = a$.

Using substitution $y^{2^a} = x^{-\alpha 2^a}$ we deduce that $x^{\alpha 2^{a+1}} = y^{2^{a+1}} = x^{-2^{a+1}(\gamma 2^f - 1)}$. Hence $\gamma 2^f + \alpha - 1 \equiv 0 \pmod{2^{m-1}}$. If $2^{m-1} || \gamma 2^f + \alpha - 1$, then by (13) we get $d = m - 1$, and hence $\beta = 1$. So (13) reduces to $\gamma 2^f + \alpha - 1 \equiv 2^{m-1} \pmod{2^m}$. Combining this with (12) we obtain the values of the parameters. On the other hand, if $2^m | \gamma 2^f + \alpha - 1$, then by (13) we have $d = m$. By (13) we have $\gamma 2^f + \alpha - 1 \equiv 0 \pmod{2^m}$. Combining this with (12) we obtain the values of the parameters.

Case 2: $c = a - 1$.

Using similar arguments as before we have $\gamma 2^{f+1} + \alpha - 1 \equiv 0 \pmod{2^m}$. By (13) we get $d = m$. Hence (13) reduces to $\gamma 2^{f+1} + \alpha - 1 \equiv 0 \pmod{2^m}$. Combining this with (12) we obtain the values of f and α, β, γ, as listed in Table 2. □

Corollary 9 *If the dessin $\mathscr{D}$ from Theorem 3 is self-Petrie-dual, then it is isomorphic to one of the following regular dessins:*

1. $\mathscr{D}_1(a, b) = (G, x, y)$ *where* $1 \le b \le a$ *and G has a presentation*

$$G = \langle x, y \mid x^{2^{a+1}} = y^{2^{a+1}} = [x, z] = [y, z] = 1, y^{2^a} = x^{2^a}, z^{2^b} = x^{2^a}, z = [x, y]\rangle.$$

2. $\mathscr{D}_2(a, b) = (G, x, y)$ *where* $1 \le b \le a$ *and G has a presentation*

$$G = \langle x, y \mid x^{2^{a+1}} = y^{2^{a+1}} = z^{2^b} = [x, z] = [y, z] = 1, y^{2^a} = x^{2^a}, z = [x, y]\rangle.$$

3. $\mathscr{D}_3(a, b) = (G, x, y)$ *where* $0 \le b \le a - 2$ *and G has a presentation*

$$G = \langle x, y \mid x^{2^{a+1}} = y^{2^{a+1}} = [x, z] = [y, z] = 1, z^{2^b} = x^{2^{a-1}} y^{2^{a-1}}, y^{2^a} = x^{2^a}, z = [x, y]\rangle.$$

Proof Since $\mathscr{D}$ is symmetric and self-Petrie-dual, it is reflexible. By Proposition 2 the multiplicity 2^m must be equal to 2, so $m = 1$. Checking Table 2 we have the listed families of self-Petrie-dual symmetric dessins. □

Remark 2 Let $\mathscr{M}$ be a regular bipartite map. It is shown [13, Theorem 25] that if the group $\mathrm{Aut}^+(\mathscr{M})$ of orientation-preserving automorphisms of $\mathscr{M}$ is nilpotent of class $c \ge 2$, then the subgroup $\mathrm{Aut}_0^+(\mathscr{M})$ of colour-preserving automorphisms of

$\mathrm{Aut}^+(\mathscr{M})$ is nilpotent of class at most $c-1$. By the correspondence between symmetric dessins and regular bipartite maps, all regular bipartite maps whose automorphism groups are 2-groups of class 3 classified in [1] are contained as a subclass in our classification of symmetric 2-dessins of class 2. The interested reader is referred to [27] for details.

Acknowledgments The main material of the paper is based on the first author's PhD thesis [27]. She acknowleges the support during the last years from the Department of Mathematics in the Faculty of Natural Sciences of Matej Bel University in Banská Bystrica. The authors are grateful to Prof. Shao-Fei Du for his generosity to share his unpublished work in [1], and to the anonymous referees for the helpful and detailed suggestions and comments which have substantially improved the presentation of the paper. The first and third author are supported by Zhejiang Provincial Natural Science Foundation of China (LY16A010010), and by Scientific Research Foundation of Zhejiang Ocean University (21065014015, 21065014115). The second author is supported by the following grants: VEGA 1/0150/14, APVV-0223-10, and the grant APVV-ESF-EC-0009-10 within the EUROCORES Programme EUROGIGA (Project GReGAS) of the European Science Foundation and the Slovak-Chinese bilateral grant APVV-SK-CN-0009-12.

References

1. Ban, Y.-F., Du, S.-F., Liu, Y., Nedela, R., Škoviera, M.: Classification of regular maps whose automorphism groups are 2-groups of class three. preprint (2012).
2. Belyĭ, G.V.: Galois extensions of a maximal cyclotomic field. Izv. Akad. Nauk. SSSR Ser. Mat., **43**, 267–276 (1979).
3. Breda d'Azevedo, A., Nedela, R.: Join and intersection of hypermaps. Acta. Univ. M. Belii., **9**, 13–28 (2001).
4. Conder, M.D.E.: All proper orientable regular hypermaps on surfaces of genus 2 to 101. URL: https://www.math.auckland.ac.nz/~conder/OrientableProperHypermaps101.txt.
5. Conder, M.D.E., Du, S.-F., Nedela, R., Škoviera, M.: Bounding the size of a regular map with nilpotent automorphism group. preprint (2014).
6. Conder, M.D.E., Jones, G.A., Streit, M., Wolfart, J.: Galois actions and regular dessins of small genera. Rev. Mat. Iberoam., **29**, 163–181 (2013).
7. Corn, D., Singerman, D.: Regular hypermaps. European J. Combin., **9**, 337–351 (1988).
8. Coste, A.D., Jones, G.A., Streit, M., Wolfart, J.: Generalised Fermat hypermaps and Galois orbits. Glasgow Math. J., **51**(2), 289–299 (2009).
9. Du, S.-F., Jones, G.A., Kwak, J.H., Nedela, R., Škoviera, M.: Regular embeddings of $K_{n,n}$ where n is a power of 2. I: Metacyclic case. European J. Combin., **28**, 1595–1609 (2007).
10. Du, S.-F., Jones, G.A., Kwak, J.H., Nedela, R., Škoviera, M.: Regular embeddings of $K_{n,n}$ where n is a power of 2. II: The non-metacyclic case. European J. Combin., **31**, 1946–1956 (2010).
11. Gonzalo, R., Rodrígues, E.R.: Riemann surfaces and abelian variaties with an automorphism of prime order, Duke Math. J. **69**, 199–217 (1993).
12. Hidalgo, R.A.: The bipartite graphs of abelian dessins d'enfants. Ars Math. Contemporanea, **6**, 301–304 (2013).
13. Hu, K., Nedela, R., Wang, N.-E.: Nilpotent dessins: Decomposition theorem and classification of the abelian dessins. arXiv:1508.04523 (2015).
14. Hu, K., Nedela, R., Wang, N.-E.: Nilpotent groups of class two which underly a unique regular dessin. Geom. Dedicata. doi:10.1007/s10711-015-0074-8 (2015).
15. Huppert, B.: Endliche Gruppen (Vol. 1). Springer-Verlag, Berlin (1967).

16. James, L.D.: Operations on hypermaps, and outer automorphisms. European J. Combin., **9**, 551–560 (1988).
17. Jones, G.A.: Regular embeddings of complete bipartite graphs: classification and enumeration. Proc. Lond. Math. Soc., **101**(3), 427–453 (2010).
18. Jones, G.A.: Regular dessins with a given automorphism group. In: Izquierdo, M., Broughton, S.A., Costa, A.F., Rodríguez, R.E., (eds) Riemann and Klein surfaces, Automorphisms, Symmeries and Moluli Spaces. American Math. Soc., Providence, Rhode Island (2014).
19. Jones, G.A., Nedela, R., Škoviera, M.: Complete bipartite graphs with a unique regular embedding. J. Combin. Theory Ser. B, **98**, 241–248 (2008).
20. Jones, G.A., Nedela, R., Škoviera, M.: Regular embeddings of $K_{n,n}$ where n is an odd prime power. European J. Combin., **28**, 1863–1875 (2007).
21. Jones, G.A., Pinto, D.: Hypermap operations of finite order. Discrete Math., **310**, 1820–1827 (2010).
22. Jones, G.A., Singerman, D.: Belyǐ functions, hypermaps and Galois groups. Bull. London Math. Soc., **28**, 561–590 (1996).
23. Kallel, S., Sjerve, D.: On the group of automorphisms of cyclic covers of the Riemann sphere. Math. Proc. Cambridge Philos. Soc. 138 , no. 2, 267C-287 (2005).
24. Lando, S. K., Zvonkin, A. K.: Graphs on Surfaces and Their Applications, Encyclopaedia of Mathematical Sciences: Lower-Dimensional Topology II 141, Berlin, New York: Springer-Verlag, (2004).
25. Malnič, A., Nedela, R., Škoviera, M.: Regular maps with nilpotent automorphism groups. European J. Combin., **33**(8), 1974–1986 (2012).
26. Schneps, L., ed.: The Grothendieck Theory of Dessins d'Enfants, London Mathematical Society Lecture Note Series, Cambridge: Cambridge University Press (1994).
27. Wang, N.-E.: Regular bipartite maps. PhD thesis, Matej Bel University, Banská Bystrica (2014).
28. Wilson, S.E.: Operators over regular maps, Pacific J. Math. **81** 559–568 (1979).
29. Wilson, S.E.: Parallel products in groups and maps. J. Algebra, **167**, 539–546 (1994).

MIX
Papier aus verantwortungsvollen Quellen
Paper from responsible sources
FSC® C105338

If you have any concerns about our products,
you can contact us on
ProductSafety@springernature.com

In case Publisher is established outside the EU,
the EU authorized representative is:
Springer Nature Customer Service Center GmbH
Europaplatz 3, 69115 Heidelberg, Germany

Printed by Libri Plureos GmbH
in Hamburg, Germany